W9-CNN-582

WITHDRAWN FROM LIBRARY

MONTGOMERY COLLEGE LIBRARY
ROCKVILLE CAMPUS

WITHDRAWN FROM LIBRARY

Applied Mathematical Sciences | Volume 25

B. Davies

Integral Transforms and Their Applications

Springer-Verlag New York Heidelberg Berlin

B. Davies
The Australian National University
Post Office Box 4
Canberra, A.C.T. 2600

AMS Classifications: 44-01, 44A10, 44A15, 44A20, 44A30

Library of Congress Cataloging in Publication Data

Davies, Brian, 1937-
Integral transforms and their applications.

(Applied mathematical sciences; v. 25)
Bibliography: p.
Includes index.
1. Integral transforms. I. Title. II. Series.
QA1.A647 vol. 25 [QA432] 510'.8s [515'.723] 77-27330

All rights reserved.
No part of this book may be translated or reproduced in any form without written permission from Springer-Verlag.
© 1978 by Springer-Verlag New York Inc.

Printed in the United States of America

9 8 7 6 5 4 3 2 1

ISBN 0-387-90313-5 Springer-Verlag New York
ISBN 3-540-90313-5 Springer-Verlag Berlin Heidelberg

Preface

This book is intended to serve as introductory and reference material for the application of integral transforms to a range of common mathematical problems. It has its immediate origin in lecture notes prepared for senior level courses at the Australian National University, although I owe a great deal to my colleague Barry Ninham, a matter to which I refer below. In preparing the notes for publication as a book, I have added a considerable amount of material additional to the lecture notes, with the intention of making the book more useful, particularly to the graduate student involved in the solution of mathematical problems in the physical, chemical, engineering and related sciences.

Any book is necessarily a statement of the author's viewpoint, and involves a number of compromises. My prime consideration has been to produce a work whose scope is selective rather than encyclopedic; consequently there are many facets of the subject which have been omitted--in not a few cases after a preliminary draft was written--because I believe that their inclusion would make the book too long. Some of the omitted material is outlined in various problems and should be useful in indicating possible approaches to certain problems. I have laid great stress on the use of complex variable techniques, an area of mathematics often unfashionable, but frequently of great power. I have been particularly severe in excising formal proofs, even though there is a considerable amount of "pure mathematics" associated with the understanding and use of generalized functions, another area of enormous utility in mathematics. Thus, for the formal aspects of the theory of integral transforms I must refer the reader to one of the many excellent books addressed to this area; I have chosen an approach which is more common in published research work in applications. I can only hope that the course which I have steered will be of great interest and help to students and research workers who wish to use integral transforms.

It was my priviledge as a student to attend lectures on mathematical physics by Professor Barry W. Ninham, now at this university. For several years it was his intention to publish a comprehensive volume on mathematical techniques in physics, and he prepared draft material on several important topics to this end. In 1972 we agreed to work on this project jointly, and continued to do so until 1975. During that period it became apparent that the size, and therefore cost, of such a large volume would be inappropriate to the current situation, and we decided to each publish a smaller book in our particular area of interest. I must record my gratitude to him for agreeing that one of his special interests--the use of the Mellin transform in asymptotics--

should be included in the present book. In addition there are numerous other debts which I owe to him for guidance and criticism.

References to sources of material have been made in two ways, since this is now a fairly old subject area. First, there is a selected bibliography of books, and I have referred, in various places, to those books which have been of particular assistance to me in preparing lectures or in pursuing research. Second, where a section is based directly on an original paper, the reference is given as a footnote. Apart from this, I have not burdened the reader with tedious lists of papers, especially as there are some comprehensive indexing and citation systems now available.

A great deal of the final preparation was done while I was a visitor at the Unilever Research Laboratories (UK) and at Liverpool University in 1975, and I must thank those establishments for their hospitality, and the Australian National University for the provision of study leave. Most of the typing and retyping of the manuscript has been done by Betty Hawkins of this department while the figures were prepared by Mrs. L. Wittig of the photographic services department, ANU. Timothy Lewis, of Applied Mathematics at Brown University, has proofread the manuscript and suggested a number of useful changes. To these people I express my gratitude and also to Professor Lawrence Sirovich for his encouragement and helpful suggestions. This book is dedicated to my respected friend and colleague, Barry Ninham.

Brian Davies

Canberra, Australia
1977

Table of Contents

Page

Part I: The Laplace Transform

§1. DEFINITION AND ELEMENTARY PROPERTIES[1]

1.1. The Laplace Transform

Let $f(t)$ be an arbitrary function defined on the interval $0 \le t < \infty$; then

$$F(p) = \int_0^\infty e^{-pt} f(t)\, dt \tag{1}$$

is the Laplace transform of $f(t)$, provided that the integral exists. We shall confine our attention to functions $f(t)$ which are absolutely integrable on any interval $0 \le t \le a$, and for which $F(\alpha)$ exists for some real α. It may readily be shown that for such a function $F(p)$ is an analytic function of p for $\mathrm{Re}(p) > \alpha$, as follows. First note that the functions

$$\phi(p,T) = \int_0^T e^{-pt} f(t)\, dt \tag{2}$$

are analytic in p, and then that $\phi(p,T)$ converges uniformly to $F(p)$ in any bounded region of the p plane satisfying $\mathrm{Re}(p) > \alpha$, as $T \to \infty$. It follows from a standard theorem on uniform convergence[2] that F(p) is analytic in the half-plane $Re(p) > \alpha$.

As simple examples of Laplace transforms, we have

(i) Heaviside unit step function

$$h(t) = \begin{cases} 1, & t > 0 \\ 0, & t < 0 \end{cases} \tag{3}$$

$$H(p) = \int_0^\infty e^{-pt}\, dt$$

$$= 1/p, \quad Re(p) > 0 , \tag{4}$$

(ii)

$$f(t) = e^{i\omega t}, \quad \omega \text{ real} \tag{5}$$

$$F(p) = \int_0^\infty e^{-pt}\, e^{i\omega t}\, dt$$

$$= \frac{1}{p-i\omega}, \quad Re(p) > 0, \tag{6}$$

(iii)

$$f(t) = t^\gamma\, e^{\beta t}, \quad \alpha \text{ real} > -1 \tag{7}$$

$$F(p) = \frac{\gamma!}{(p-\beta)^{\gamma+1}}, \quad Re(p) > Re(\beta). \tag{8}$$

An important feature of these examples, and indeed of many of the Laplace transforms which occur in applications, is that the analytic function defined by (1) in the half-plane $Re(p) > \alpha$ can be analytically continued into the remainder of the plane once the singularity structure has been elucidated. Thus the functions defined by (4) and (6) exhibit only a simple pole; in the case of (8) there is a branch point at $p = \beta$ except for the special case that γ is an integer, when we get a pole.

1.2. Important Properties

There are a number of simple properties which are of recurring importance in the application of the Laplace transform to specific problems. In order to simplify somewhat the statement of these results, we introduce the notation

$$\mathscr{L}[f] = F(p) = \int_0^\infty e^{-pt} f(t)\, dt \tag{9}$$

which emphasizes the operator nature of the transform.

Linearity: If we consider the linear combination

$$f(t) = \sum_{k=1}^{n} a_k f_k(t) \tag{10}$$

where the a_k are arbitrary constants, then

$$\mathscr{L}[f] = \sum_{k=1}^{n} a_k \mathscr{L}[f_k] \ . \tag{11}$$

One immediate consequence of this is that if f depends on a variable x which is independent of t, we have

$$\mathscr{L}[\partial f/\partial x] = \partial \mathscr{L}[f]/\partial x, \tag{12}$$

$$\mathscr{L}\left[\int_a^b f\, dx\right] = \int_a^b \mathscr{L}[f]\, dx. \tag{13}$$

These results follow by trivial manipulation of the integrals in the half-space $Re(p) > \alpha$ in which all the integrals converge absolutely and uniformly (in x). But then they must also hold over the entire region of the complex p plane to which the transforms may be analytically continued.

Derivatives and Integrals: If we apply integration by parts to (1), we obtain

$$\mathscr{L}[f'(t)] = p\mathscr{L}[f] - f(0+), \tag{14}$$

$$f(0+) = \lim_{t \to 0} f(t).$$

The distinction which we have made between the value of $f(0)$ and the limit of $f(t)$ as $t \to 0$ is of importance in problems where there are discontinuities at $t = 0$. In many problems, initial values of functions are specified with the implied meaning that they are limiting values for small t, and the distinction becomes unimportant and may be neglected. By repeating the procedure of integration by parts, we can derive the general result

$$\mathscr{L}[f^{(n)}(t)] = p^n\mathscr{L}[f] - \sum_{k=1}^{n} p^{n-k} f^{(k-1)}(0+). \tag{15}$$

A similar result holds for differentiation of the Laplace transform with respect to p. By differentiating under the integral sign, we obtain

$$\frac{d^n}{dp^n} F(p) = (-1)^n \mathscr{L}[t^n f(t)]. \tag{16}$$

Suppose now that we define $g(t)$ by

$$g(t) = \int_0^t f(\tau)\, d\tau. \tag{17}$$

Then by interchanging orders of integration, we get

$$\begin{aligned} G(p) &= \int_0^\infty e^{-pt}\, dt \int_0^t f(\tau)\, d\tau \\ &= \int_0^\infty f(\tau)\, d\tau \int_\tau^\infty e^{-pt}\, dt \\ &= \frac{1}{p} \int_0^\infty e^{-p\tau} f(\tau)\, d\tau \\ &= \frac{1}{p} F(p), \end{aligned} \tag{18}$$

where the real part of p must be sufficiently large to

ensure that all of the integrals converge. A complementary result can be obtained by considering

$$\begin{aligned}\mathscr{L}[t^{-1} f(t)] &= \int_0^\infty e^{-pt} \frac{f(t)}{t}\, dt \\ &= \int_0^\infty f(t)\, dt \int_p^\infty e^{-qt}\, dq \qquad (19) \\ &= \int_p^\infty F(q)\, dq,\end{aligned}$$

which is valid provided the integrals exist. Both of these procedures may be iterated to give more general results, which we will not list here.

Translations: Let $\tau > 0$ and suppose that $f(t) = 0$ for $t < 0$, then

$$\begin{aligned}\mathscr{L}[f(t-\tau)] &= \int_\tau^\infty e^{-pt} f(t-\tau)\, dt \\ &= \int_0^\infty e^{-p(t'+\tau)} f(t')\, dt' \qquad (20) \\ &= e^{-p\tau} \mathscr{L}[f(t)].\end{aligned}$$

This result applies to translations to the right; in particular the inverse Laplace transform of $\exp(-p\tau)F(p)$, where $F(p) = \mathscr{L}[f(t)]$, will give $f(t-\tau)$ for $t > \tau$ and zero for $t < \tau$. For translations to the left we have

$$\begin{aligned}\mathscr{L}[f(t+\tau)] &= \int_0^\infty e^{-pt} f(t+\tau) dt \\ &= \int_\tau^\infty e^{-p(t'-\tau)} f(t')\, dt' \qquad (21) \\ &= e^{p\tau} \mathscr{L}[f(t)] - \int_0^\tau e^{p(\tau-t')} f(t')\, dt'.\end{aligned}$$

The finite integral cannot be neglected unless $f(t) = 0$

for $t < \tau$, as it accounts for the part of the function which has been 'lost' by translation to negative t values where the Laplace transform does not operate.

Convolutions: The convolution of two functions $f_1(t)$ and $f_2(t)$ is defined by

$$g(t) = \int_0^t f_1(\tau)\, f_2(t-\tau)\, d\tau. \tag{22}$$

Now we take the Laplace transform of $g(t)$, and by changing the order of integration and writing $t' = t-\tau$, we obtain

$$\begin{aligned} G(p) &= \int_0^\infty e^{-pt}\, dt \int_0^t f_1(\tau)\, f_2(t-\tau)\, d\tau \\ &= \int_0^\infty f_1(\tau)\, d\tau \int_0^\infty e^{-pt}\, f_2(t-\tau)\, dt \\ &= \int_0^\infty e^{-p\tau}\, f_1(\tau)\, d\tau \int_0^\infty e^{-pt'}\, f_2(t')dt' \\ &= F_1(p)\, F_2(p). \end{aligned} \tag{23}$$

Thus the transform of a convolution is simply the product of the individual transforms--a result which is of considerable importance. Obviously this result can be iterated to obtain a connection between an n-fold convolution of n functions and the product of the transforms of these functions.

Simple Applications:

(i)

$$\begin{aligned} \mathscr{L}[\sin \omega t] &= \frac{1}{2i}\mathscr{L}[e^{i\omega t}] - \frac{1}{2i}\mathscr{L}[e^{-i\omega t}] \\ &= \frac{\omega}{p^2+\omega^2} \end{aligned} \tag{24}$$

(ii)

$$\mathscr{L}[\cos \omega t] = \frac{1}{2}\mathscr{L}[e^{i\omega t}] + \frac{1}{2}\mathscr{L}[e^{-i\omega t}]$$
$$= \frac{p}{p^2+\omega^2} \qquad (25)$$

(iii)

$$\mathscr{L}[t \sin \omega t] = -\frac{d}{dp}\mathscr{L}[\sin \omega t]$$
$$= \frac{2p\omega}{(p^2+\omega^2)} \qquad (26)$$

(iv)

$$\mathscr{L}[e^{-at} \sin \omega t] = \frac{\omega}{(a+p)^2+\omega^2} \qquad (27)$$

[by replacing p by $a+p$ in (24)].

(v)

$$\mathscr{L}[t^{-1} \sin \omega t] = \int_p^\infty \mathscr{L}[\sin \omega t]\, dq$$
$$= \int_p^\infty \frac{\omega\, dq}{q^2+\omega^2} \qquad (28)$$
$$= \text{arc tan } (\omega/p)$$

(vi) Let

$$\text{Si}(z) = \int_0^z \frac{\sin t}{t}\, dt, \qquad (29)$$

then

$$\mathscr{L}[\text{Si}(z)] = p^{-1}\mathscr{L}[t^{-1} \sin t]$$
$$= p^{-1} \text{ arc cot } (p). \qquad (30)$$

Less trivial applications of the properties of the transform, particularly in the solution of differential equations and integral equations of convolution form, are the subject matter of Sections 3-5.

1.3. Asymptotic Properties: Watson's Lemma

Consider equation (1) for large p. By inspection, it seems reasonable to assume that the only significant region of integration is $0 \leq t < 1/p$, so that we could write as an approximation

$$\begin{aligned} F(p) &\simeq f(0) \int_0^\infty e^{-pt}\, dt \\ &= f(0)/p, \qquad p \gg 1. \end{aligned} \tag{31}$$

Such information, linking properties of functions and their transforms directly, may be very useful in application. However, the example given in equations (7) and (8), where

$$F(p) \simeq \frac{\gamma!}{p^{\gamma+1}}, \quad p \gg \beta \tag{32}$$

shows that we need a sharper result than (31).

Definitions: If two functions $f(x)$ and $g(x)$ satisfy the relation

$$\lim_{x \to x_0} [f(x)/g(x)] = 1, \tag{33}$$

then we say that they are asymptotically equal as $x \to x_0$, and write

$$f(x) \sim g(x), \quad x \to x_0. \tag{34}$$

In the event that x is a complex variable, we may need to add some restriction about the way in which x approaches x_0, for example

$$\begin{aligned} 1 + e^{-z} \sim 1, \quad z \to \infty, \\ |\arg(z)| < \pi/2. \end{aligned} \tag{35}$$

If now (33) is replaced by the condition that

$$\lim_{x \to x_0} [f(x)/g(x)] = 0, \tag{36}$$

then we write

$$f(x) = o(g(x)), \tag{37}$$

and if $|f(x)/g(x)|$ is bounded as x approaches x_0, then we write

$$f(x) = \mathcal{O}(g(x)). \tag{38}$$

In this book we shall frequently use the notations (34) and (38); the small o notation (37) will not occur often.

Asymptotic Expansion: An expansion of the form

$$f(x) \sim \sum_{\nu=1}^{\infty} g_\nu(x), \quad x \to x_0 \tag{39}$$

is called an asymptotic expansion if

$$g_{\nu+1}(x) = o(g_\nu(x)). \tag{40}$$

The meaning of such an expansion is that

$$f(x) = \sum_{\nu=1}^{n} g_\nu(x) + \mathcal{O}(g_{n+1}(x)) \tag{41}$$

so that a finite number of terms of the series gives an approximation to the function $f(x)$ of "order" $g_{n+1}(x)$ when x approaches x_0. Viewed as an infinite series (39) may be convergent or it may be divergent.

Watson's Lemma: We will now state and prove an important result, of which (31) is a special case, linking the asymptotic expansion of a function $f(t)$ about $t = 0$ with the asymptotic expansion of $F(p)$ as $p \to \infty$. Suppose that $f(t)$ has the asymptotic expansion

$$f(t) \sim \sum_{\nu=1}^{\infty} a_\nu t^{\lambda_\nu}, \qquad t \to 0,$$
$$-1 < \mathrm{Re}(\lambda_1) < \mathrm{Re}(\lambda_2) < \mathrm{Re}(\lambda_3) < \ldots; \tag{42}$$

then $F(p)$ has the corresponding asymptotic expansion

$$F(p) \sim \sum_{\nu=1}^{\infty} \frac{a_\nu \lambda_\nu!}{p^{\lambda_\nu+1}}, \qquad |p| \to \infty,$$
$$-\pi/2 < \arg(p) < \pi/2. \tag{43}$$

Note the effect of the restriction $|\arg(p)| < \pi/2$ is to ensure that $\mathrm{Re}(p)$ becomes infinite as $|p|$ does in this sector. To derive the stated result, we introduce the function

$$f_n(t) = f(t) - \sum_{\nu=1}^{n} a_\nu t^{\lambda_\nu}, \tag{44}$$

in terms of which $F(p)$ is given by

$$F(p) = \sum_{\nu=1}^{n} \frac{a_\nu \lambda_\nu!}{p^{\lambda_\nu+1}} + \mathscr{L}[f_n(t)]. \tag{45}$$

To compute bounds on $\mathscr{L}[f_n(t)]$ we choose positive numbers t_n and K_n so that

$$|f_n(t)| < K_n\, t^{\mathrm{Re}(\lambda_\nu)}, \qquad 0 \le t \le t_n. \tag{46}$$

Also, we know that there must be some real value α for which the integral defining $F(p)$ converges, and we use this constant to define the functions

$$\phi_n(t) = \int_{t_n}^{t} e^{-\alpha s} f_n(s)\, ds. \tag{47}$$

The importance of the choice of α is that the functions ϕ_n will be bounded, and we write A_n for the maximum value of $|\phi_n(t)|$, $t_n \le t < \infty$. Using these definitions, we can break up the integral defining $\mathscr{L}[f_n(t)]$ into two parts, and calculate the following bounds:

$$\left|\int_0^{t_n} e^{-pt} f_n(t)\, dt\right| < \frac{\mathrm{Re}(\lambda_n)!\ K_n}{[\mathrm{Re}(p)]^{\mathrm{Re}(\lambda_n)+1}} = \mathscr{O}(p^{-\lambda_n-1}) \tag{48}$$

and

$$\left|\int_{t_n}^{\infty} e^{-pt} f_n(t)\, dt\right| = \left|(p-\alpha)\int_{t_n}^{\infty} e^{-(p-\alpha)t} \phi_n(t)\, dt\right| < \frac{A_n|p-\alpha|}{\mathrm{Re}(p)-\alpha} \exp\{-(\mathrm{Re}(p)-\alpha)t_n\}. \tag{49}$$

This latter integral tends to zero exponentially as p tends to infinity in the given sector, consequently (48) shows that

$$\mathscr{L}[f_n(t)] = \mathscr{O}(p^{-\lambda_n-1}) \tag{50}$$

and Watson's lemma (43) is proved. The result must be used with caution. It gives information about the behavior of $F(p)$ for large p which is consequent upon the behavior of $f(t)$ for small t. The question of a converse implication is discussed in Sections 2 and 6.

Problems

Deduce the following general relationships.[3]

1. If $f(t+T) = f(t)$, $t > 0$ where $T > 0$ is a constant,

$$F(p) = (1-e^{-Tp})^{-1} \int_0^T e^{-pt} f(t)\, dt.$$

2. If $f(t) = \frac{1}{t}\frac{d}{dt} g(t)$,

$$F(p) = \int_p^{\infty} q\, G(q)\, dq.$$

3. If $f(t) = \int_0^t u^{-1}\, g(u)\, du,$

$$F(p) = p^{-1} \int_p^{\infty} G(q)\, dq.$$

4. If $f(t) = \int_t^{\infty} u^{-1}\, g(u)\, du,$

$$F(p) = p^{-1} \int_0^{p} G(q)\, dq.$$

5. $\int_0^{\infty} t^{-1}\, f(t)\, dt = \int_0^{\infty} F(p)\, dp$

Find the Laplace transforms of the following functions.[4]

6. $\sinh(at)$

7. $\cosh(at)$

8. $\cos(at)\cosh(bt)$

9. $t^{-1}\sinh(at)$

10. $\sin(at^{1/2})$

11. $t^{1/2}\cos(at^{1/2})$

12. $t^{\nu}\cos(at)$, $\nu > -1$

13. $\int_t^{\infty} u^{-1}\, e^{-u}\, du$

Using Problem 5, evaluate

14. $\int_0^{\infty} t^{-1}\sin(\omega t)\, dt$

15. $\int_0^{\infty} t^{-1}\{e^{-at} - e^{-bt}\}\, dt$, $a > 0,\ b > 0.$

By taking the Laplace transform with respect to t, evaluate the integrals

16. $\int_0^{\infty} \frac{x\sin(xt)}{1+x^2}\, dx$

17. $\int_0^{\infty} \exp\left(-x^2 - \frac{t^2}{x^2}\right) dx.$

18. Let

$$f(t) = \begin{cases} t^{p-1}, & t > 0, \quad \mathrm{Re}(p) > 0 \\ 0, & t < 0, \end{cases}$$

$$g(t) = \begin{cases} t^{q-1}, & t > 0, \quad \mathrm{Re}(q) > 0 \\ 0, & t < 0, \end{cases}$$

$$h(t) = \begin{cases} \frac{(p-1)!(q-1)!}{(p+q-1)!} t^{p+q-1}, & t > 0 \\ 0, & t < 0. \end{cases}$$

Show that the Laplace transform of the convolution of $f(t)$ with $g(t)$ is equal to $\mathscr{L}[h(t)]$. Hence derive the formula[5]

$$\int_0^1 x^{p-1} (1-x)^{q-1} dx = \frac{(p-1)!(q-1)!}{(p+q-1)!} .$$

19. Show that

$$\mathscr{L}[\ln(1+t^{1/2})] \sim \sum_{\nu=1}^{\infty} \frac{(-1)^{\nu-1}(\frac{1}{2}\nu-1)!}{2p^{(1/2)\nu+2}}, \quad p \to \infty.$$

20. Show that

$$\mathscr{L}[t^{-1/2}(t+2)^{-1/2}] \sim$$

$$(\frac{\pi}{2p})^{1/2} \sum_{\nu=0}^{\infty} \frac{(-1)^{\nu}\, 1^2 \cdot 3^2 \cdot 5^2 \cdot \ldots \cdot (2\nu-1)^2}{\nu!\ (8p)^{\nu}}, \quad p \to \infty.$$

Footnotes

1. The results given in this section may be found in many places. We mention in particular DITKIN & PRUDNIKOV (1965), DOETSCH (1971), and WIDDER (1944).
2. AHLFORS (1966), Ch. 5.
3. Many more general relationships may be found in ERDELYI, et al. (1954), Ch. 4.
4. Extensive tables of Laplace transforms are available; for instance, ERDELYI, et. al. (1954).
5. Anticipating the result that the Laplace transform has a unique inverse.

§2. THE INVERSION THEOREM

2.1. The Riemann-Lebesgue Lemma

As necessary preliminaries to a statement and proof of the inversion theorem, which together with its elementary properties makes the Laplace transform a powerful tool in applications, we must first take note of some results from classical analysis.[1] Suppose that $f(x)$ is a function continuous on the closed interval $a \leq x \leq b$ (and hence uniformly continuous). We will investigate the asymptotic properties of the integral

$$I(\omega) = \int_a^b f(x)\, e^{i\omega x}\, dx \tag{1}$$

for large real ω. By some trivial changes of variable we can write

$$\begin{aligned} I(\omega) &= \int_a^{a+\pi/\omega} f(x)\, e^{i\omega x}\, dx + \int_{a+\pi/\omega}^{b} f(x)\, e^{i\omega x}\, dx \\ &= \int_a^{a+\pi/\omega} f(x)\, e^{i\omega x}\, dx - \int_a^{b-\pi/\omega} f(x+\pi/\omega)\, e^{i\omega x} dx \end{aligned} \tag{2}$$

and

$$I(\omega) = \int_a^{b-\pi/\omega} f(x)\, e^{i\omega x}\, dx + \int_{b-\pi/\omega}^{b} f(x)\, e^{i\omega x}\, dx, \tag{3}$$

and thus

$$\begin{aligned} I(\omega) &= \frac{1}{2}\int_a^{a+\pi/\omega} f(x)\, e^{i\omega x}\, dx + \frac{1}{2}\int_{b-\pi/\omega}^{b} f(x)\, e^{i\omega x}\, dx \\ &\quad + \frac{1}{2}\int_a^{b-\pi/\omega} [f(x)-f(x+\pi/\omega)]\, e^{i\omega x}\, dx. \end{aligned} \tag{4}$$

It is easily seen, by a mean value theorem for integrals, that the first two integrals in (4) are functions of asymptotic order ω^{-1}; furthermore, since $f(x)$ is uniformly continuous, we can make the integrand in the third integral arbitrarily small by choosing ω sufficiently large. Thus we have proved that

$$\lim_{\omega\to\infty} \int_a^b f(x)\, e^{i\omega x}\, dx = 0, \tag{5}$$

which is known as the Riemann-Lebesgue lemma.

Infinite Interval: The extension of (5) to the case where one limit or both may be infinite will also be needed. For example, if $f(x)$ is a function defined on the interval $0 \le x < \infty$, for which

$$\int_0^\infty |f(x)|\, dx \tag{6}$$

converges, then we can write

$$\int_0^\infty f(x)\, e^{i\omega x}\, dx = \int_0^a f(x)\, e^{i\omega x}\, dx + \varepsilon, \qquad |\varepsilon| \le \int_a^\infty |f(x)|\, dx, \tag{7}$$

and because of the absolute convergence (6), it is possible to make $|\varepsilon|$ arbitrarily small by a suitable choice of a. Using (5) on the finite integral, we have its extension to the infinite integral, i.e.,

$$\lim_{\omega\to\infty} \int_0^\infty f(x)\, e^{i\omega x}\, dx = 0. \tag{8}$$

Dirichlet Conditions: We say that a function $f(x)$ satisfies Dirichlet's conditions in the interval $a \le x \le b$ if it has at most a finite number of maxima, minima, and points of discontinuity in the interval, and takes only a finite jump at any discontinuity. The importance of the Dirichlet conditions to the theorems which we need is that they enable the interval $a \le x \le b$ to be divided into subintervals, in each of which the function is both uniformly continuous and monotonic. This latter property allows us to use the second mean value theorem for integrals, which states that if $f(x)$ is a monotonic function and $g(x)$ a continuous function on the interval $a \le x \le b$, then there is a point c in the interval such that

$$\int_a^b f(x)\, g(x)\, dx = f(a) \int_a^c g(x)\, dx + f(b) \int_c^b g(x)\, dx. \quad (9)$$

Returning now to equation (1), if $f(x)$ satisfies Dirichlet's conditions, then we can take the interval $a \le x \le b$ to be one of the subintervals in which it is monotonic and continuous; then the integral is equal to

$$f(a) \int_a^c e^{i\omega x}\, dx + f(b) \int_c^b e^{i\omega x}\, dx \quad (10)$$

$$= \mathcal{O}(\omega^{-1}), \quad \omega \to \infty.$$

For an arbitrary interval, we must add up a finite number of such results, and so equation (5) is replaced by the much stronger condition

$$\int_a^b f(x)\, e^{i\omega x}\, dx = \mathcal{O}(\omega^{-1}), \quad \omega \to \infty. \quad (11)$$

Note however, that we may not set $a = -\infty$ or $b = \infty$ in this result without imposing restrictions on $f(x)$ in addition to the convergence of equation (6).

2.2. Dirichlet Integrals

In addition to integrals of the form (1), we must consider what are known as Dirichlet integrals, viz.,

$$\int_a^b f(x)\, \frac{\sin(\omega x)}{x}\, dx \quad (12)$$

in the limit that ω tends to infinity. Suppose now that $a = 0$, $b > 0$, and $f(x)$ satisfies Dirichlet's conditions on $0 \le x \le b$. Choose c so that $f(x)$ is continuous and monotonic on the interval $0 \le x \le c$; then an application of the Riemann-Lebesgue lemma shows that

$$\lim_{\omega\to\infty} \int_0^b f(x) \frac{\sin(\omega x)}{x} dx = \lim_{\omega\to\infty} \int_0^c f(x) \frac{\sin(\omega x)}{x} dx, \tag{13}$$

and in addition, we can use the second mean value theorem to write

$$\begin{aligned}\int_0^c f(x) \frac{\sin(\omega x)}{x} dx &= f(0+) \int_0^c \frac{\sin(\omega x)}{x} dx \\ &\quad + \int_0^c [f(x)-f(0+)]\frac{\sin(\omega x)}{x} dx \\ &= f(0+) \int_0^c \frac{\sin(\omega x)}{x} dx \\ &\quad + [f(c)-f(0+)] \int_h^c \frac{\sin(\omega x)}{x} dx,\end{aligned} \tag{14}$$

where $0 < h < c$. It is a standard result that the integral

$$\int_0^\infty \frac{\sin x}{x} dx \tag{15}$$

is convergent, and has the value $\pi/2$. Consequently, the expression

$$\left| \int_h^c \frac{\sin(\omega x)}{x} dx \right| \tag{16}$$

is bounded by some constant M, so that we can write

$$\begin{aligned}&\lim_{\omega\to\infty} \int_0^c f(x) \frac{\sin(\omega x)}{x} dx = \frac{\pi}{2} f(0+) + \varepsilon \\ &|\varepsilon| < |f(c) - f(0+)|M\end{aligned} \tag{17}$$

We can make ε arbitrarily small by choosing c so that

$|f(c)-f(0+)|M$ is sufficiently small, and this does not affect the restriction previously placed on c. Consequently

$$\lim_{\omega\to\infty} \int_0^b f(x) \frac{\sin(\omega x)}{x}\, dx = \frac{\pi}{2} f(0+). \tag{18}$$

By a similar argument, it can be shown that

$$\lim_{\omega\to\infty} \int_{-b}^0 f(x) \frac{\sin(\omega x)}{x}\, dx = \frac{\pi}{2} f(0-). \tag{19}$$

Finally we note that these results are unchanged if b is set to infinity, since the added integral tends to zero in the limit $\omega \to \infty$ by the Riemann-Lebesgue lemma.

2.3. The Inversion Integral

Let $f(x)$ be a function with the Laplace transform $F(p)$ for which the defining integral

$$F(p) = \int_0^\infty f(x)\, e^{-px}\, dx \tag{20}$$

converges in the half plane $\mathrm{Re}(p) > c$. Consider the integral

$$I_R(x) = \frac{1}{2\pi i} \int_{\gamma-iR}^{\gamma+iR} e^{px}\, F(p)\, dp, \quad \gamma > c. \tag{21}$$

We substitute for $F(p)$ the integral (20), and interchange the orders of integration (an operation which is valid because (20) is uniformly convergent with respect to p when $\gamma > c$), to transform (21) to the formula

$$\begin{aligned} I_R(x) &= \int_0^\infty f(y)\, dy\, \frac{1}{2\pi i} \int_{\gamma-iR}^{\gamma+iR} e^{p(x-y)}\, dy \\ &= \frac{1}{\pi} \int_0^\infty f(y)\, e^{\gamma(x-y)} \frac{\sin R(x-y)}{x-y}\, dy \\ &= \frac{1}{\pi} \int_{-x}^\infty f(x+u)\, e^{-\gamma u} \frac{\sin Ru}{u}\, du. \end{aligned} \tag{22}$$

If we break the integral into two, from $-x$ to zero, and zero to infinity, and allow R to become infinite, there are three possibilities; namely,

$$I_R(x) \to \begin{cases} 0 , & x < 0 \\ \frac{1}{2}f(0+), & x = 0 \\ \frac{1}{2}[f(x-0) + f(x+0)], & x > 0. \end{cases} \tag{23}$$

This result is generally known as the inversion theorem for Laplace transforms, and is expressed by the reciprocal pair of equations

$$F(p) = \int_0^\infty f(x)\, e^{-px}\, dx, \quad \mathrm{Re}(p) > c, \tag{24}$$

$$f(x) = \frac{1}{2\pi i} \int_{\gamma-i\infty}^{\gamma+i\infty} F(p)\, e^{px}\, dp, \quad \gamma > c, \tag{25}$$

where $f(x)$ is taken as $\frac{1}{2}[f(x-0) + f(x+0)]$ at a point of discontinuity.

2.4. Inversion of Rational Functions.

In many situations it is necessary to calculate the inverse Laplace transform of a rational function

$$F(p) = \frac{A(p)}{B(p)} , \tag{26}$$

where

$$A(p) = \sum_{i=0}^{m} a_i p^i,$$

$$B(p) = \sum_{i=0}^{n} b_i p^i, \quad (b_n \neq 0), \tag{27}$$

and $n > m$. The need for such inversions arises particularly in the solution of equations with constant coefficients (Section 3), and in techniques of rational approximation (Section 22). We commence with the integral

$$\frac{1}{2\pi i}\int_C \frac{A(p)}{B(p)}\, e^{px}\, dp, \tag{28}$$

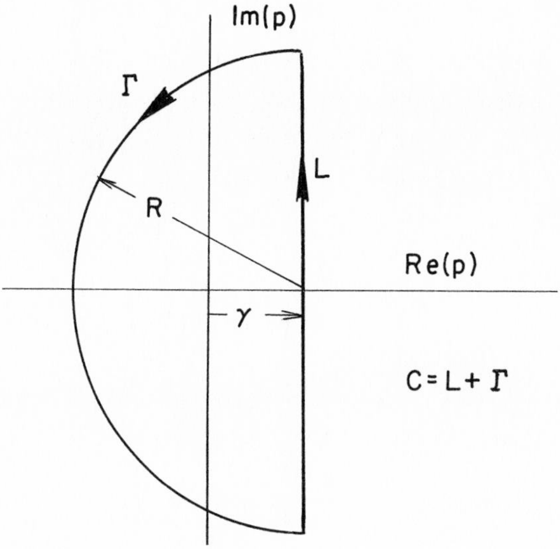

Figure 2.1

where the contour is shown in Figure 1. We are only concerned with the case $x > 0$, and since $n > m$, we can obtain a bound for the integral around the semi-circle by writing

$$p = \gamma - Re^{i\theta}, \quad -\pi/2 \leq \theta \leq \pi/2 \tag{29}$$

and taking R sufficiently large so that $|F(p)| < K/R$. Then

$$\left| \int_\Gamma \frac{A(p)}{B(p)}\, e^{px}\, dp \right| < K \int_{-\pi/2}^{\pi/2} e^{-Rx\cos\theta} d\theta$$

$$= \mathcal{O}(R^{-1}), \quad R \to \infty. \tag{30}$$

Thus the integral (28) is the inversion integral in the limit $R \to \infty$. Since the only singularities of the integrand are poles produced by the zeros of the denominator, we find that the original function $f(x)$ is the sum of the residues at these poles.

The evaluation of these residues can be calculated by decomposing $F(p)$ into partial fractions and then inverting each term. In particular, if the roots $\alpha_1, \alpha_2, \alpha_3, \ldots, \alpha_n$ of $B(p)$ are distinct then we can write immediately

$$F(p) = \sum_{i=1}^{n} \frac{A(\alpha_i)}{B'(\alpha_i)(p-\alpha_i)}$$
$$f(x) = \sum_{i=1}^{n} \frac{A(\alpha_i)}{B'(\alpha_i)} e^{\alpha_i x} \tag{31}$$

obtaining an original function which is a sum of exponentials.

A similar, but more complicated analysis can be made if $B(p)$ has multiple zeros. For simplicity, we concentrate here on the contribution from one such root, as the extension to the general case is trivial in principle but tedious in practice. If α is a root of multiplicity m, then the partial fraction expansion of $F(p)$ will contain the terms

$$\frac{\gamma_1}{(p-\alpha)} + \frac{\gamma_2}{(p-\alpha)^2} + \ldots + \frac{\gamma_m}{(p-\alpha)^m}, \tag{32}$$

and the corresponding contribution to the original function $f(x)$ is

$$\left[\gamma_1 + \frac{\gamma_2}{1!} x + \frac{\gamma_3}{2!} x^2 + \ldots + \frac{\gamma_m}{(m-1)!} x^{m-1}\right] e^{\alpha x}. \tag{33}$$

An Example: If

$$f(p) = \frac{1}{1-p^4} \tag{34}$$

then a straightforward partial fraction decomposition leads to

$$F(p) = \frac{-1}{4(p-1)} + \frac{1}{4(p+1)} - \frac{i}{4(p-i)} + \frac{i}{4(p+i)} \tag{35}$$

and

$$f(x) = \frac{1}{2}\sin x - \frac{1}{2}\sinh x. \tag{36}$$

2.5. Taylor Series Expansion

For small values of x, it may be more appropriate to have an expansion for $f(x)$ as a Taylor series, instead of a cumbersome expression involving a complete knowledge of the roots of the polynomial $B(p)$. Returning to the integral (28), we assume that R is so large that all of the poles lie inside the contour, and then deform the contour to be a circle, centered at the origin, with all of the zeros of $B(p)$ still on the inside. On this circle we may expand $F(p)$ in a convergent power series in inverse powers of p; term by term evaluation of the integral will then give the Taylor series[2] for $f(x)$. An example will demonstrate the technique. Suppose that

$$F(p) = \frac{1}{1+p^2}, \tag{37}$$

then

$$F(p) = \sum_{n=0}^{\infty} \frac{(-1)^n}{p^{2n+2}}, \quad |p| > 1 \tag{38}$$

and term by term inversion on a contour which is a circle of radius more than unity gives

$$f(x) = \sum_{n=0}^{\infty} \frac{(-1)^n x^{2n+1}}{(2n+1)!} \tag{39}$$

$$= \sin x.$$

For rational functions, then, there is a converse to Watson's lemma: an expansion in inverse powers of p which is both asymptotic and convergent for large p implies an expansion of the original function in powers of x, again both asymptotic and convergent. In the example chosen here (37-39) all of the series are elementary and can be written out in full; however, knowledge of the first few terms of the one expansion is sufficient to construct the first few terms of the other in more difficult problems.

Problems

1. Show that if $f(x)$ has derivatives up to $f^{(n)}(x)$, and if $f^{(n)}(x)$ is absolutely integrable, then

$$\lim_{\omega\to\infty} \omega^n \int_a^b f(x)\, e^{i\omega x}\, dx = 0$$

where a and/or b may be infinite.

Find the inverse Laplace transforms of the following functions using the inversion integral.

2. $\dfrac{\lambda p+\mu}{(p+a)^2}$

3. $\dfrac{\lambda p+\mu}{(p+a)(p+b)}$

4. Show that if $F(p)$ has the expansion

$$F(p) = \sum_{n=0}^{\infty} \frac{a_n}{p^{n+1}}$$

which is convergent for $|p| > R$, then the inverse function has the power series expansion

$$f(t) = \sum_{n=0}^{\infty} \frac{a_n}{n!} t^n .$$

Footnotes

1. For a thorough treatment of the material in Sections 2.1-2.3, see, for example, APOSTOL (1957), Ch. 15.

2. Often known as the Heaviside series expansion. See Section 6.5 for the general case.

§3. ORDINARY DIFFERENTIAL EQUATIONS[1]

3.1. First and Second Order Differential Equations

Linear differential equations with constant coefficients are an important area of application of the Laplace transform. As a prelude to the discussion of such problems we discuss first two particularly simple examples, since the connection with the classical methods of solution is readily apparent in these cases.

First Order Equations: Consider the initial value problem

$$\begin{aligned} y'(t) + by(t) &= f(t), \quad t > 0, \\ y(0) &= y_0, \end{aligned} \tag{1}$$

which can be solved by using the integrating factor $\exp(bt)$ to give

$$y(t) = y_0 \, e^{-bt} + \int_0^t e^{b(\tau - t)} \, f(\tau) \, d\tau. \tag{2}$$

Now we take the Laplace transform of (1); after applying (1.15) we have

$$[p \, Y(p) - y_0] + b \, Y(p) = F(p), \tag{3}$$

which is an algebraic equation. $Y(p)$ is found immediately, viz.,

$$\begin{aligned} Y(p) &= G(p)[y_0 + F(p)], \\ G(p) &= [p + b]^{-1}, \end{aligned} \tag{4}$$

and this is obviously equivalent to the classical solution (2) since $G(p)F(p)$ is the transform of a convolution, and $[p + b]^{-1}$ is the transform of $\exp(-bt)$. The advantage of the Laplace transform over the classical method is not apparent from this simple example, however, it is interesting to see how (4) gives a different emphasis from (2). In particular, the function $G(p)$, which contains information about the analytical behavior of the solution, plays a prominent role, while the initial value, which is no more important than the function $f(t)$, enters on an equal footing with that function and is incorporated from the outset.

Second Order Equations: Now we consider the second order initial value problem

$$\begin{aligned} y''(t) + by'(t) + cy(t) &= f(t), \quad t > 0, \\ y(0) &= y_0, \\ y'(0) &= v_0. \end{aligned} \tag{5}$$

This equation arises in many elementary applications which may be found in standard texts.[2] If we take the Laplace transform, and again use (1.15), we obtain

$$[p^2Y(p) - py_0 - v_0] + b[pY(p) - y_0] + cY(p) = F(p). \tag{6}$$

The equation for $Y(p)$ is algebraic, and can be solved immediately to give

$$\begin{aligned} Y(p) &= G(p)[(p+b)y_0 + v_0 + F(p)], \\ G(p) &= [p^2+bp+c]^{-1}. \end{aligned} \tag{7}$$

Once again, inversion gives the solution as a term depending on the initial conditions plus a convolution integral. In some cases it is more convenient to invert the function $G(p)F(p)$ directly, rather than write it as a convolution and evaluate the latter; nevertheless, the general form of the solution is important for understanding the role of $G(p)$. An analysis of (7) depends on factoring the quadratic expression p^2+bp+c; two different cases emerge:

<u>Unequal Roots</u>: If $p^2 + bp + c = (p-\alpha_1)(p-\alpha_2)$ with $\alpha_1 \neq \alpha_2$, then we can write

$$G(p) = \frac{1}{\alpha_1-\alpha_2}\left[\frac{1}{p-\alpha_1} - \frac{1}{p-\alpha_2}\right],$$

$$Y(p) = \frac{1}{\alpha_1-\alpha_2}\left[\frac{(\alpha_1+b)y_0+v_0}{p-\alpha_1} - \frac{(\alpha_2+b)y_0+v_0}{p-\alpha_2}\right] + G(p)F(p). \tag{8}$$

Inversion of the various terms then gives

$$y(t) = [(\alpha_1+b)y_0 + v_0]g_1(t) + [(\alpha_2+b)y_0 + v_0]g_2(t) + \int_0^t g(t-\tau)\, f(\tau)\, d\tau,$$

$$g_1(t) = \frac{1}{\alpha_1-\alpha_2} e^{\alpha_1 t},$$

$$g_2(t) = \frac{1}{\alpha_2-\alpha_1} e^{\alpha_2 t}, \tag{9}$$

$$g(t) = g_1(t) + g_2(t).$$

<u>Equal Roots</u>: If $p^2 + bp + c = (p-\alpha)^2$, then we have

$$G(p) = \frac{1}{(p-\alpha)^2},$$

$$Y(p) = \frac{y_0}{p-\alpha} + \frac{(\alpha+b)y_0 + v_0}{(p-\alpha)^2} + G(p)F(p), \tag{10}$$

and inversion gives

$$y(t) = y_0\, g(t) + [(\alpha+b)y_0 + v_0]\, h(t) + \int_0^t h(t-\tau)\, f(\tau)\, d\tau,$$

$$g(t) = e^{\alpha t}, \quad h(t) = t\, e^{\alpha t}. \tag{11}$$

General Features: Two important features are implicit in these results. First, information on the behavior of $f(t)$ for large t is contained in the function $G(p)$, since its poles determine both the functions which determine the dependence on initial conditions, and the influence function which is used in the convolution integral. If both poles of $G(p)$ have negative real part, then the influence of the initial conditions dies out for increasing time. Also in this case the influence of the function f at time τ on y at time t $(t > \tau)$ diminishes as t becomes large. The second feature of note is that $y(t)$ is continuous even if $f(t)$ has discontinuities. This property follows from the fact that $y(t)$ depends on $f(t)$ via an integral with a finite integrand, so that it is automatically continuous.

3.2. Higher Order Differential Equations

The analysis of the nth order differential equation

$$a_n y^{(n)}(t) + a_{n-1} y^{(n-1)}(t) + \dots + a_0 y(t) = f(t), \quad t > 0 \tag{12}$$

proceeds in a similar fashion. The Laplace transform of (12) gives the algebraic equation

$$\sum_{k=1}^{n} a_k [p^k\, Y(p) - \sum_{\ell=0}^{k-1} p^{k-\ell-1}\, y^{(\ell)}(0)] + a_0 Y(p) = F(p) \tag{13}$$

which can be reversed to give

$$
\begin{aligned}
Y(p) &= G(p)\,[H(p) + F(p)], \\
G(p) &= \left[\sum_{k=0}^{n} a_k p^k\right]^{-1}, \\
H(p) &= \sum_{\ell=0}^{n-1} y^{(p)}(0) \sum_{k=\ell+1}^{n} a_k\, p^{k-\ell-1},
\end{aligned}
\tag{14}
$$

where we have defined a polynomial $H(p)$ which contains all the information about initial conditions. A formal solution to (12) can now be found by inverting the functions $G(p)$ and $G(p)H(p)$, namely

$$
\begin{aligned}
y(t) &= \mathscr{L}^{-1}[GH] + \int_0^t g(t-\tau)\, f(\tau)\, d\tau, \\
g(t) &= \mathscr{L}^{-1}[G],
\end{aligned}
\tag{15}
$$

which exactly parallels the solutions given in (2), (9), and (11) for first and second order equations.

Stability: The inversions involved in (15) are of rational functions, and have been considered in Section 2.4. A most important question is that of the stability of the solution, that is, whether the function $y(t)$ increases without bound for large time without a corresponding increase in the driving function $f(t)$ to cause this behavior. This asymptotic behavior depends solely on the position of the poles of $G(p)$ in the complex plane, for if we turn off the driving force at some time $T > 0$, then we can write (15) as

$$
y(t) = \mathscr{L}^{-1}[GH] + \int_0^T g(t-\tau)\, f(\tau)\, d\tau. \tag{16}
$$

We know from Section 2.4 that the inversions of $G(p)$ and $G(p)H(p)$ have the general form

$$\mathscr{L}^{-1}[G] = \sum_{i,j} g_{ij}\, t^i\, e^{\alpha_j t},$$
$$\mathscr{L}^{-1}[GH] = \sum_{i,j} h_{ij}\, t^i\, e^{\alpha_j t}, \qquad (17)$$

where the poles of $G(p)$ are at $p = \alpha_j$, and on substitution into (16) it is readily seen that the large time behavior of the contribution of the pole at α_k is determined as follows:

$\mathrm{Re}\,(\alpha_k) < 0$,	exponentially damped;
$\mathrm{Re}\,(\alpha_k) > 0$,	exponentially growing;
$\mathrm{Re}\,(\alpha_k) = 0$,	bounded if the root is simple, otherwise unbounded.

For stability, we want the solution to remain bounded, hence all the poles must be in the left-hand half-plane, except possibly for simple poles on the imaginary axis.

Transfer Function: Our analysis of (12) shows that $G(p)$ plays a central role in determining $y(t)$. In many physical applications, the function $f(t)$ represents an input to a system, and the corresponding response is measured by $y(t)$. The relation

$$Y(p) = G(p)\, F(p) \qquad (18)$$

can be represented diagrammatically as

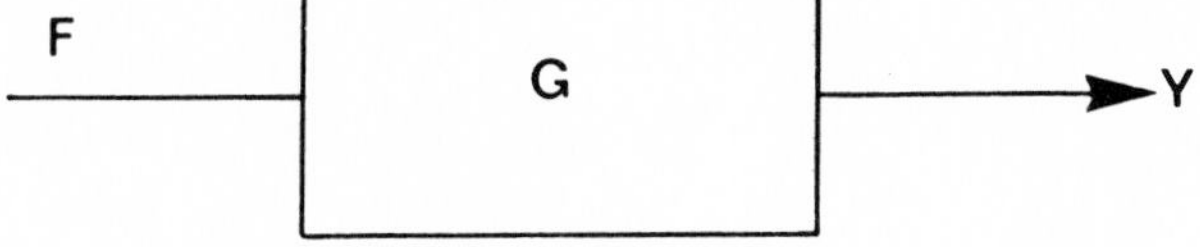

Figure 3.1

where the box labelled G represents a linear system described mathematically by G. G is known as the transfer function in this context, and this is a most important concept in the analysis of linear systems. Suppose now that the function $y(t)$ is used as the forcing function in another system. Then we have, for the second system,

$$\begin{aligned} Y_1(p) &= G_1(p)\ Y(p) \\ &= G_1(p)\ G(p)\ F(p), \end{aligned} \tag{19}$$

and this may be represented by the diagram

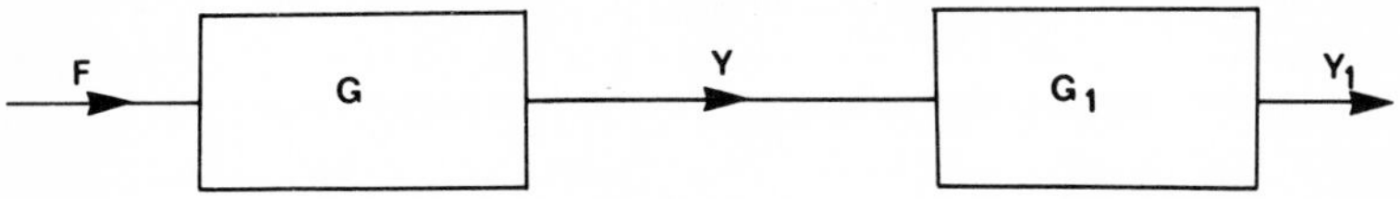

Figure 3.2

In the analysis of more complicated systems, this method of representation leads to a very clear formulation of the overall problem. For details the reader should consult specialized books on the subject.[3]

<u>An Example</u>: If

$$\begin{aligned} &y^{(4)}(t) + 4y(t) = \sin t, \\ &y^{(3)}(0) = y^{(2)}(0) = y^{(1)}(0) = y(0) = 0, \end{aligned} \tag{20}$$

then we get

$$\begin{aligned} Y(p) &= G(p)\ \frac{1}{p^2+1}\ , \\ G(p) &= \frac{1}{p^4+4}\ . \end{aligned} \tag{21}$$

Now we can factor $Y(p)$ as

$$Y(p) = \frac{-i}{4(p-i)} + \frac{i}{4(p+i)} + \frac{3-5i}{32(p-1-i)} + \frac{3+5i}{32(p-1+i)} - \frac{3+5i}{32(p+1-i)} - \frac{3-5i}{32(p+1+i)}, \tag{22}$$

from which we obtain $y(t)$ as

$$y(t) = \frac{1}{2}\sin t + \frac{1}{16}e^{t}(3\cos t + 5\sin t) - \frac{1}{16}e^{-t}(3\cos t - 5\sin t). \tag{23}$$

Note the fact that $y(t)$ grows exponentially for large t, a fact which is evident from the factorization

$$\begin{aligned}(p^4+4) &= (p^2 + 2i)(p^2 - 2i) \\ &= (p-1+i)(p+1-i)(p-1-i)(p+1+i).\end{aligned} \tag{24}$$

3.3. Simultaneous Differential Equations

As we have shown, the Laplace transform is an effective method for dealing with the solution of a single differential equation with constant coefficients. However, the full power and elegance of the method only become apparent when it is applied to a system of simultaneous differential equations. Moreover, it is possible to gain an insight into problems with a comparatively small amount of calculation, especially as compared to the classical methods. We consider first two examples to illustrate what can happen.

Example 1: Consider the mechanical system shown schematically in Figure 3. Two springs of negligible mass and spring constant k support two masses of equal mass m in the manner shown. The bottom mass is attached to a linear damping device which applies a resistance proportional to velocity.

If we denote the (vertical) displacements of the upper and lower masses by $y_1(t)$ and $y_2(t)$, respectively, taking a downward displacement as positive, we can write the equations of motion in the form

$$\begin{aligned} my_1'' + ky_1 - k(y_2-y_1) &= 0, \\ my_2'' + cy_2' - k(y_2-y_1) &= 0. \end{aligned} \tag{25}$$

After taking the Laplace transform and rearranging, the equations become

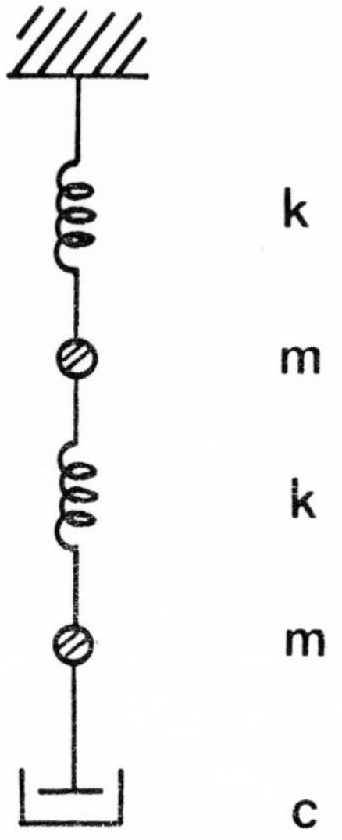

Figure 3.3

$$\begin{aligned} (p^2+2\omega^2)Y_1(p) - \omega^2 Y_2(p) &= py_1(0) + y_1'(0), \\ -\omega^2 Y_1(p) + (p^2+\gamma p+\omega^2)Y_2(p) &= (p+\gamma)y_2(0) + y_2'(0), \end{aligned} \tag{26}$$

where $\omega^2 = k/m$ and $\gamma = c/m$. Solutions for $Y_1(p)$ and $Y_2(p)$ are readily found by elimination, viz.,

$$\begin{aligned} Y_1(p) &= G(p)H_1(p), \\ Y_2(p) &= G(p)H_2(p), \\ G(p) &= [(p^2+\omega^2)(p^2+\gamma p+\omega^2)-\omega^4]^{-1}, \\ H_1(p) &= [p^2+\gamma p+\omega^2][py_1(0) + y_1'(0)] \\ &\qquad + \omega^2[(p+\ )y_2(0) + y_2'(0)], \\ H_2(p) &= \omega^2[py_1(0) + y_1'(0)] \\ &\qquad + [p^2+2\omega^2][(p+\gamma)y_2(0) + y_2'(0)]. \end{aligned} \tag{27}$$

Now the analysis of this expression may proceed analogously to that of the single fourth order equation; for simplicity we assume here that the damping is small, so that we can write

$$
\begin{aligned}
G(p) &= (p^2+\Gamma_1 p+\Omega_1{}^2)^{-1}(p^2+\Gamma_2 p+\Omega_2{}^2)^{-1}, \\
\Omega_1{}^2 &\simeq \tfrac{1}{2}\,\omega^2\,(3+\sqrt{5}) \simeq 2.62\,\omega^2, \\
\Omega_2{}^2 &\simeq \tfrac{1}{2}\,\omega^2\,(3-\sqrt{5}) \simeq 0.38\,\omega^2, \\
\Gamma_1 &\simeq \tfrac{1}{2}\,\gamma\,(1-1/\sqrt{5}) \simeq 0.27\,\gamma, \\
\Gamma_2 &\simeq \tfrac{1}{2}\,\gamma\,(1+1/\sqrt{5}) \simeq 0.72\,\gamma.
\end{aligned}
\tag{28}
$$

Each of the quadratic factors has a pair of complex conjugate roots with negative real part, hence the system is stable. Also, the polynomials $H_1(p)$ and $H_2(p)$ are of lower degree than $1/G(p)$, so we can find partial fraction decompositions for $Y_1(p)$ and $Y_2(p)$. As a simple example, we consider the solution for $y_1(t)$ when the initial conditions are $y_1(0) = y_1'(0) = y_2(0) = 0$. Then we have

$$
\begin{aligned}
Y_1(p) &= \omega^2 y_2'(0)G(p) \\
&\simeq \frac{\omega^2 y_2'(0)}{\Omega_2{}^2 - \Omega_1{}^2}\left[\frac{i}{2(p+\frac{1}{2}\Gamma_1 - i\Omega_1)} - \frac{i}{2(p+\frac{1}{2}\Gamma_1 + i\Omega_1)}\right. \\
&\qquad \left. - \frac{i}{2(p+\frac{1}{2}\Gamma_2 - i\Omega_2)} + \frac{i}{2(p+\frac{1}{2}\Gamma_2 + i\Omega_2)}\right],
\end{aligned}
\tag{29}
$$

$$
y_1(t) \simeq \frac{y_2'(0)}{2\sqrt{5}}\,[e^{-\frac{1}{2}\Gamma_2 t}\sin(\Omega_2 t) - e^{-\frac{1}{2}\Gamma_1 t}\sin(\Omega_1 t)],
$$

where we have again used the conditions $\Gamma_1 \ll \Omega_1$, $\Gamma_2 \ll \Omega_2$ to find the poles of $G(p)$. Solutions for arbitrary initial conditions can be found by similar algebraic manipulations.

<u>Example 2</u>: We consider the electrical circuit shown

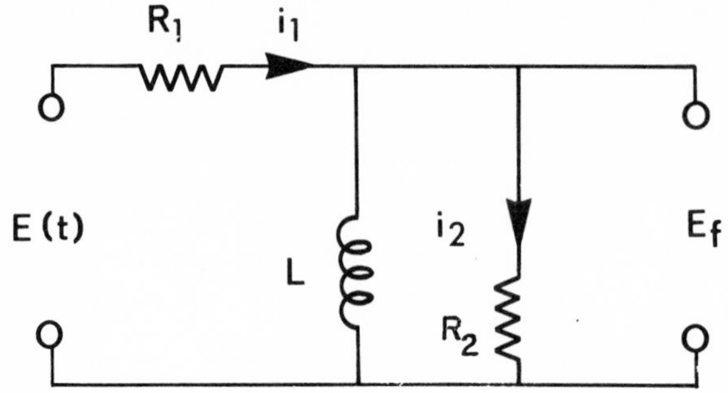

Figure 3.4

in Figure 4. Here we want to determine the voltage E_f across R_2 from a knowledge of the input voltage $E(t)$ and initial conditions. E_f is equal to $i_2 R_2$, and must also equal the voltage across L, which is $L(i_1' - i_2')$, since $i_1 - i_2$ is the current through L. A further equation comes from the fact that E is the sum of E_f and $i_1 R_1$. Putting these facts down, we have

$$\begin{aligned} E &= i_1 R_1 + E_f, \\ E_f &= L(i_1' - i_2') \\ &= i_2 R_2. \end{aligned} \tag{30}$$

Now we introduce the notation $y_1 = i_1$, $y_2 = i_2$, and eliminate E_f from the problem in favor of i_2. Then (30) becomes the simultaneous pair of equations

$$\begin{aligned} R_1 y_1 + R_2 y_2 &= E, \\ L y_1' - L y_2' - R_2 y_2 &= 0. \end{aligned} \tag{31}$$

On taking the Laplace transform, and writing $\bar{E}(p) = \mathscr{L}[E]$, we have

$$\begin{aligned} R_1 Y_1(p) + R_2 Y_2(p) &= \bar{E}(p), \\ pL\, Y_1(p) - (pL+R_2)\, Y_2(p) &= L[y_1(0) - y_2(0)]. \end{aligned} \tag{32}$$

These equations can be solved for Y_1 and Y_2 to give

$$\begin{aligned} Y_1(p) &= G(p)[H_1(p) + (pL+R_2)\bar{E}(p)], \\ Y_2(p) &= G(p)[H_2(p) + pL\,\bar{E}(p)], \\ G(p) &= [pL(R_1+R_2) + R_1R_2]^{-1}, \\ H_1(p) &= R_2L[y_1(0) - y_2(0)], \\ H_2(p) &= -R_1L[y_1(0) - y_2(0)]. \end{aligned} \tag{33}$$

This time we have a solution in which new features appear, which becomes evident if we try to write y_1 or y_2 as the inversion of the initial value term plus a convolution involving $E(t)$. For y_1, this convolution would be

$$\begin{aligned} &\int_0^t g_1(t-\tau)\,E(\tau)\,d\tau, \\ &g_1(t) = \mathscr{L}^{-1}[(pL + R_2)\,G(p)], \end{aligned} \tag{34}$$

but the inversion integral for $g_1(t)$ does not exist in the classical sense. The most straightforward way out of this problem is to rearrange (33) by writing $(pL+R_2)G(p)$ and $pLG(p)$ as quotients plus remainders, where each remainder is a fraction with the denominator of higher order in p than the numerator. Explicitly,

$$\begin{aligned} Y_1(p) &= \frac{\bar{E}(p)}{R_1+R_2} + G(p)\left[H_1(p) + \frac{R_2^{\,2}}{R_1+R_2}\,\bar{E}(p)\right], \\ Y_2(p) &= \frac{\bar{E}(p)}{R_1+R_2} + G(p)\left[H_2(p) - \frac{R_1R_2}{R_1+R_2}\,\bar{E}(p)\right]. \end{aligned} \tag{35}$$

Inserting the functions G, H_1, and H_2, we can now write the formal solutions

$$
\begin{aligned}
y_1(t) &= \frac{E(t)}{R_1+R_2} + \frac{R_2[y_1(0) - y_2(0)]}{R_1+R_2} e^{-\alpha t} \\
&\quad + \frac{R_2^{\,2}}{(R_1+R_2)^2 L} \int_0^t e^{-\alpha(t-\tau)} E(\tau)\, d\tau, \\
y_2(t) &= \frac{E(t)}{R_1+R_2} - \frac{R_1[y_1(0) - y_2(0)]}{R_1+R_2} e^{-\alpha t} \\
&\quad - \frac{R_1 R_2}{(R_1+R_2)^2 L} \int_0^t e^{-\alpha(t-\tau)} E(\tau)\, d\tau, \\
\alpha &= \frac{R_1 R_2}{(R_1+R_2) L}.
\end{aligned} \tag{36}
$$

This solution differs from previous solutions which we have discussed in two important respects. First, it will not necessarily satisfy the initial conditions which we try to impose. In fact, if we put $t = 0$ in (36), we have the relation $R_1 y_1(0) + R_2 y_2(0) = E(0)$, which is (31a). So there is no real contradiction; the basic equations imply that the possible initial values of y_1, y_2, and E are related, and the solution is consistent with this restriction. The other interesting new feature is the appearance of $E(t)$ as a component of the solution. This means that if $E(t)$ has a discontinuity at some time, then $y_1(t)$ and $y_2(t)$ will also be discontinuous at that time since the other terms in these functions cannot be discontinuous if $E(t)$ is finite. But this raises another problem, since we have assumed that y_1 and y_2 are differentiable in writing down (31). It is a remarkable fact, which we will consider in more detail in Section 9, that the Laplace transform still handles the situation correctly.

Alternative Formulation: We consider the last example again, this time formulating the problem so that we do not differentiate a discontinuous function. If we define variables $u_1 = i_1$, $u_2 = i_1 - i_2$, then (31) becomes

$$\begin{aligned} R_1 u_1(t) + R_2 u_2(t) &= E(t), \\ L\, u_2'(t) - R_2[u_1(t) - u_2(t)] &= 0. \end{aligned} \tag{37}$$

We can now eliminate u_1 algebraically to get the first order differential equation

$$u_2'(t) + \alpha u_2(t) = \frac{R_2}{(R_1+R_2)L} E(t). \tag{38}$$

However, the output voltage $E_f = R_2 i_2$ is given by

$$E_f(t) = \frac{R_2}{R_1+R_2} E(t) - \frac{R_1 R_2}{R_1+R_2} u_2(t) \tag{39}$$

and this will exhibit the phenomenon of being discontinuous wherever $E(t)$ is discontinuous.

<u>Normal and Anomalous Systems</u>: We now consider the system of differential equations

$$\sum_{j=1}^{n} a_{ij}\, y_j'(t) + \sum_{j=1}^{n} b_{ij}\, y_j(t) = f_i(t), \quad i = 1,2,\ldots,n, \tag{40}$$

where the a_{ij} and b_{ij} are constants. Any set of differential equations with constant coefficients can be reduced to this form without making assumptions of differentiability beyond those implicit in the original set. For example, (25) can be written

$$\begin{aligned} my_3' + 2ky_1 - ky_2 &= 0, \\ cy_2' + my_4' - ky_1 + ky_2 &= 0, \\ y_1' - y_3 &= 0, \\ y_2' - y_4 &= 0, \end{aligned} \tag{41}$$

by introducing two extra variables y_3 and y_4. Now we take the Laplace transform of (40), and get

$$\sum_j (a_{ij}p + b_{ij})\, Y_j(p) = F_i(p) + H_i, \qquad H_i = \sum_j a_{ij}\, y_j(0). \tag{42}$$

The equations can be solved for $Y_i(p)$:

$$Y_i(p) = \sum_j G_{ij}(p)[F_j(p) + H_j], \tag{43}$$

where the functions $G_{ij}(p)$ are the elements of the inverse of the matrix C with elements $a_{ij}p + b_{ij}$. Using Cramer's rule, we can express $G_{ij}(p)$ as a ratio of determinants,

$$G_{ij}(p) = (-1)^{i+j}\, |C_{ji}(p)| \,/\, |C|, \tag{44}$$

where C_{ji} is obtained from C by deleting row j and column i. In practice, Cramer's rule is unlikely to prove a viable method of constructing the functions G_{ij}; our interest here, however, is simply to discover their analytic structure. In particular, we are interested to know the order of the polynomials $|C(p)|$ and $|C_{ji}(p)|$. Since they are sums of products of linear factors in p, the maximum order is n for $|C|$ and $n-1$ for $|C_{ji}|$. In fact the coefficient of p^n in $|C|$ is obviously $|A|$, where A is the matrix formed by a_{ij}. This leads to the following distinction: if $|A| \neq 0$, the system is said to be normal; if $|A| = 0$ it is said to be anomalous. For a normal system, the functions $G_{ij}(p)$ are therefore rational functions which can decomposed as partial fractions, exactly as $G(p)$ was decomposed in Section 2.4. Consequently, the inverse transforms $g_{ij}(t)$ may be defined, and the formal solution to (40) is

$$y_i(t) = -\sum_{j=1}^{n} H_j\, g_{ij}(t) + \sum_{j=1}^{n} \int_0^t g_{ij}(t-\tau)\, f_j(\tau)\, d\tau. \qquad (45)$$

From this we see that the solutions are continuous for finite inputs $f_j(t)$, and that no restrictions are placed on the possible initial values by the solution.

Anomalous systems are different, as we have seen in Example 2 above. Some of the functions $G_{ij}(p)$ cannot be inverted as ordinary functions, consequently the solutions may be discontinuous for discontinuous inputs. Moreover, the solution will place certain restrictions on the initial values. The reason for this is not hard to find; if $|A| = 0$ then there are non-trivial solutions of the homogeneous equation

$$\sum_{i=1}^{n} \alpha_i\, a_{ij} = 0, \qquad j = 1,2,\ldots,n. \qquad (46)$$

If we multiply (40) by these coefficients and sum, the derivative terms vanish and we obtain

$$\sum_{i,j=1}^{n} \alpha_i\, b_{ij}\, y_j(t) = \sum_{i=1}^{n} \alpha_i\, f_i(t), \qquad (47)$$

which is a linear relationship between the unknown functions $y_i(t)$ and the inputs $f_i(t)$. One possibility is to use this relationship to eliminate one unknown from (40), which will give a new system of $n-1$ equations. If necessary, this can be repeated until eventually we obtain a normal set.

3.4. Equations with Polynomial Coefficients

The Laplace transform can sometimes be used to obtain solutions of ordinary differential equations with non-constant coefficients, as we now show in connection with Bessel functions. Bessel's equation for functions of order ν is

$$J_\nu''(x) + \frac{1}{x} J_\nu'(x) + \left(1 - \frac{\nu^2}{x^2}\right) J_\nu(x) = 0. \tag{48}$$

Near the origin, the two linearly independent solutions of this equation have the asymptotic form $x^{\pm\nu}$, except when $\nu = 0$, in which case[4] the second solution behaves like $\ell n\ x$. We will consider only the functions of the first kind, defined by

$$J_\nu(x) \sim \frac{(x/2)^\nu}{\nu!}, \quad x \to 0, \tag{49}$$

and for this purpose we make the substitution $J_\nu(x) = x^{-\nu} f_\nu(x)$, leading to the new differential equation

$$x\, f_\nu''(x) - (2\nu-1)\, f_\nu'(x) + x\, f_\nu(x) = 0. \tag{50}$$

On taking the Laplace transform, this second order equation for $f_\nu(x)$ is converted into a first order differential equation for $F_\nu(p)$, namely

$$(1+p^2)F_\nu'(p) + (2\nu+1)p\, F_\nu(p) = 0. \tag{51}$$

The point of the substitution of $f_\nu(x)$ for $J_\nu(x)$ is that this equation does not depend on $f_\nu(0)$ or $f_\nu'(0)$. The general solution of (51) is

$$F_\nu(p) = \frac{A_\nu}{(1+p^2)^{\nu+1/2}}. \tag{52}$$

The constants A_ν can be fixed by appealing to the relationship between the asymptotic forms of $f_\nu(x)$ for small x, and $F_\nu(p)$ for large p [equations (1.42) and (1.43)], giving for $F_\nu(p)$

$$F_\nu(p) = \frac{2^\nu(\nu-\frac{1}{2})!}{\sqrt{\pi}(1+p^2)^{\nu+1/2}}. \tag{53}$$

Some discussion of the inversion of this Laplace transform may be found in Section 6.

More General Applications: Direct use of the Laplace transform on equations such as (48) depends on first finding the asymptotic form of the solution near the origin, and then using this information in such a way that unknown quantities such as $f(0)$ are eliminated from the transformed equation. Since these steps are only intermediate, the final result being a particular integral representation of the desired solution, it is better to write down the solution as an integral from the outset. This approach is generally known as Laplace's method, and is the subject of Section 20.

Problems

In Problems 1-11, solve the given differential equations. If initial conditions are not stated, take them as arbitrary.

1. $y' + y = 1, \quad y(0) = 2.$
2. $y'' + \omega^2 y = \cos(\nu t), \quad \nu \neq \omega.$
3. $y'' + y = \sin(\omega t).$
4. $y'' + 4y' + 8y = 1, \quad y(0) = y'(0) = 0.$
5. $y''' + y = 1, \quad y(0) = y'(0) = y''(0) = 0.$
6. $y''' + y = t, \quad y(0) = y'(0) = y''(0) = 0.$

7. $y' - z = -t^2,$

$y - z' = 2t\, e^{-t}.$

8. $x' = y,$

$y' = z,$

$z' = x.$

9. $y'' + 2z = 0,$

$y - 2z' = 0.$

10. $x'' + ay' - bx = 0,$

$y'' - ax' - by = 0.$

11. $x_n' + \alpha(x_n + x_{n-1}) = 0, \quad n \geq 1,$

$x_0' + \alpha x_0 = 0,$

$x_n(0) = 0, \quad n \geq 1,$

$x_0(0) = 1.$

12. A constant voltage E_0 is applied from time $t = 0$ to a circuit consisting of a resistor R and capacitor C in series. The charge on C is initially zero. Find an expression for the current $i(t)$. Is it possible to specify $i(0)$ arbitrarily?

13. A constant voltage E_0 is applied from time $t = 0$ to a circuit consisting of a resistor R, a capacitor C, and an inductance L in series. Find an expression for the charge on C for $t > 0$.

14. For the same circuit as in Problem 13, find the charge on C if the applied voltage is $E_0 \sin(\omega t)$.

15. Express the solution to Problem 13 as a convolution if the applied voltage is an arbitrary function $E(t)$.

16. Two flywheels of moment of inertia I_1 and I_2 are coupled by an elastic shaft of stiffness k. The first flywheel is coupled to a motor which applies a constant torque from $t = 0$, the second drives a load which applies a torque proportional to the angular velocity. Describe the motion of the system. (Neglect the moment of inertia of the shaft.)

17. We wish to find the motion of a particle projected from a fixed point on the earth's surface. Let the origin of the coordinates be 0, the x-axis point east, the y-axis north, and the z-axis opposite to the acceleration due to gravity. Then the equations of motion are

$$x'' = 2\omega(y' \sin \lambda - z' \cos \lambda),$$

$$y'' = -2\omega\, x' \sin \lambda,$$

$$z'' = g + 2\omega\, x' \cos \lambda,$$

where λ is the latitude of 0, and ω the angular velocity of the earth. Find the solution for arbitrary initial values of x', y', and z'.

18. A motor produces a torque proportional to a control voltage fed into a regulator. It is coupled to a flywheel of moment of intertia I which drives a load exerting a constant retarding torque N. If the regulator control voltage is proportional to $\omega - \omega_0$, where ω is the angular speed and ω_0 a constant, find an expression for $\omega(t)$. Assume that $\omega(0) = 0$.

19. Suppose that in Problem 16 the torque supplied by the motor is proportional to a control voltage fed into a regulator. Let the voltage be proportional to $\omega-\omega_0$, where ω is the angular velocity of the second flywheel. Discuss the stability of the system. (Hint: If the polynomial $a_n p^n + a_{n-1}p^{n-1} + \dots + a_0$ has real coefficients, then the ratio $-a_{n-1}/a_n$ is equal to the sum of the real parts of the roots).

Footnotes

1. A thorough treatment of the material in this section may be found in DOETSCH (1971), Ch. 3.

2. A very large number of applications may be found in THOMPSON (1957), Ch. 3.

3. See, for example, KAPLAN (1962) and VAN DER POL & BREMMER (1955), Ch. 8.

4. Logarithmic terms appear in the second solution whenever ν is an integer. See Section 20.6.

§4. PARTIAL DIFFERENTIAL EQUATIONS

4.1. The Diffusion Equation

As an example to show how the Laplace transform may be applied to the solution of partial differential equations, we consider the diffusion of heat in an isotropic solid body. For temperature gradients which are not too large, it is a good approximation to assume that the heat flow is proportional to the temperature gradient, so that the temperature $u(\underset{\sim}{r},t)$ must satisfy the partial differential equation

$$\kappa \frac{\partial u}{\partial t} = \nabla^2 u, \tag{1}$$

where κ is a constant given by $\kappa = \rho c/k$, and ρ is the density, c the specific heat, and k the thermal conductivity of the material. (We have assumed that ρ, c, and k are all constants.) A method of solution of (1) for particular problems is illustrated by the following two examples.[1]

Semi-infinite Region: Suppose that the body fills the region $x \geq 0$, and that u is a function only of x and t. We want to find $u(x,t)$ if the temperature is initially equal to T_0, and the plane $x = 0$ is held at the temperature T_1. For this problem the diffusion equation (1) reduces to the form

$$\frac{\partial u}{\partial t} = \frac{1}{\beta^2}\frac{\partial^2 u}{\partial x^2}, \quad \beta^2 = \kappa. \tag{2}$$

By analogy with Section 3, where we showed that the Laplace transform reduces the problem of solving ordinary differential equations with constant coefficients to an algebraic problem, we might expect that a similar method applied to a partial differential equation would afford some simplification. We introduce the Laplace transform with respect to time

$$U(x,p) = \int_0^\infty e^{-pt} u(x,t)\, dt \tag{3}$$

and take the transform of (2), to get

$$pU(x,p) - T_0 = \frac{1}{\beta^2} \frac{d^2}{dx^2} U(x,p). \tag{4}$$

Thus the problem will be solved if we solve the ordinary differential equation (4) subject to appropriate boundary conditions. The boundary conditions on $u(x,t)$ are: (i) $u(0,t) = T_1$ and (ii) $u(x,t)$ remains finite as x goes to infinity. The boundary conditions on $U(x,p)$ are obtained by taking the transforms of these, so we have

$$U(0,p) = \frac{T_1}{p}, \tag{5}$$

$$U(x,p) \text{ finite}, \quad x \to \infty,$$

and the unique solution of (4) subject to these restrictions is

$$U(x,p) = \frac{T_0}{p} + \frac{T_1 - T_0}{p} e^{-\beta x\sqrt{p}}. \tag{6}$$

To complete the solution we need the inverse Laplace transform of (6), which we obtain in Section 6.2. The result is

$$u(x,t) = T_0 + (T_1 - T_0) \operatorname{erfc} (\beta x/2\sqrt{t}), \tag{7}$$

where the function erfc (x), known as the complementary error function, is given by the integral

$$\operatorname{erfc}(x) = \frac{2}{\sqrt{\pi}} \int_x^\infty e^{-u^2} du. \tag{8}$$

Infinite Slab: We consider the case when the body fills the region $0 \leq x \leq \ell$ and is initially at temperature T_0. One face $(x = 0)$ is maintained at this temperature, while the second face is supplied with heat from time $t = 0$ at a constant rate H. We want to find the heat flow through the first face as a function of time. Our partial differential equation is again (2), and the Laplace transform is again (4). The general solution of (4) is

$$U(x,p) = \frac{T_0}{p} + A \sinh (\beta x\sqrt{p}) + B \cosh (\beta x\sqrt{p}), \tag{9}$$

where the constants A and B will be determined by the boundary conditions. Since $u(0,t) = 0$ we immediately have $B = 0$. To determine the other constant, and also the heat flow through the face $x = 0$, we must consider the function $q(x,t) = -\kappa\, \partial u/\partial x$, which represents the heat flow at an arbitrary point. From (9), we can write for the transform of q

$$Q(x,p) = C \cosh (\beta x\sqrt{p}), \tag{10}$$

where $C = -\kappa\beta A\sqrt{p}$. The boundary condition at $x = \ell$ is $q(\ell,t) = -H$ or $Q(\ell,p) = -H/p$, which determines C. We require the heat flow through the face $x = 0$; the Laplace transform of this quantity is

$$Q(0,p) = \frac{H}{p \cosh (\beta\ell\sqrt{p})}. \tag{11}$$

This function may be inverted using the methods of Sections 6.2 or 6.1. The results are, respectively,

$$\begin{aligned} q(0,t) &= 2H \sum_{r=1}^{\infty} (-1)^{r-1} \operatorname{erfc} \{\beta\ell(r-\tfrac{1}{2})/\sqrt{t}\}, \\ &= H[1+ \frac{4}{\pi} \sum_{r=1}^{\infty} \frac{(-1)^r}{(2r-1)} \exp \{-(r-\tfrac{1}{2})^2\, \pi^2 t/\beta^2\ell^2\}]. \end{aligned} \tag{12}$$

The first expansion is useful for computation for small t, while the second is a good expansion for large t. In particular, we see that $q(0,t) \to H$ for large t, as it must.

4.2. Wave Propagation

The simplest continuous vibrational system is a uniform flexible string of mass ρ per unit length, stretched to a tension T. If the string executes small transverse vibrations in a plane, then the displacement $u(x,t)$ must satisfy the partial differential equation

$$\frac{\partial^2 u}{\partial t^2} = a^2 \frac{\partial^2 u}{\partial x^2} + \frac{f(x,t)}{\rho}, \tag{13}$$

where $a^2 = T/\rho$ and $f(x,t)$ is the external force per unit length. In addition, $u(x,t)$ must satisfy boundary conditions, which depend on the manner in which the string is supported. We consider two simple problems involving (13) to further illustrate how the Laplace transform may be applied to such equations.

Semi-infinite String: We consider (13) with $f \equiv 0$ for the region $x \geq 0$, with u and $\partial u/\partial t$ initially zero. For $t \geq 0$, the end $x = 0$ is subjected to the time varying displacement $\phi(t)$. The Laplace transform of (13), together with the boundary condition at $x = 0$, give the equations

$$\begin{aligned} a^2 \frac{d^2U}{dx^2} - p^2U &= 0, \\ U(0,p) &= \Phi(p), \end{aligned} \tag{14}$$

and the solution which is bounded for $x \to \infty$ is simply

$$U(x,p) = \Phi(p)\, e^{-px/a}. \tag{15}$$

The corresponding displacement is easily found using the translation properties of the transform, viz.,

$$u(x,t) = \begin{cases} \phi(t - x/a), & x < at \\ 0 & , \quad x > at. \end{cases} \tag{16}$$

Hence the displacement which is imposed on the end propagates down the string at velocity a.

Finite String: Suppose now we fix the string at $x = \ell$, while still subjecting the end $x = 0$ to an arbitrary time varying displacement. Then we must solve (14) subject to the additional boundary condition $U(\ell,p) = 0$. The solution is easily found to be

$$U(x,p) = \Phi(p) \frac{\sinh[p(\ell-x)/a]}{\sinh[p\ell/a]} . \tag{17}$$

This function may be inverted to find $u(x,t)$ in a manner which describes the physical picture very well: By replacing the hyperbolic sine functions by exponentials and expanding the denominator, we get

$$\begin{aligned} U(x,p) &= \Phi(p)[e^{-px/a} - e^{p(x-2\ell)/a} + e^{-p(x+2\ell)/a} \\ &\qquad - e^{-p(x-4\ell)/a} + \ldots], \\ u(x,t) &= \phi(t-x/a) - \phi(t+(x-2\ell)/a) + \phi(t-(x+2\ell)/a) \\ &\qquad - \phi(t+(x-4\ell)/a) + \ldots, \end{aligned} \tag{18}$$

where we assume in writing the expression for $u(x,t)$ that $\phi(t) = 0$ for $t < 0$. This result represents the propagation of $\phi(t)$ at velocity a while $t < \ell/a$; however, the wave reaches $x = \ell$ at this time and a second term, $-\phi(t+(x-2\ell)/a)$, begins to contribute. This represents the reflection of ϕ at the boundary, with the same amplitude but opposite sign. Further reflections occur at $t = 2\ell/a$, $3\ell/a$, etc., as the disturbance travels up and down the string.

Infinite Transmission Line: Coaxial cables and other electrical transmission lines are systems which are amenable to simple analysis via the Laplace transform. We consider here a rather simple example: a line which has no resistive losses. The line is described by two parameters, the inductance L and capacitance C per unit length. Consider a small length δx of the line (Figure 1).

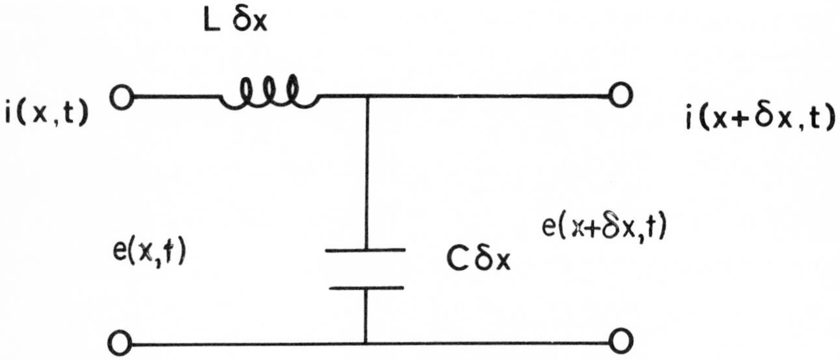

Figure 4.1

By equating the currents and voltages at x and $x + \delta x$ and taking the limit $\delta x \to 0$, we find that they must satisfy the simultaneous partial differential equations

$$\frac{\partial e(x,t)}{\partial x} = - L \frac{\partial i(x,t)}{\partial t} ,$$
$$\frac{\partial i(x,t)}{\partial x} = - C \frac{\partial e(x,t)}{\partial t} . \tag{19}$$

Suppose now that we connect a voltage source $\phi(t)$ at $x = 0$, commencing at $t = 0$. For initial conditions we take $e(x,0) = i(x,0) = 0$. Taking the Laplace transform of

(19), we get

$$\frac{\partial E(x,p)}{\partial x} = - pLI(x,p),$$
$$\frac{\partial I(x,p)}{\partial x} = - pCE(x,p), \tag{20}$$

which must be solved subject to $E(0,p) = \Phi(p)$. The solution which is bounded as $x \to \infty$ is

$$E(x,p) = \Phi(p)\ e^{-px/v},$$
$$v^2 = 1/LC, \tag{21}$$

so that the disturbance $\phi(x,t)$ propagates at velocity v exactly as for an infinite string.

Termination of a Finite Line: Suppose now that the line is of finite length ℓ, terminated by a resistance R (Figure 2).

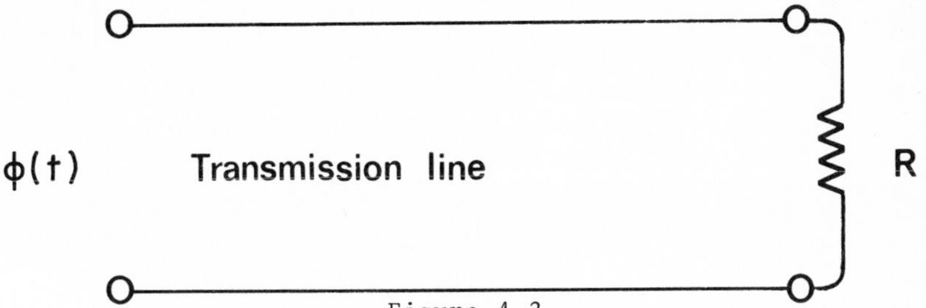

Figure 4.2

Again we apply a voltage $\phi(t)$ at $x = 0$, and look at the way the signal propates from the source to the load R. Equation (20) must be solved subject to the boundary condition $E(\ell,p) = RI(\ell,p)$, which is Ohm's law for the load. The solution now becomes

$$E(x,p) = \Phi(p) \frac{R \cosh [p(\ell-x)/v] + Lv \sinh [p(\ell-x)/v]}{R \cosh [p\ell/v] + Lv \sinh [p\ell/v]}. \tag{22}$$

Now if $R = 0$, we recover (17); the physical interpretation is the same as before, the electric signal being reflected back and forth along the line. The solution for $R = \infty$ is similar; we will not pause to consider the details. From a practical point of view, the purpose of a transmission line is to transfer energy from the source to the load. We therefore ask if R can be chosen so as to eliminate any reflection at $x = \ell$, and it is evident from the form of (22) that the choice $R = Lv = \sqrt{(L/C)}$ is the only one which achieves this aim, since we then have

$$E(x,p) = \Phi(p)\, e^{-px/v}, \tag{23}$$

which is equivalent to (21). It is of interest to calculate in this case the ratio of voltage to current at $x = 0$. Using (23) in (20) we have $E(0,p)/I(0,p) = R$, but only if $R = \sqrt{(L/C)}$. Thus for this particular choice, the system, transmission line plus load, appears to the voltage source to be the load without an intervening line. The input signal is transmitted at velocity v without change of form or loss of energy, and delivered to the load R without reflection. The quantity $R = \sqrt{(L/C)}$ is known as the impedance of the line, and the line is usually referred to as a Heaviside distortionless line.

Problems

1. The plane boundary $(x = 0)$ of a semi-infinite body is maintained at temperature $f(t)$ from $t = 0$. The body is initially at a uniform temperature T_0. Find an expression for the subsequent temperature $u(x,t)$ at each point in the body.

2. Obtain explicit solutions to Problem 1 for the following special cases:

(i) $f(t) = \begin{cases} T_1, & 0 < t < t_0, \\ T_0, & t_0 < t, \end{cases}$

(ii) $f(t) = At$.

3. Solve Problem 1 if, instead of maintaining the boundary $x = 0$ at a fixed temperature, we supply heat at a rate $Q(t)$ per unit area. Give an explicit solution if $Q(t)$ is a constant. Find also an expression for the temperature at $x = 0$.

4. The plane boundary $(x = 0)$ of a semi-infinite body radiates heat according to Newton's law (Q proportional to ΔT, where ΔT is the temperature difference between the body and the medium in contact with it). If the body is initially at uniform temperature T_1 and the other medium is at T_0, find an expression for the temperature distribution of the body at subsequent times.

5. Two semi-infinite bodies, initially at uniform temperatures T_1 and T_2 respectively, are brought into thermal contact at $t = 0$. Describe the subsequent equalization of temperature.

6. A slab of thickness ℓ is initially at temperature T_0. From $t = 0$, one face $(x = 0)$ is held at temperature T_1, the other $(x = \ell)$ at T_2. Find the subsequent temperature distribution. Give forms useful both for small t and large t.

7. Solve Problem 6 if the face $x = 0$ radiates according to Newton's law, while the face $x = \ell$ is held at temperature T_1.

8. Solve Problem 7 if the face $x = \ell$ is supplied with heat at a constant rate Q.

9. A solid sphere of radius a is initially at uniform temperature T_0. From $t = 0$ the surface is kept at temperature T_1. Find an expression for the temperature distribution at subsequent times.

10. An infinite solid has in it a circular cavity of radius a. It is initially at temperature T_0; from $t = 0$ the surface of the cavity is held at temperature T_1. Find the subsequent temperature distribution.

11. Solve Problem 9 if the sphere is surrounded by a medium of temperature T_1 into which it radiates according to Newton's law.

12. A stretched string, fixed at $x = 0$ and $x = \ell$, is plucked at its mid-point and released (from rest) at $t = 0$. Find an infinite series solution for the subsequent motion of the string.

13. Show that

$$\frac{1}{2\pi i}\int_{\gamma-i\infty}^{\gamma+i\infty} e^{a\lambda}\,\frac{d\lambda}{\lambda} = \begin{cases} a, & a > 0 \\ 0, & a \le 0 \end{cases}$$

where $\gamma > 0$. Hence obtain from the Laplace transform solution of Problem 12 a direct picture of the displacement of the string in the time interval $0 \le t \le 2\ell/c$.

14. A capacitor C_0, initially charged to potential E_0, is connected at $t = 0$ to a semi-infinite transmission line with inductance L and capacitance C per unit length. Find the distribution of current at subsequent times.

15. A finite line of length ℓ and parameters L and C is terminated by a resistance R. If the end $x = 0$ is connected to a constant potential E_0 from $t = 0$, show that the potential across the load at subsequent times is given by

$$e(\ell,t) = \begin{cases} 0, & 0 < t < \ell/v \\ E_0\left[1 - \left(\frac{z-R}{z+R}\right)\right]^n, & (2n-1)\ell/v < t < (2n+1)\ell/v \end{cases}$$

$$z = \sqrt{L/C}\,,$$

$$v = 1/\sqrt{LC}\,.$$

16. A finite line of length ℓ and parameters L and C, is connected at $x = 0$ to a potential source $E(t)$ in series with a resistance R. The end $x = \ell$ is open. Find the potential at $x = \ell$ for $t > 0$. Is there any value of R for which transmission takes place without distortion?

17. A line of inductance L and capacitance C per unit length also has resistance R and leakage conductance G (G = the inverse of the leakage resistance) per unit length. If the line is infinite, and if a potential $\phi(t)$ is applied at $x = 0$ from $t = 0$, find an expression for the voltage distribution for $t > 0$. In particular, study the case $RC = LG$.

18. Using residues, find a series solution for problem 15.

19. A line with $RC = LG$, of length ℓ, is short-circuited at $x = \ell$. From $t = 0$ a potential $\phi(t) = A \sin(\omega t)$ is applied at $x = 0$. Find the voltage distribution in a form which explicitly shows the role of reflections at the ends.

20. A constant potential E_0 is applied from $t = 0$ to the end of a semi-infinite cable (a line with $L = G = 0$). Find the voltage at subsequent times.

21. Solve Problem 20 if the cable is finite and terminated at $x = \ell$ by a resistance R_0.

Footnotes

1. Many more examples may be found in CARSLAW & JAEGER (1941), Chs. 5-10.

§5. INTEGRAL EQUATIONS

5.1. Convolution Equations of Volterra Type

Integral equations in which the unknown function appears in a convolution occur in some important situations. The equation

$$g(x) = f(x) + \lambda \int_a^b k(x-y)\, g(y)\, dy, \tag{1}$$

where $f(x)$ and $k(x)$ are given functions and λ a given constant, is an example of a Fredholm integral equation of the second kind. (An equation of the first kind is one in which the unknown function g does not appear outside the integral.) If the upper limit of integration b is replaced by the variable x, then (1) is said to be of Volterra, rather than Fredholm, type. By the change of variables $x' = x-a$, $y' = y-a$, (1) may then be written

$$g(x') = f(x') + \lambda \int_0^{x'} k(x'-y')\, g(y')\, dy'. \tag{2}$$

General Method of Attack: Applying the Laplace transform to (2) leads immediately to the algebraic equation

$$G(p) = F(p) + \lambda\, K(p)\, G(p) \tag{3}$$

with the immediate consequence

$$G(p) = \frac{F(p)}{1-\lambda\, K(p)}, \tag{4}$$

and inversion yields the solution. Another problem of frequent interest in connection with (2) is the determination of the resolvent kernel, that is, the function $\Gamma(t)$ such that

$$g(x) = f(x) + \int_0^x \Gamma(x-y)\, f(y)\, dy. \tag{5}$$

Now (4) may be written

$$G(p) = F(p) + \frac{\lambda K(p)}{1-\lambda K(p)} F(p), \tag{6}$$

so that $\Gamma(t)$ is the inverse Laplace transform of $\lambda K/[1-\lambda K]$. We illustrate these few comments with some examples.

Example 1: Consider the equation

$$s = \int_0^s e^{s-t} g(t)\, dt. \tag{7}$$

The Laplace transformation yields

$$\frac{1}{p^2} = \frac{1}{p-1} G(p), \tag{8}$$

hence

$$G(p) = \frac{p-1}{p^2} = \frac{1}{p} - \frac{1}{p^2}, \qquad g(t) = 1-t. \tag{9}$$

Example 2: Consider the equation

$$g(x) = 1 - \int_0^x (x-y)\, g(y)\, dy. \tag{10}$$

Then

$$G(p) = \frac{1}{p} - \frac{1}{p^2} G(p), \tag{11}$$

which gives for the solution

$$G(p) = \frac{p}{1+p^2}, \qquad g(x) = \cos x . \tag{12}$$

Example 3: Let Γ be the resolvent of the equation

$$g(s) = f(s) + \lambda \int_0^s e^{s-t} g(t)\, dt. \tag{13}$$

Then, denoting the Laplace transform of Γ by $\Omega(p)$, we can easily obtain

$$\Omega(p) = \frac{\lambda}{p-\lambda-1},$$
$$\Gamma(s) = \lambda\, e^{-(\lambda+1)s}. \tag{14}$$

Equation (5) reads in this case

$$g(s) = f(s) + \lambda \int_0^s e^{(\lambda+1)(s-t)} f(t)\, dt. \tag{15}$$

Example 4: A less trivial example is furnished by the equation

$$g(t) = \frac{1}{(1+t)^2} + a \int_0^t \frac{g(s)\, ds}{(1+t-s)^2}, \tag{16}$$

which occurs in the solution of a semi-empirical model for subsidence caused by mining operations.[1] We introduce the notation $k(t) = (1+t)^{-2}$; then for $\operatorname{Re}(p) \geq 0$ we have

$$\begin{aligned} K(p) &= \int_0^\infty \frac{e^{-pt}\, dt}{(1+t)^2} \\ &= 1 - p \int_0^\infty \frac{e^{-pt}\, dt}{(1+t)} \\ &= 1 - p\, e^p \int_p^\infty \frac{e^{-u}\, du}{u}, \end{aligned} \tag{17}$$

where the last step follows from the substitution $u = p(1+t)$. The last integral is the exponential integral (see Appendix C), so

$$K(p) = 1 - p\, e^p E_1(p), \tag{18}$$

which defines K in the entire complex plane cut along the negative real axis. Laplace transformation of (16) now gives

$$G(p) = K(p) + a\, K(p)\, G(p). \tag{19}$$

Evaluation of $g(t)$ will involve numerical approximations to the inversion integral;[2] however, we can deduce some important information, particularly about the behavior of $g(t)$ for large t, from simple analytic information. Consider the inversion integral for $g(t)$,

$$g(t) = \frac{1}{2\pi i} \int_{\gamma - i\infty}^{\gamma + i\infty} \frac{e^{pt} \; K(p)}{1 - a \; K(p)} \, dp, \tag{20}$$

where $\gamma > 0$ is chosen so that the contour is to the right of all zeros of the denominator $1 - a\,K(p)$. Now we know from Appendix C that the exponential integral has a logarithmic branch point at the origin. Furthermore, it may be shown (Problem 6) that the imaginary part of $K(p)$ is not zero except for real positive p, so that $1 - a\,K(p)$ can only have zeros for $p \geq 0$. Again, for real p we have

$$\begin{aligned} K(0) &= 1, \\ K'(p) &= -\int_0^\infty t\, e^{-pt}\, k(t)\, dt < 0, \end{aligned} \tag{21}$$

so that if $a < 1$, $1 - a\,K(p)$ has no zeros; if $a = 1$ the origin is a zero; and if $a > 1$ there is one simple zero for some real positive p_0. We assign the analysis of the special case $a = 1$ to the problems. For $a < 1$, we can deform the contour in (20) to encircle the negative real axis. Using the superscripts $\pm$ to denote the values of a function at $p = -\xi \pm i\varepsilon$, $\xi > 0$, $\varepsilon \to 0$, we obtain in this case

$$\begin{aligned} g(t) &= I(t) \\ &= \frac{1}{2\pi i} \int_0^\infty e^{-\xi t} \, [G^-(-\xi) - G^+(-\xi)] \, d\xi, \end{aligned} \tag{22}$$

and using the properties

$$\begin{aligned} K^+(-\xi) &= [K^-(-\xi)]^*, \\ K^-(-\xi) - K^+(-\xi) &= 2\pi i \; \xi \; e^{-\xi}, \end{aligned} \tag{23}$$

we have

$$I(t) = \int_0^\infty \frac{\xi \, e^{-(1+t)\xi} \, d\xi}{|1 - a\,K^+(-\xi)|^2} > 0. \tag{24}$$

This integral can easily be bounded, since $|1 - a\,K^+|$ must have a minimum value, and replacing the denominator by this value we find that there is some constant A for which

$$I(t) < \frac{A}{(1+t)^2}. \tag{25}$$

Hence the solutions to (16) tend to zero for $a < 1$. If $a > 1$, a similar analysis can be made, except that when we deform the contour we must pick up the residue at the simple pole $p = p_0$. (See Figure 1.) This gives

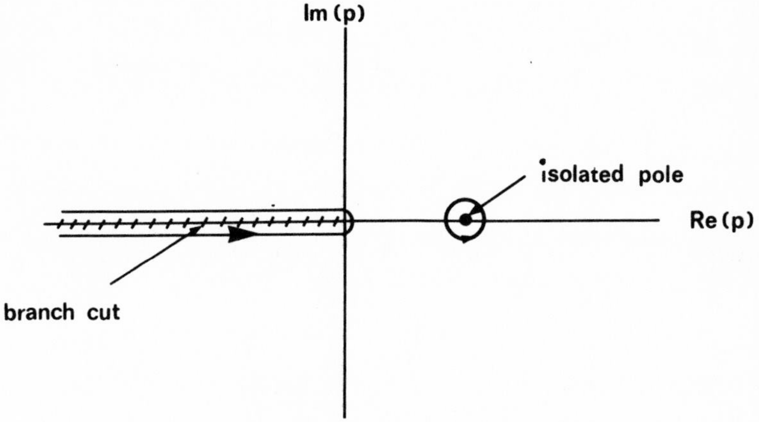

Figure 5.1

$$g(t) = \frac{p_0 \, e^{p_0 t}}{a(p_0 - a - 1)} + I(t), \quad a > 1 \tag{26}$$

showing that the solution is exponentially growing in this case.

5.2. Convolution Equations over an Infinite Range.[3]

We consider the integral equation[4]

$$\begin{aligned} f(x) &= \lambda \int_0^\infty k(|x-y|)\, g(y)\, dy, \quad x \geq 0 \\ k(s) &= e^{-\alpha s} \qquad\qquad\qquad\quad , \quad \alpha > 0. \end{aligned} \tag{27}$$

Taking the Laplace transform of both sides and splitting the integral over y into two, over $y \leq x$ and $y \geq x$, we obtain

$$F(p) = \lambda[G(p)K(p) + \int_0^\infty e^{-px}\, dx \int_x^\infty k(y-x)g(y)\, dy]. \tag{28}$$

The double integral can be rearranged as

$$\int_0^\infty g(y)\, dy \left[\int_{-\infty}^{y} k(y-x)e^{-px}\, dx - \int_{-\infty}^{0} k(y-x)e^{-px}\, dx\right]$$

$$= G(p)\ K(-p) - \int_0^\infty g(y)\, dy \int_0^\infty k(x+y)e^{px}\, dx \tag{29}$$

$$= G(p)\ K(p) + (p-\alpha)^{-1}\ G(\alpha),$$

providing all the integrals converge.[5] Substituting this result into (28) and solving for $G(p)$, we obtain

$$\begin{aligned} G(p) &= \frac{\lambda^{-1}F(p) - (p-\alpha)^{-1}G(\alpha)}{K(p) + K(-p)} \\ &= \frac{\lambda(p+\alpha)G(\alpha) - (p^2-\alpha^2)F(p)}{2\alpha\lambda}. \end{aligned} \tag{30}$$

The appearance of $p^2F(p)$ shows that the solution for $g(x)$ will involve $f''(x)$; hence we rearrange (30) as

$$\begin{aligned} 2\alpha\lambda\ G(p) &= -[p^2F(p) - pf(0) - f'(0)] + \alpha^2F(p) \\ &\quad - p[f(0) - \lambda\ G(\alpha)] - [f'(0) - \alpha\lambda\ G(\alpha)]. \end{aligned} \tag{31}$$

This expression has a convergent inverse only when the terms of the form $Ap + B$ are identically zero, i.e., only if

$$\begin{aligned} f(0) &= \lambda\, G(\alpha), \\ f'(0) &= \alpha\lambda\, G(\alpha). \end{aligned} \tag{32}$$

These restrictions may be obtained directly from the integral equation, so it is not surprising that they occur as necessary conditions for the convergence of the inversion integral. Subject to these restrictions, it follows from (31) that the integral equation has the explicit solution

$$g(x) = \frac{\alpha^2 f(x) - f''(x)}{2\alpha\lambda}, \tag{33}$$

although it must be noted that the solutions to this differential equation are not necessarily related by an integral equation of the type (27); the initial conditions (32) are also needed.

General Considerations: We consider an integral equation of the form

$$a\, g(x) = bf(x) + \lambda \int_0^\infty k(|x-y|)\, g(y)\, dy, \quad x \geq 0 \tag{34}$$

where either a or b may be chosen to be zero. Laplace transformation, followed by rearrangement of the double integral, yields[6]

$$\begin{aligned} a\, G(p) = b\, F(p) + \lambda \Big[& G(p)K(p) + G(p)K(-p) \\ & - \int_0^\infty g(y)\, dy \int_0^\infty k(x+y) e^{px}\, dx \Big]. \end{aligned} \tag{35}$$

Now we suppose that the kernel function is a linear combination of exponential functions with polynomial coefficients, namely

$$k(s) = \sum_{i=1}^{n} \phi_i(s)\, e^{-\alpha_i s},$$
$$\phi_i(s) = \sum_{j=0}^{m_i} a_{ij}\, s^j. \qquad (36)$$

Substituting a representative term from (36) into the double integral occuring in (35) yields

$$\int_0^\infty g(y)\, dy \int_0^\infty a_{ij}(x+y)^j\, e^{-\alpha_i(x+y) + px}\, dx$$

$$= a_{ij} \sum_{k=0}^{j} \frac{j!}{k!(j-k)!} \int_0^\infty g(y) y^{j-k} e^{-\alpha_i y} \int_0^\infty x^k e^{(p-\alpha_i)x}\, dx$$

$$= a_{ij} \sum_{k=0}^{j} \frac{j!}{(j-k)!} \frac{G^{(j-k)}(\alpha_i)}{(p-\alpha_i)^{k+1}}, \qquad (37)$$

that is, we obtain a partial fraction expansion whose coefficients are constants, which are related to the Laplace transform $G(p)$ at the points $p = \alpha_i$. To solve (35), we temporarily regard these constants as arbitrary. Denoting the double integral in (35) by $N(p)$, we obtain the explicit solution for $G(p)$

$$G(p) = \frac{\lambda\, N(p) - b\, F(p)}{\lambda[K(p) + K(-p)] - a}. \qquad (38)$$

In general the inversion integral for this $G(p)$ will not converge, and this will restrict the constants, which were temporarily assumed to be arbitrary, to certain fixed values. Consequently (34) will lead not only to a functional relationship between $f(x)$ and $g(x)$, but may also give a set of subsidiary conditions, as we saw in the example discussed above.

A Further Example: We solve the equation

$$g(x) = f(x) + \lambda \int_0^\infty |x-y|\ e^{-\alpha|x-y|}\ g(y)\ dy, \qquad x \geq 0,\ \alpha > 0. \tag{39}$$

The Laplace transform gives

$$G(p) = F(p) + \frac{\alpha\lambda(p^2+\alpha^2)}{(p^2-\alpha^2)^2}\ G(p) + \frac{\gamma_1}{p-\alpha} + \frac{\gamma_2}{(p-\alpha)^2}\ . \tag{40}$$

Solving for $G(p)$, we obtain

$$G(p) = \left[F(p) + \frac{\gamma_1}{p-\alpha} + \frac{\gamma_2}{(p-\alpha)^2}\right][1 + \Psi(p)],$$

$$\Psi(p) = \frac{2\lambda(p^2+\alpha^2)}{(p^2-\alpha^2)^2 - 2\lambda(p^2+\alpha^2)}\ . \tag{41}$$

Inversion of (41) is possible for any values of γ_1 and γ_2, hence we have

$$g(x) = f(x) + (\gamma_1+\gamma_2 x)\ e^{\alpha x} - \int_0^x [f(y) + (\gamma_1+\gamma_2 x)e^{\alpha y}]\psi(x-y)\ dy. \tag{42}$$

It is not difficult to show that the constants γ_1 and γ_2 are indeed arbitrary in this case (see Problem 18).

5.3. The Percus-Yevick Equation for Hard Rods

One of the central problems of statistical mechanics is the determination of the pair distribution function.[7] Several approximate integral equations have been proposed to determine this function; the most successful appears to be the Percus-Yevick equation. We consider here the one-dimensional form which can be written

$$\phi(x) = Q^2 + n\int_{-\infty}^{\infty} \phi(x')\ f(x')\phi(x-x')\ e(x-x')\ dx',$$
$$Q^2 = 1 - n\int_{-\infty}^{\infty} \phi(x')\ f(x')\ dx', \tag{43}$$

where $e(x) = \exp[-\beta V(x)]$, $f(x) = e(x)-1$, $\beta = 1/kT$, n is the density, $V(x)$ the interaction energy between a pair of particles, k Boltzmann's constant, and T the temperature. For hard rods of length a, we put $V(x) = \infty$, $|x| < a$; $V(x) = 0$, $|x| > a$, so that $e(x)$ and $f(x)$ are step functions.

Equation (43) determines a function $\phi(x)$ which is related to the pair distribution function $g(x)$ by

$$g(x) = \phi(x)\ e(x). \tag{44}$$

In the ensuing treatment, we shall also employ the function

$$h(x) = \phi(x)\ f(x), \tag{45}$$

known as the direct correlation function.[8] Substituting these definitions into (43) and taking the Laplace transform, we have[9]

$$G(p) - H(p) = \frac{Q^2}{p} + n\,\mathscr{L}\left[\int_{-\infty}^{\infty} g(x')\ h(x-x')\ dx'\right]. \tag{46}$$

The integral in (46) is subjected to manipulations similar to, but more complicated than, those which we employed on (34). We split it up into three regions: (i) $x' < 0$, (ii) $0 < x' < x$, (iii) $x' > x$, and deal with each in turn.

(i)

$$\int_0^{\infty} e^{-px}\ dx \int_{-\infty}^{0} g(x')\ h(x-x')\ dx$$
$$= \int_0^{\infty} e^{-px}\ dx \int_x^{\infty} h(x'')\ g(x-x'')\ dx'' \tag{47}$$
$$= 0,$$

where we have used the variable change $x'' = x-x'$, and the properties

$$g(s) = 0, \quad |s| < a, \qquad h(s) = 0, \quad |s| > a. \tag{48}$$

(ii) If $0 < x' < x$, we have a convolution, giving the contribution

$$G(p)\, H(p). \tag{49}$$

(iii) By interchanging orders of integration,

$$\begin{aligned}\int_0^\infty e^{-px}\, dx \int_x^\infty g(x')\, h(x-x')\, dx \\ = \int_0^\infty g(x')\, dx' \left[\int_{-\infty}^{x'} e^{-px}\, h(x-x')\, dx \right. \\ \left. - \int_{-\infty}^{0} e^{-px}\, h(x-x')\, dx\right].\end{aligned} \tag{50}$$

In dealing with both of these integrals we need to note that $g(x)$ and $h(x)$ are even functions. In the first integral, the substitution $x'' = x-x'$ yields the contribution $G(p)\, H(-p)$; in the second a change of sign of both variables gives (47) with p replaced by $-p$. Hence (46) becomes

$$G(p) - H(p) = \frac{Q^2}{p} + n\, G(p)\, [H(p) + H(-p)]. \tag{51}$$

The problem with this result is the occurrence of $H(-p)$; we now show how this can be circumvented.[10] Solving for $G(p)$, we obtain

$$G(p) = \frac{Q^2 p^{-1} + H(p)}{1 - n\, H(p) - n\, H(-p)}\,. \tag{52}$$

The function $H(p)$ is an entire function of p, since $h(x) = 0$ for $|x| > a$, and we assume it to be finite for $|x| < a$. Also $G(p)$ is regular for $\mathrm{Re}(p) \geq 0$ (except at

$p = 0$), so that the denominator cannot have zeros for $\mathrm{Re}(p) \geq 0$. But the denominator is an even function of p, hence the function $G(p)$ is entire except for a simple pole at $p = 0$. Consequently, the function

$$p^2\, G(p)\, [Q^2 p^{-1} - H(-p)] \tag{53}$$

is an entire function; it is not difficult to show (Problem 21) that it is also bounded as $|p| \to \infty$. It follows that the function is a constant; evaluating it at $p = 0$ we have

$$p^2\, G(p)\, [Q^2 p^{-1} - H(-p)] = Q^2. \tag{54}$$

We use this result to eliminate $H(-p)$ from (51), obtaining

$$G(p) - H(p) = Q^2\left[\frac{1}{p} - \frac{n}{p^2}\right] + n\, G(p)[H(p) + Q^2 p^{-1}]. \tag{55}$$

The function $h(x)$ can be obtained by simple considerations, using (48). First note that the inversion of (55) gives

$$g(x) - h(x) = Q^2[1-nx] + n\int_0^x g(x')[h(x-x') + Q^2]dx', \tag{56}$$

a considerable simplification on the original equation (43). For $|x| < a$, the convolution is zero because $g(x') = 0$ for $|x'| < a$, hence

$$h(x) = \begin{cases} -Q^2(1 - nx), & |x| < a \\ 0, & |x| > a. \end{cases} \tag{57}$$

The constant Q^2 can be evaluated by inserting this result into the definition of Q^2 (43b); this gives $Q^2 = (1-na)^{-2}$. Eq. (56) is now a convolution equation of Volterra type, and is amenable to analysis using the methods of Section 5.1. Details are left as a problem.

Problems

1. Show that Abel's integral equation

$$\int_0^t \frac{\phi(\tau)}{(t-\tau)^\alpha}\, d\tau = f(t), \quad 0 < \operatorname{Re}(\alpha) < 1$$

has the solution

$$\phi(t) = \frac{\sin(\alpha\pi)}{\pi}\left[\frac{f(0)}{t^{1-\alpha}} - \int_0^t \frac{f'(\tau)}{(t-\tau)^{1-\alpha}}\, d\tau\right].$$

2. Solve the integral equation

$$\sin s = \int_0^s J_0(s-t)\, g(t)\, dt.$$

3. By introducing the change of variables $s = x^{1/2}$, $u = y^{1/2}$, show how the solutions of the equations

$$f(s) = \int_0^s k(s^2-u^2)\, g(u)\, du$$

and

$$\phi(x) = \int_0^x k(x-y)\, \psi(y)\, dy$$

are related.

4. Solve the integral equation

$$f(s) = \int_0^s \frac{g(t)}{(s^2-t^2)^\alpha}\, dt.$$

5. Find the resolvent for the equation

$$g(s) = f(s) + \lambda \int_0^s \frac{g(t)}{(s-t)^\alpha}\, dt.$$

6. Define

$$\phi(x) = \lim_{\varepsilon\to 0} \operatorname{Im}\,[K(x+i\varepsilon)]$$

where $K(p)$ is given by (17). Show that

$$\phi(x) = \begin{cases} 0, & x \geq 0 \\ \pi x e^x, & x \leq 0. \end{cases}$$

Since it may also be shown that

$$\lim_{|p|\to\infty} \text{Im}\,[K(p)] = 0,$$

deduce the fact that $\text{Im}\,[K(p)] > 0$ for all p satisfying $\text{Im}(p) > 0$.

7. Analyze (16) when $a = 1$ to determine the behavior for large t.

8. Consider the integral equation[11]

$$f(x) = \int_0^x k(x-t)\, g(t)\, dt.$$

Under what condition (onthe kernel function k) may the solution be written

$$g(x) = \int_0^x k(x-t)\, P_n(d/dt)\, f(t)\, dt,$$

where k is the same kernel function, P_n is a polynomial of degree n, and $f(0) = f'(0) = \ldots = f^{(n-1)}(0)=0$.

Solve the following integral equations in the form given in Problem 8:

9. $f(x) = \int_0^x e^{-(x-t)}\, g(t)\, dt$

10. $f(x) = \int_0^x [\sinh(x-t) - \sin(x-t)]\, g(t)\, dt$

11. $f(x) = \int_0^x \sin^2(x-t)\, g(t)\, dt$

12. $f(x) = \int_0^x e^{x-t}\, \text{erf}(\sqrt{x-t})\, g(t)\, dt$

13. $f(x) = \int_0^x J_0(\sqrt{x-t})\, g(t)\, dt$

Solve the integral equations:

14. $$f(s) = 2\int_s^1 \frac{t\,g(t)\,dt}{\sqrt{t^2-s^2}}$$

15. $$f(s) = s\int_s^\infty \frac{g'(t)}{\sqrt{t-s}}\,dt$$

16. $$g(x) = \frac{1}{\sqrt{x}}\,e^{-a/4x} + \frac{i}{\sqrt{\pi}}\int_0^x \frac{g(y)}{\sqrt{x-y}}\,dy$$

17. Obtain (27) from (32) and (33).

18. Prove that the constants γ_1 and γ_2 in (42) are arbitrary. [Hint: examine the equation

$$\phi(x) = \int_0^x \psi(x-y)\phi(y)\,dy.]$$

19. Solve

$$g(x) = \int_0^\infty \sin(|x-y|)\,g(y)\,dy.$$

20. Solve

$$g(x) = 1 + \int_0^\infty |x-y|\,\cos(|x-y|)\,g(y)\,dy.$$

21. Show that if the pair distribution function $g(x)$ is bounded as $x \to \infty$, then the function defined in (53) is bounded as $|p| \to \infty$.

22. Show that (57) implies

$$Q^2 = (1-na)^{-2}$$

where Q^2 is defined in (43).

23. Investigate the Volterra integral equation (56) for $g(x)$.

24. In three dimensions, the Percus-Yevick equation may be written

$$\phi(\underset{\sim}{r}) = 1 - n \int \phi(\underset{\sim}{r})\ f(\underset{\sim}{r})\ d\underset{\sim}{r}$$

$$+ n \int \phi(\underset{\sim}{r}')\ f(\underset{\sim}{r}')\ \phi(\underset{\sim}{r}-\underset{\sim}{r}')\ e(\underset{\sim}{r}-\underset{\sim}{r}')\ d\underset{\sim}{r}'.$$

Assume a spherically symmetrical solution, and introduce variables $|\underset{\sim}{r}'|$, $|\underset{\sim}{r}-\underset{\sim}{r}'|$, and the angle θ between $\underset{\sim}{r}'$ and $\underset{\sim}{r}-\underset{\sim}{r}'$ into the integrals. By considering the function $x\,\phi(x)$, in the case of hard spheres $[V(r) = \infty$ for $r \leq a$, $V(r) = 0$ for $r > a]$ find the direct correlation function explicitly and derive a convolution integral equation for the pair distribution function $g(x) = \phi(x)\ e(x)$.

Footnotes

1. See J. H. Giese, SIAM Review (1963), 5, 1.

2. Some numerical values for the case $a = -2$ computed by Padé approximation may be found in L. Fox and E. J. Goodwin, Phil. Trans. Roy. Soc. Lond. (1953), A245, 501.

3. N. Mullineux and J. R. Reed, Q. Appl. Math. (1967), 25, 327.

4. Equations of this type may be solved by the Weiner-Hopf technique (see Section 18). However, we are interested here in a class of problems which can be solved by more elementary methods.

5. We must first take $\mathrm{Re}(p) < \alpha$, and then use analytic continuation on the final result to extend it to $\mathrm{Re}(p) > \alpha$.

6. As with (29), a process of analytic continuation may be involved.

7. This is the probability of finding two particles at the stated positions. For an infinite uniform system it is a function only of the relative positions of the two.

8. This identification is only valid in the Percus-Yevick approximation.

9. M. S. Wertheim, J. Math. Phys. (1964), 5, 643. The more general case where $V(x) \neq 0$ for $a \leq |x| \leq \ell$ is also analyzed using Laplace transforms.

10. The ensuing procedure is a simple example of the type of argument which is used in the Wiener-Hopf technique (Section 18).

11. Problems 8-13 and some related material may be found in D. O. Reudink, SIAM Review (1967), 9, 4.

§6. THE INVERSION INTEGRAL

6.1. Inversion of Meromorphic Functions

Analytic information about the inversion integral is usually obtained by "closing the contour", as in Section 2.4 for rational functions. More generally, we may consider a Laplace transform $F(p)$ which is meromorphic, that is, a function having no singularities other than poles, and investigate the integral

$$\frac{1}{2\pi i}\int_{L+\Gamma} F(p)\, e^{pt}\, dp, \tag{1}$$

where the contour is shown in Figure 1.

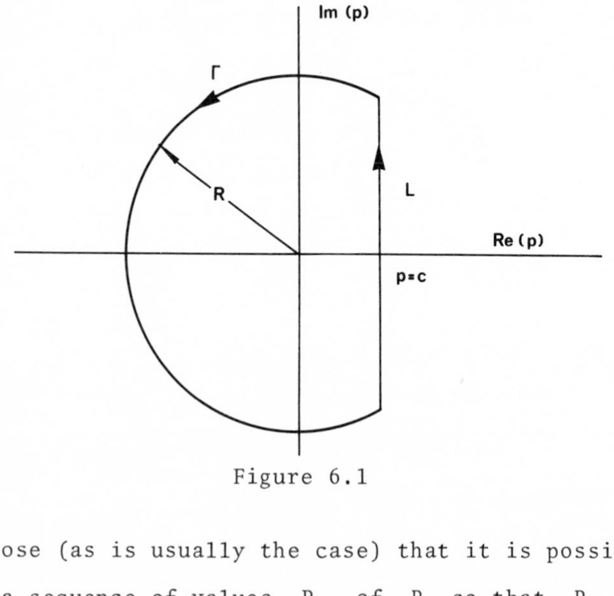

Figure 6.1

We suppose (as is usually the case) that it is possible to choose a sequence of values R_n of R so that $R_n \to \infty$ as $n \to \infty$, while on the corresponding contours Γ_n the integrand satisfies the inequality

$$|e^{pT}\, F(p)| < A|p|^{-k}, \qquad k > 0 \tag{2}$$

for some T. (The reason for choosing a discrete sequence of contours rather than allowing R to vary continuously is to avoid having poles lying on the contours.) It can readily be shown from (2) that

$$\lim_{n\to\infty} \frac{1}{2\pi i} \int_{\Gamma_n} F(p)\, e^{pt}\, dp = 0, \qquad t > T, \tag{3}$$

so that as we take the limit $n \to \infty$ of (1) we recover the inversion integral. The only singularities enclosed by these contours are poles, hence the inversion integral is given by the sum of the residues at these poles.

Heaviside Expansion Theorem: Suppose that the poles of $F(p)$, at $p = \alpha_k$, are all simple; then the function $H(p) = 1/F(p)$ has simple zeros at $p = \alpha_k$, and the residues of $F(p)$ are given by $1/H'(\alpha_k)$. More generally, if there is a convenient factorization

$$F(p) = G(p)/H(p), \tag{4}$$

where $G(p)$ is an entire function and $H(p)$ has only simple zeros, the inversion integral is given by the series

$$f(t) = \sum_{k=1}^{\infty} \frac{G(\alpha_k)}{H'(\alpha_k)}\, e^{\alpha_k t}. \tag{5}$$

This result, first formulated by Heaviside in relation to his operational calculus, is known as the Heaviside expansion theorem.

Examples:

(i) $$F(p) = \frac{1}{p \cosh(\gamma\sqrt{p})}, \qquad c > 0 \tag{6}$$

This transform was derived in Section 4.1 in connection with the solution of a partial differential equation. Notice that the Taylor series for the cosh function has only even powers of its argument, so that $F(p)$ does not have a branch point at the origin. For negative t, we can close the contour in the right-hand half-plane (Figure 2), and readily show that the integral along Γ goes to zero for large R.

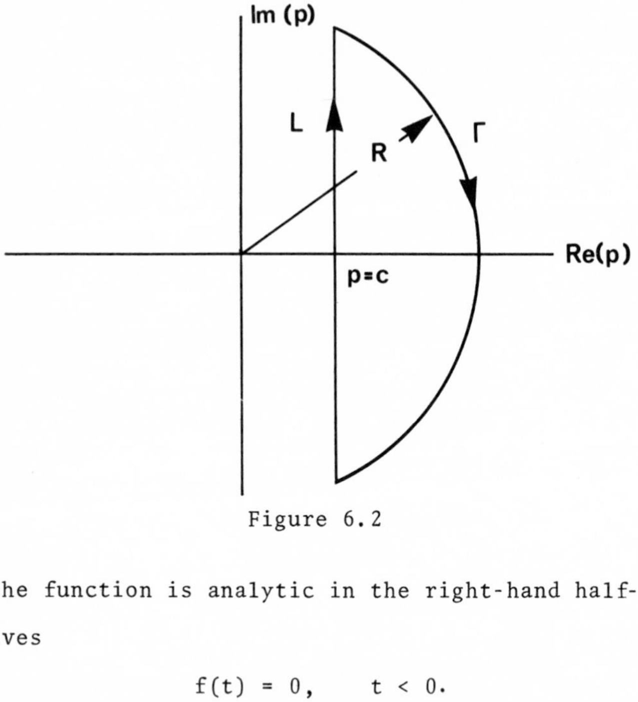

Figure 6.2

Since the function is analytic in the right-hand half-plane, this gives

$$f(t) = 0, \qquad t < 0. \tag{7}$$

This feature is a general one in the inversion of Laplace transforms as is shown by the inversion theorem [equation (2.23)].

For $t > 0$, we may close the contour in the left-hand half-plane, and the Heaviside expansion theorem gives

$$f(t) = 1 + \frac{2}{\pi}\sum_{k=1}^{\infty} \frac{(-1)^k}{(k-\frac{1}{2})} e^{-(k-1/2)^2\pi^2 t/\gamma^2}. \tag{8}$$

(ii)

$$F(p) = \frac{\cosh(\delta\sqrt{p})}{p\cosh(\gamma\sqrt{p})}, \qquad c > 0 \tag{9}$$

Again the function does not have a branch point, and the Heaviside expansion theorem gives

$$f(t) = 1 + \frac{2}{\pi}\sum_{k=1}^{\infty} \frac{(-1)^k \cos[\delta(k-\frac{1}{2})\pi/\gamma]}{(k-\frac{1}{2})} e^{-(k-\frac{1}{2})^2\pi^2 t/\gamma^2}. \tag{10}$$

6.2. Inversions Involving a Branch Point

If the Laplace transform has a branch point, possibly in addition to singularities in the form of poles, then it is appropriate to consider the integral (1) with a new contour as shown in Figure 3. For convenience, we have assumed that the branch point is at the origin, and that there is only one. Extension of the following techniques to more general

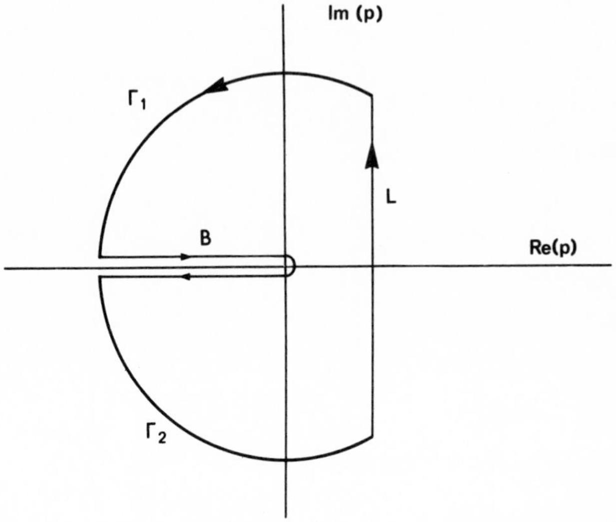

Figure 6.3

situations is not difficult (in principle). Assuming that we can again make the contribution from Γ vanish by taking R sufficiently large, we have

$$f(t) = \Sigma \text{ residues at poles} + \frac{1}{2\pi i}\int_B F(p)\, e^{pt}\, dp. \tag{11}$$

Special Case: The treatment of the loop integral depends on the behavior of the integrand near the branch point. If $F(p) \sim p^{\gamma}$, with $\text{Re}(\gamma) > -1$, then we can "shrink" the contour onto the branch cut. Writing $p = u\exp(\pm i\pi)$ according as $\text{Im}(p)$ is positive or negative, this leads to

$$\int_B F(p)\, e^{pt}\, dp = \int_0^\infty [F(ue^{-i\pi}) - F(ue^{i\pi})]\, e^{-ut}\, du\,. \tag{12}$$

In some cases it may be possible to evaluate the integral explicitly; in other cases an asymptotic series for large t follows immediately by the use of Watson's lemma. As an example, consider the function

$$F(p) = \frac{1}{\sqrt{p}}\, e^{-\gamma\sqrt{p}}\,. \tag{13}$$

Substituting into (12) leads to

$$f(t) = \frac{1}{\pi}\int_0^\infty u^{-1/2}\, e^{-ut}\cos(\gamma u^{1/2})\, du, \qquad t > 0. \tag{14}$$

The integral can be reduced to a more standard form by the substitution $ut = s^2$, giving for $f(t)$ the expression

$$f(t) = (\pi t)^{-1/2}\, e^{-\gamma^2/4t}. \tag{15}$$

We leave it to the reader to verify that the application of Watson's lemma to (14) yields the Taylor series of (15) in ascending powers of $(\gamma^2/4t)$, which is the asymptotic series for large t.

<u>More General Case</u>: The Laplace transform

$$F(p) = \frac{1}{p} e^{-\gamma\sqrt{p}}, \tag{16}$$

which occurred in Section 4.1, cannot be treated by substitution into (12) because it diverges too strongly at $p = 0$. We consider the loop integral more carefully, breaking it up into three parts (Figure 4). For the integral around the

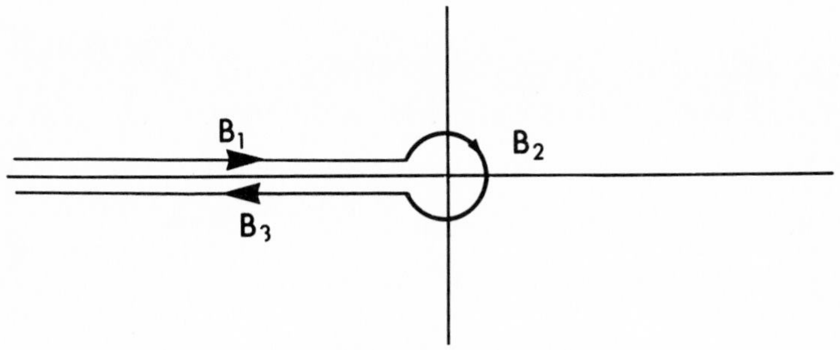

Figure 6.4

small circle, we write $p = \varepsilon \exp(i\theta)$, and readily show that

$$\frac{1}{2\pi i} \int_{B_2} F(p)\, e^{pt}\, dp = 1 + \mathcal{O}(\varepsilon). \tag{17}$$

For the other two contributions, we can use (12) with the lower limit replaced by $u = \varepsilon$; subsequently setting ε to zero gives

$$f(t) = 1 - \frac{1}{\pi} \int_0^\infty e^{-ut} \sin(\gamma u^{1/2}) \frac{du}{u}. \tag{18}$$

Comparing this result with (14), we see immediately that

$$
\begin{aligned}
-\frac{1}{\pi}\int_0^\infty e^{-ut}\sin(\gamma u^{1/2})\,\frac{du}{u} \\
&= -(\pi t)^{-1/2}\int_0^\gamma e^{-\mu^2/4t}\,d\mu \\
&= -2\,\pi^{-1/2}\int_0^{\gamma/2\sqrt{t}} e^{-u^2}\,du \\
&= -\operatorname{erf}(\gamma/2\sqrt{t}),
\end{aligned}
\tag{19}
$$

and hence the complete inverse of (16) is

$$
f(t) = \operatorname{erfc}(\gamma/2\sqrt{t}), \qquad t > 0. \tag{20}
$$

6.3. Watson's Lemma for Loop Integrals

The above examples involving a branch cut were reduced quite readily to real integrals to which Watson's lemma could be applied. We consider here an extension of Watson's lemma to loop integrals which has the advantages of being direct in application to the inversion integral, and of working for a wider range of integrals than may be treated by (12). Specifically, we will show that if $F(p)$ has the asymptotic expansion

$$
\begin{aligned}
F(p) \sim \sum_{\nu=1}^{\infty} a_\nu p^{\lambda_\nu}, \quad p \to 0, \\
-\pi \le \arg(p) \le \pi,
\end{aligned}
\tag{21}
$$

where $\mathrm{Re}(\lambda_1) < \mathrm{Re}(\lambda_2) < \mathrm{Re}(\lambda_3) < \ldots$, and $\mathrm{Re}(\lambda_\nu)$ increases without bound as $\nu \to \infty$, then the loop integral

$$
f(t) = \frac{1}{2\pi i}\int_{-\infty}^{0+} F(p)\,e^{pt}\,dp \tag{22}
$$

has the asymptotic expansion

$$f(t) \sim \sum_{\nu=1}^{\infty} \frac{a_\nu}{(-\lambda_\nu - 1)!\, t^{\lambda_\nu + 1}}, \qquad t \to \infty,$$

$$-\pi/2 < \arg(t) < \pi/2 . \tag{23}$$

In (22) the notation $-\infty, 0+$ means that the contour begins and ends at $p = -\infty$, and circles the origin once in the positive direction, as in Figure 4.

The proof is quite simple. Define a set of functions $F_n(p)$ by

$$F_n(p) = F(p) - \sum_{\nu=1}^{n} a_\nu P^{\lambda_\nu} \tag{24}$$

and substitute into (23) to get

$$f(t) = \sum_{\nu=1}^{n} a_\nu \frac{1}{2\pi i} \int_{-\infty}^{0+} p^{\lambda_\nu} e^{pt}\, dp$$

$$+ \frac{1}{2\pi i} \int_{-\infty}^{0+} F_n(p)\, e^{pt}\, dp. \tag{25}$$

If $\mathrm{Re}(t) > 0$, it is permissible to make the substitution $u = pt$ in the first integral while using the same contour for u as for p; the Hankel integral representation of $1/z!$ (Appendix A) then gives

$$\frac{1}{2\pi i} \int_{-\infty}^{0+} p^\lambda e^{pt}\, dp = \frac{1}{(-\lambda-1)!\, t^{\lambda+1}} . \tag{26}$$

To deal with the remainder term, choose n sufficiently large so that $\mathrm{Re}(\lambda_{n+1}) > -1$; then we can shrink the contour onto the branch cut--which we could not do with (22) because there is no restriction on $\mathrm{Re}(\lambda_1)$. Equation (12) followed by Watson's lemma for real integrals then yields the estimate

$$\frac{1}{2\pi i} \int_{-\infty}^{0+} F_n(p)\, e^{pt}\, dp = \mathcal{O}(t^{-\lambda_{n+1}-1}). \tag{27}$$

This completes the proof.

6.4. Asymptotic Forms for Large t

The information gleaned above may be applied to many inverse Laplace transforms (and, as we shall see in Part II, to Fourier transforms also) to recover asymptotic information for large values of the time. If the singularities of $F(p)$ all take the form of isolated poles and/or isolated branch points, then by a suitable deformation of the inversion contour we may reduce the integral to a sum of residues at the poles plus a sum of loop integrals around the branch points. These latter may usually be estimated for large t by Watson's lemma for loop integrals. For example, if there is a branch point at $p = a$, and if $F(p)$ has the asymptotic expansion

$$F(p) \sim \sum_{\nu=1}^{\infty} a_\nu (p-a)^{\lambda_\nu}, \qquad (28)$$

then the substitution $p' = p-a$ reduces the loop integral to

$$\frac{1}{2\pi i} \int_{-\infty}^{0+} F(p'+a) e^{(p'+a)t} \, dp'$$

$$\sim e^{at} \sum_{\nu=1}^{\infty} \frac{a_\nu}{(-\lambda_\nu - 1)! \, t^{\lambda_\nu + 1}}, \qquad t \to \infty . \qquad (29)$$

Formulas appropriate for asymptotic expansions involving logarithmic functions may also be derived--see Problem 9. For large t, the contribution from each pole and each branch point is dominated by the exponential factor, and the asymptotic form of the complete inversion integral will be governed by the singularity whose position $p = a$ has the most positive real part.

Examples: (i) Consider the function

$$F(p) = \frac{1}{\sqrt{p(p+a)}} \quad , \; a > 0 \tag{30}$$

which has branch points at $p = 0$ and $p = -a$. Since $a > 0$, we need consider only the origin for large t, so that

$$F(p) \sim \frac{1}{\sqrt{pa}} \left[1 - \frac{p}{2a} + \frac{3p^2}{8a^2} + \ldots\right], \tag{31}$$

with the corresponding asymptotic expansion

$$f(t) \sim \frac{1}{\sqrt{\pi at}} \left[1 + \frac{1}{4at} + \frac{9}{32a^2t^2} + \ldots\right]. \tag{32}$$

(ii) The Bessel function $J_0(t)$ has the Laplace transform [(3.53)]

$$F(p) = \frac{1}{\sqrt{p^2+1}} . \tag{33}$$

There are two branch points, both on the imaginary axis, and consequently of equal importance for large t. The necessary asymptotic expansions of $F(p)$ are

$$F(p) \sim \begin{cases} \dfrac{e^{-i\pi/4}}{\sqrt{2(p-i)}} \left[1 - \dfrac{p-i}{4i} - \dfrac{3(p-i)^2}{32} + \ldots\right], & p \to i \\[2ex] \dfrac{e^{i\pi/4}}{\sqrt{2(p+i)}} \left[1 + \dfrac{p+i}{4i} - \dfrac{3(p+i)^2}{32} + \ldots\right], & p \to -i, \end{cases} \tag{34}$$

from which it follows that

$$\begin{aligned} J_0(t) &\sim \sqrt{2/\pi t} \cos (t-\pi/4) \, [1 - 9/128t^2 + \ldots] \\ &+ \sqrt{2/\pi t} \sin (t-\pi/4) \, [1/8t - \ldots], \quad t \to \infty . \end{aligned} \tag{35}$$

Because there are no other singularities in this case, there are no neglected terms which are exponentially small.[1]

6.5. Heaviside Series Expansion

For small values of the time, it is often possible to extend the technique of Section 2.5 to derive an expansion in ascending powers of t. Sometimes this expansion will be a convergent Taylor series, but more often it will be an asymptotic expansion. We deal with the latter, since it includes the former as a special case. Suppose then that the Laplace transform $F(p)$ has an asymptotic expansion[2]

$$F(p) \sim \sum_{\nu=1}^{\infty} a_\nu p^{-\lambda_\nu} ; \tag{36}$$

then for any n we can define the function $F_n(p)$ in the usual way by

$$F_n(p) = F(p) - \sum_{\nu=1}^{n} a_\nu p^{-\lambda_\nu} \tag{37}$$

and deform the contour into the right-hand half-plane so that

$$|F_n(p)| < A_n |p|^{-\mathrm{Re}(\lambda_n)} \tag{38}$$

Some elementary considerations, the details of which we omit, then lead to the Heaviside series expansion, namely,

$$f(t) \sim \sum_{\nu=1}^{\infty} \frac{a_\nu t^{\lambda_\nu - 1}}{(\lambda_\nu - 1)!} . \tag{39}$$

<u>An Example</u>: We consider again the Bessel function $J_0(t)$. Expanding (33) in descending powers of p gives

$$F(p) \sim \sum_{k=1}^{\infty} \frac{\sqrt{\pi}}{(-k-\frac{1}{2})!\ k!\ p^{(2k+1)}} , \tag{40}$$

with the corresponding Heaviside series expansion

$$J_0(t) = \sum_{k=0}^{\infty} \frac{(-1)^k (t/2)^{2k}}{k!\ k!} . \tag{41}$$

Since the expansion of $F(p)$ is a convergent series for

$|p| > 1$, the series (41) is also convergent.

Problems

Find the inverse Laplace transforms of the following functions using the inversion integral.

1. $\dfrac{1}{p\sqrt{p+1}}$

2. $\dfrac{1}{a+\sqrt{p}}$

3. $\dfrac{1-e^{-ap}}{p}$, $a > 0$

4. $\dfrac{e^{-ap}-e^{-bp}}{p}$, $0 \le a < b$

5. $\dfrac{e^{-ap}-e^{-bp}}{p^2}$, $0 \le a < b$

6. $\ln\left(\dfrac{p+b}{p+a}\right)$

7. $\ln\left(\dfrac{p^2+b^2}{p^2+a^2}\right)$

8. Show that if $F(p)$ has the asymptotic expansion

$$F(p) \sim \sum_{\nu=1}^{\infty} a_\nu p^{\lambda_\nu} \ln p, \quad p \to \infty,$$

$$-\pi \le \arg(p) \le \pi,$$

where $\operatorname{Re}(\lambda_1) < \operatorname{Re}(\lambda_2) < \operatorname{Re}(\lambda_3) < \ldots$, and $\operatorname{Re}(\lambda_\nu)$ increases without bound, then the loop integral (22) has the asymptotic expansion

$$f(t) \sim \sum_{\nu=1}^{\infty} \frac{a_\nu}{(-\lambda_\nu-1)!} \frac{\psi(-\lambda_\nu) - \ln t}{t^{\lambda_\nu+1}},$$

where

$$\psi(\alpha+1) = \frac{d}{d\alpha}[\ell n\ \alpha!].$$

10. Invert

$$F(p) = p\ \ell n\ (1 - a^2/p^2).$$

11. Find power series for the functions whose inverses were found in Problems 1, 2, 6, and 7.

12. Find an asymptotic expansion for the inverse of

$$F(p) = \frac{e^{-b\sqrt{p^2+\alpha^2}}}{\sqrt{p^2+\alpha^2}}.$$

Footnotes

1. For a discussion of the possible importance of exponentially small terms, see OLVER (1974), pp. 76-78.

2. If the expansion is convergent, then so is the inverse (39). See CARSLAW & JAEGER (1941), pp. 271-273.

Part II: The Fourier Transform

§7. DEFINITIONS AND ELEMENTARY PROPERTIES

7.1. The Exponential, Sine, and Cosine Transforms

Let $f(t)$ be an arbitrary function; then the (exponential) Fourier transform of $f(t)$ is the function defined by the integral

$$F(\omega) = \int_{-\infty}^{\infty} e^{i\omega t} f(t)\, dt \tag{1}$$

for those values of ω for which the integral exists. We shall usually refer to (1) as the Fourier transform, omitting any reference to the term exponential. The Fourier transform is related to the Laplace transform; indeed, on denoting by $\tilde{f}_{\pm}(p)$ the following Laplace transforms:

$$\tilde{f}_{\pm}(p) = \int_{0}^{\infty} e^{-pt} f(\pm t)\, dt, \qquad \mathrm{Re}(p) > \alpha_{\pm}, \tag{2}$$

we have

$$F(\omega) = \tilde{f}_{+}(-i\omega) + \tilde{f}_{-}(i\omega). \tag{3}$$

Furthermore, we see that (1) will converge for values of ω in the strip $\alpha_+ < \mathrm{Im}(\omega) < -\alpha_-$, corresponding to the regions of convergence of (2).

Inversion: Consider the inversion integrals

$$\frac{1}{2\pi i}\int_{\gamma-i\infty}^{\gamma+i\infty} e^{pt}\,\tilde{f}_{\pm}(\pm p)\,dp. \tag{4}$$

If $\gamma > \alpha_+$, the first integral gives $f(t)$ for $t > 0$ and zero for $t < 0$. Similarly, on making the substitution $p \to -p$, we see that if $-\gamma > -\alpha_-$, the second integral gives $f(-t)$ for $t < 0$ and zero for $t > 0$. Adding these two results we have

$$\begin{aligned} f(t) &= \frac{1}{2\pi i}\int_{\gamma-i\infty}^{\gamma+i\infty} e^{pt}\,[\tilde{f}_+(p) + \tilde{f}_-(-p)]\,dp \\ &= \frac{1}{2\pi}\int_{i\gamma-\infty}^{i\gamma+\infty} e^{-i\omega t}\,F(\omega)\,d\omega, \end{aligned} \tag{5}$$

where the last step follows from the substitution $p \to i\omega$, and the use of (3). Hence we have the reciprocal transform pair

$$\begin{aligned} F(\omega) &= \int_{-\infty}^{\infty} e^{i\omega t}\,f(t)\,dt, \qquad \alpha < \mathrm{Im}(\omega) < \beta \\ f(t) &= \frac{1}{2\pi}\int_{i\gamma-\infty}^{i\gamma+\infty} e^{-i\omega t}\,F(\omega)\,d\omega, \qquad \alpha < \gamma < \beta. \end{aligned} \tag{6}$$

Sine and Cosine Transforms: Consider the functions defined by the integrals

$$\begin{aligned} F_s(\omega) &= 2\int_0^{\infty} \sin(\omega t)\,f(t)\,dt, \\ F_c(\omega) &= 2\int_0^{\infty} \cos(\omega t)\,f(t)\,dt, \end{aligned} \tag{7}$$

known as the Fourier sine and Fourier cosine transforms. They may be related to the Laplace transforms (2) by

$$F_s(\omega) = -i[\tilde{f}_+(-i\omega) - \tilde{f}_+(i\omega)],$$
$$F_c(\omega) = [\tilde{f}_+(-i\omega) + \tilde{f}_+(i\omega)], \tag{8}$$

so that the integrals (7) converge (if at all) in the strip $\alpha_+ < \text{Im}(\omega) < -\alpha_-$, which includes the real axis. Employing the Laplace inversion theorem with $\gamma = 0$, we may readily show that the inversions of (7) are

$$f(t) = \frac{1}{\pi}\int_0^\infty \sin(\omega t)\, F_s(\omega)\, d\omega,$$
$$f(t) = \frac{1}{\pi}\int_0^\infty \cos(\omega t)\, F_c(\omega)\, d\omega. \tag{9}$$

Examples:

(i) $$f(t) = e^{-\alpha|t|}, \quad \text{Re}(\alpha) > 0$$
$$F(\omega) = \frac{2\alpha}{\alpha^2+\omega^2} \tag{10}$$

Here the region of convergence of the integral defining $F(\omega)$ is $|\text{Im}(\omega)| < \text{Re}(\alpha)$. Inserting $F(\omega)$ into the inversion integral (5) with $\gamma = 0$, we can easily evaluate the latter by residues. If $t > 0$, we must close the contour in the lower half-plane; if $t < 0$, we must close in the upper half-plane. The effect of switching from one pole to the other gives the result[1]

$$f(t) = \begin{cases} e^{-\alpha t}, & t > 0 \\ e^{+\alpha t}, & t < 0 \end{cases}$$
$$= e^{-\alpha|t|}. \tag{11}$$

(ii) $$f(t) = e^{-\alpha t^2}, \quad \text{Re}(\alpha) > 0 \tag{12}$$

$$\begin{aligned} F(\omega) &= \int_{-\infty}^{\infty} e^{-\alpha t^2+i\omega t}\,dt \\ &= e^{-\omega^2/4a^2} \int_{-\infty}^{\infty} e^{-\alpha u^2}\,du \\ &= (\pi/a)^{1/2}\, e^{-\omega^2/4a^2} \end{aligned} \tag{13}$$

In (13) we have written $u = t - i\omega/2\alpha$, and changed the contour from $-\infty < t < \infty$ to $-\infty < u < \infty$. In this example, the integral defining $F(\omega)$ converges for all ω, and the Fourier transform is an entire function of ω as a consequence. Since $F(\omega)$ has the same functional form as $f(t)$, the inversion integral is evaluated by a trivial modification of (13).

(iii)
$$f(t) = \begin{cases} 1, & -1 < t < 1 \\ 0, & |t| > 1 \end{cases} \tag{14}$$

$$\begin{aligned} F(\omega) &= \int_{-1}^{1} e^{i\omega t}\,dt \\ &= \frac{2\sin\omega}{\omega} \end{aligned} \tag{15}$$

To invert (15), we write the inversion integral as

$$I = \frac{1}{\pi i}\int_C \frac{e^{-i\omega(t-1)}}{\omega}\,d\omega - \frac{1}{\pi i}\int_C \frac{e^{-i\omega(t+1)}}{\omega}\,d\omega \tag{16}$$

where the contour is shown in Figure 1. If $t < -1$, we can close the contour in the upper half plane for both integrals, giving $I = 0$. If $-1 < t < 1$, we close in the lower half-plane for the second integral, obtaining $I = 1$. Finally, if $t > 1$, we close in the lower half-plane for both integrals, and the residues cancel. Hence $I = f(t)$.

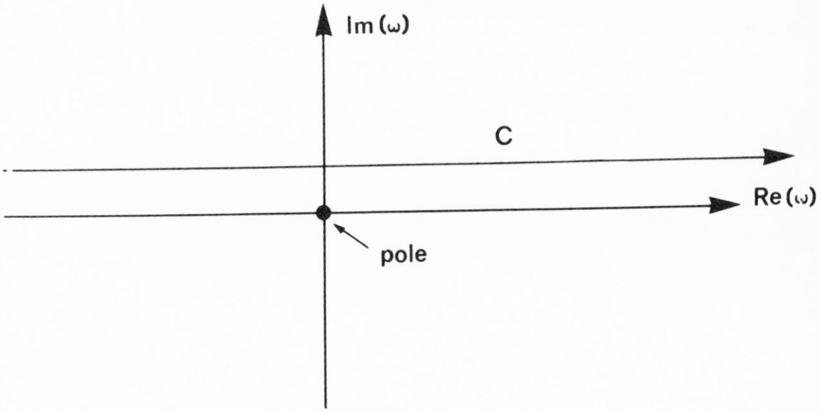

Figure 7.1

(iv)
$$f(t) = J_0(at) \tag{17}$$

For this function, the integral (1) will diverge unless ω is real, so there is no strip $\alpha < \mathrm{Im}(\omega) < \beta$ for convergence. For real ω we can evaluate (1) by writing[2]

$$\begin{aligned} F(\omega) &= \lim_{\varepsilon\to 0} \int_{-\infty}^{\infty} e^{i\omega t}\, J_0(at)\, e^{-\varepsilon|t|}\, dt \\ &= \lim_{\varepsilon\to 0} \{[a^2-(\omega-i\varepsilon)^2]^{-1/2} + [a^2-(\omega+i\varepsilon)^2]^{-1/2}\}, \end{aligned} \tag{18}$$

where we have used the Laplace transform of $J_0(at)$ in (3). On taking the limit $\varepsilon \to 0$, we have

$$F(\omega) = \begin{cases} 2(a^2-\omega^2)^{-1/2}, & |\omega| < a \\ 0 \quad , & |\omega| > a \end{cases} \tag{19}$$

where the result for $|\omega| > a$ depends on a consideration of the phases of the two terms in (18). The inversion integral may be transformed into Bessel's integral (20.50) by the substitution $\omega = a \sin\theta$, viz.

$$J_0(at) = \frac{1}{\pi}\int_{-a}^{a} \frac{e^{-i\omega t}\,d\omega}{\sqrt{a^2-\omega^2}}$$

$$= \frac{2}{\pi}\int_0^a \frac{\cos(\omega t)\,d\omega}{\sqrt{a^2-\omega^2}} \tag{20}$$

$$= \frac{2}{\pi}\int_0^{\pi/2} \cos(at\sin\theta)\,d\theta.$$

(v)
$$f(t) = \begin{cases} t^{-1/2}, & t > 0 \\ 0, & t < 0 \end{cases} \tag{21}$$

$$F(\omega) = \int_0^\infty t^{-1/2} e^{i\omega t}\,dt$$

$$= \frac{e^{i\pi/4}\sqrt{\pi}}{\sqrt{\omega}}, \qquad \mathrm{Im}(\omega) > 0 \tag{22}$$

The inversion integral is (see Figure 2 for details)

$$f(t) = \frac{e^{-i\pi/4}}{2\sqrt{\pi}}\int_C \frac{e^{-i\omega t}}{\sqrt{\omega}}\,d\omega. \tag{23}$$

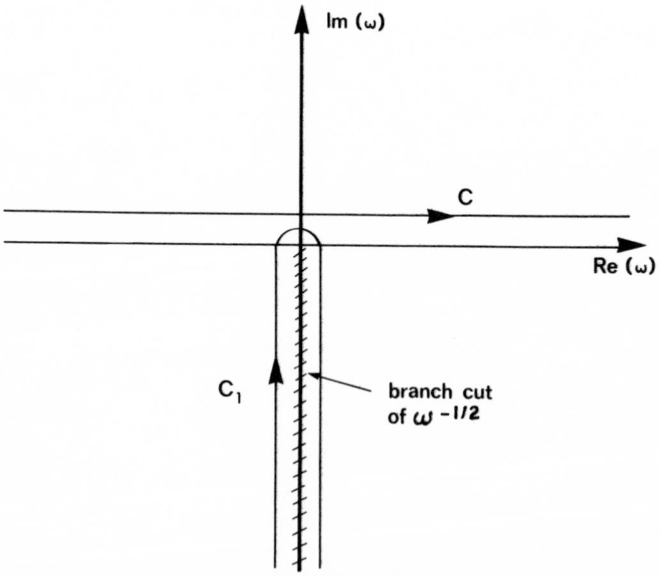

Figure 7.2

If $t < 0$, we can close the contour in the upper half-plane, giving zero. If $t > 0$, we can close C in the lower half-plane (giving C_1), and 'shrink' the contour about the branch cut. With $\xi = i\omega$, this gives

$$\frac{e^{-i\pi/4}}{2\sqrt{\pi}} \int_0^\infty e^{-\xi t} \left[\frac{e^{i\pi/4}}{\sqrt{\xi}} - \frac{e^{-3i\pi/4}}{\sqrt{\xi}} \right] d\xi$$

$$= \frac{1}{\sqrt{\pi}} \int_0^\infty \xi^{-1/2} e^{-\xi t} \, d\xi \tag{24}$$

$$= t^{-1/2}.$$

Hence we recover $f(t)$ for all t.

7.2. Important Properties

We will derive a number of simple but important properties of the exponential Fourier transform; the corresponding properties for the sine and cosine transforms, which are also simple, are given in the problems.

Derivatives: Suppose that $g(t) = f'(t)$; then

$$G(\omega) = \int_{-\infty}^{\infty} f'(t)\, e^{i\omega t}\, dt$$

$$= [f(t)\, e^{i\omega t}]_{-\infty}^{\infty} - i\omega \int_{-\infty}^{\infty} f(t)\, e^{i\omega t}\, dt \tag{25}$$

$$= -i\omega\, F(\omega),$$

provided[3] $f(t) \to 0$ as $|t| \to \infty$. Similarly if $\phi(t) = t\, f(t)$, then differentiation of the integral which defines $F(\omega)$ yields

$$\Phi(\omega) = -i \frac{d}{d\omega} F(\omega), \tag{26}$$

provided the order of integration and differentiation may be changed. Equations (25) and (26) represent a duality between operations on a function and the corresponding operation on its Fourier transform; symbolically we may express this by the correspondence

$$\begin{aligned} \frac{d}{dt} &\leftrightarrow -i\omega, \\ \frac{d}{d\omega} &\leftrightarrow \ it. \end{aligned} \tag{27}$$

Translations: Similarly, there is a duality between translations of a function and multiplication by an exponential factor. Denoting by $\mathscr{F}[f]$ the Fourier transform of $f(t)$, we have

$$\begin{aligned} \mathscr{F}[f(t-\tau)] &= \int_{-\infty}^{\infty} e^{i\omega t} f(t-\tau)\, dt \\ &= e^{i\omega\tau} \int_{-\infty}^{\infty} e^{i\omega u} f(u)\, du \\ &= e^{i\omega\tau} F(\omega), \end{aligned} \tag{28}$$

$$\begin{aligned} \mathscr{F}[e^{iat} f(t)] &= \int_{-\infty}^{\infty} e^{i(a+\omega)t} f(t)\, dt \\ &= F(a+\omega). \end{aligned} \tag{29}$$

Convolutions: A convolution integral of the type given in (1.22) has a particularly simple Laplace transform. The corresponding result for the Fourier transform stems from replacing the integration limits by $\pm\infty$; that is, we consider the Fourier transform of the function defined by the convolution integral

$$g(t) = \int_{-\infty}^{\infty} k(t-\tau)\, f(\tau)\, d\tau. \tag{30}$$

Assuming that the necessary changes of orders of integration are valid, the application of (28) gives

$$\begin{aligned} G(\omega) &= \int_{-\infty}^{\infty} \mathscr{F}[k(t-\tau)]\ f(\tau)\ d\tau \\ &= K(\omega) \int_{-\infty}^{\infty} e^{i\omega\tau}\ f(\tau)\ d\tau \\ &= K(\omega)\ F(\omega). \end{aligned} \tag{31}$$

There is a similar result, which again reveals a duality between operations on functions and their Fourier transforms, for the Fourier transform of the product of two functions. By replacing one of the functions by its inverse Fourier transform and using (29) we obtain

$$\begin{aligned} \mathscr{F}[f(t)\ g(t)] &= \frac{1}{2\pi} \int_{-\infty}^{\infty} F(\omega')\ d\omega'\ \mathscr{F}[e^{-i\omega' t}\ g(t)] \\ &= \frac{1}{2\pi} \int_{-\infty}^{\infty} F(\omega')\ G(\omega-\omega')\ d\omega'. \end{aligned} \tag{32}$$

Parseval Relations: One immediate and important consequence of (32) is obtained by putting $\omega = 0$. The resulting equation, which involves the function $G(-\omega)$, may be made more symmetrical by replacing $g(t)$ by its complex conjugate function $g^*(t)$, and noting that

$$\mathscr{F}[g^*(t)] = G^*(-\omega). \tag{33}$$

Hence on writing $\mathscr{F}[f(t)\ g^*(t)]_{\omega=0}$ as an integral we have

$$\int_{-\infty}^{\infty} f(t)\ g^*(t)\ dt = \frac{1}{2\pi} \int_{-\infty}^{\infty} F(\omega)\ G^*(\omega)\ d\omega, \tag{34}$$

which is Parseval's relation.

7.3. Spectral Analysis

Suppose that $f(t)$ represents the value of some physical quantity at time t. Then if the Fourier transform exists for real ω, the representation

$$f(t) = \frac{1}{2\pi}\int_{-\infty}^{\infty} F(\omega)\, e^{-i\omega t}\, d\omega \tag{35}$$

is a way of expressing $f(t)$ as a linear combination of simple harmonic functions $\cos(\omega t) \pm i \sin(\omega t)$. This means that the frequency content of the signal $f(t)$ is spread over a continuous range of frequencies ω, the amplitude of a given frequency being proportional to $F(\omega)$. If $|f(t)|^2$ is a measure of the intensity of the quantity $f(t)$ at time t, then we may regard the function $|F(\omega)|^2$ as a measure of the intensity at the frequency ω. Parseval's relation allows us to given these two statements a consistent quantitative meaning: if $|f(t)|^2\delta t$ is the power content of $f(t)$ in the time interval from t to $t + \delta t$, then we may interpret $|F(\omega)|^2\delta\omega/2\pi$ as the power content in the frequency range ω to $\omega + \delta\omega$, for then the relation

$$\int_{-\infty}^{\infty} |f(t)|^2 dt = \frac{1}{2\pi}\int_{-\infty}^{\infty} |F(\omega)|^2\, d\omega \tag{36}$$

gives an unambiguous meaning to the concept of the total power content of the quantity $f(t)$.

Illustrative Example: Consider the following simple mechanical problem. A mass m is suspended by a spring with force constant k, subject to a linear damping force proportional to its velocity and driven by an external force $f(t)$. The equation for the displacement of the particle from equilibrium is

$$m\, x''(t) + \gamma\, x'(t) + k\, x(t) = f(t). \tag{37}$$

For simplicity we put $m = 1$ and $k = 1$, and consider the case of light damping, $\gamma \ll 1$. First we make the driving force a periodic function, $f(t) = \sin(\omega t)$, and look for steady-state solutions $x(t) = B\sin(\omega t + \beta)$, where B and β are functions of ω but not t. Direct substitution into (37) gives us two relations for these quantities, namely

$$\begin{aligned} B[(1-\omega^2)\cos\beta - \omega\gamma\sin\beta] &= 1, \\ B[(1-\omega^2)\sin\beta + \omega\gamma\cos\beta] &= 0, \end{aligned} \tag{38}$$

from which we obtain

$$|B|^2 = \frac{1}{(1-\omega^2)^2 + \omega^2\gamma^2} . \tag{39}$$

The rate at which energy is dissipated by friction is $\gamma|x'(t)|^2$, hence the energy dissipated per cycle is

$$\int_0^{2\pi/\omega} \gamma|x'(t)|^2\, dt = \frac{\pi\gamma}{\omega}|\omega B|^2. \tag{40}$$

A graph of this quantity is shown in Figure 3.

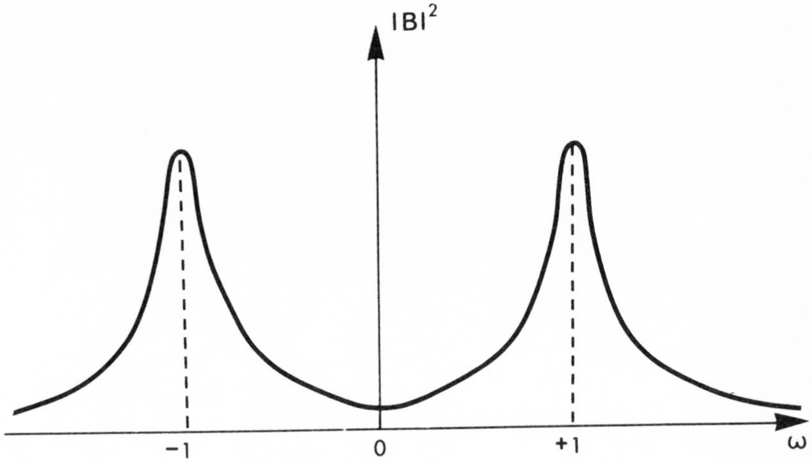

Figure 7.3

Notice that for small γ the peaks at $\omega \simeq \pm 1$ are high and narrow.

Now we apply the force

$$f(t) = \begin{cases} 0 \ , & t < 0 \\ \sin t, & 0 \le t \le 2\pi \\ 0 \ , & t > 2\pi \end{cases} \tag{41}$$

which is one cycle of a sine wave at the resonant frequency. The solution of (37) for $\gamma << 1$ and this force is approximately given by

$$x(t) = \begin{cases} \frac{1}{2}(\sin t - t \cos t), & 0 \le t \le 2\pi \\ -\pi e^{-\gamma t/2} \cos t, & t \ge 2\pi. \end{cases} \tag{42}$$

Let us calculate the total energy dissipated by friction as the result of this 'one-cycle' signal. There may seem to be two methods, viz.:

(i) Use the solution (42) to calculate the integral of $\gamma|x'(t)|^2$. Explicitly, this gives

$$E = \int_0^\infty \gamma|x'(t)|^2 dt \simeq \frac{\pi^2}{2} . \tag{43}$$

(ii) Use (40) with $\omega = 1$, and multiply by the period 2π during which the force is applied. This gives π/γ as the energy, a result which disagrees completely with (43).

The resolution of this problem is quite easy if we apply the concept of spectral analysis to the force $f(t)$. Writing

$$f(t) = \frac{1}{2\pi} \int_{-\infty}^{\infty} F(\omega)\ e^{-i\omega t}\ d\omega \tag{44}$$

and applying the steady-state result (38) to each harmonic component separately, we obtain for $x(t)$

$$x(t) = \frac{1}{2\pi} \int_{-\infty}^{\infty} F(\omega)\ B(\omega)\ e^{-i[\omega t+\beta(\omega)]}\ d\omega. \tag{45}$$

To compute the total energy, we apply Parseval's relation to the integral of $\gamma|x'(t)|^2$, giving

$$E = \frac{\gamma}{2\pi} \int_{-\infty}^{\infty} |F(\omega)|^2 \; |\omega B(\omega)|^2 \, d\omega \\ \simeq \frac{\pi^2}{2} \tag{46}$$

which agrees with (43). This illustrates the fact that the energy is spread out over a wide range of frequencies.

7.4. Kramers-Krönig Relations

Consider a linear physical system with input $x(t)$ and output (response) $y(t)$. We suppose that the law of cause and effect holds, that is, that the output $y(t)$ depends only on values of the input $x(t')$ for times $t' \leq t$. Then the most general linear relation which we may write is a convolution integral involving an influence function $k(t-t')$; viz.

$$y(t) = \int_{-\infty}^{\infty} k(t-\tau)\, x(\tau)\, d\tau, \tag{47}$$

where $k(s) = 0$ for $s < 0$ because of causality. We suppose, further, that the system is unconditionally stable, so that if $x(t) = 0$ for $t > t_0$, then $y(t) \to 0$ as $t \to \infty$. This means that the Fourier transform $K(\omega)$ has no poles in the half-plane $\mathrm{Im}(\omega) \geq 0$. Finally, suppose that a real input results in a real output; then we may show that the real and imaginary parts of $K(\omega)$ are, respectively, even and odd functions of the real variable ω.

Now consider the contour integral

$$\int_C \frac{K(\omega)\, d\omega}{\omega - \Omega} \tag{48}$$

where the contour is shown in Figure 4.

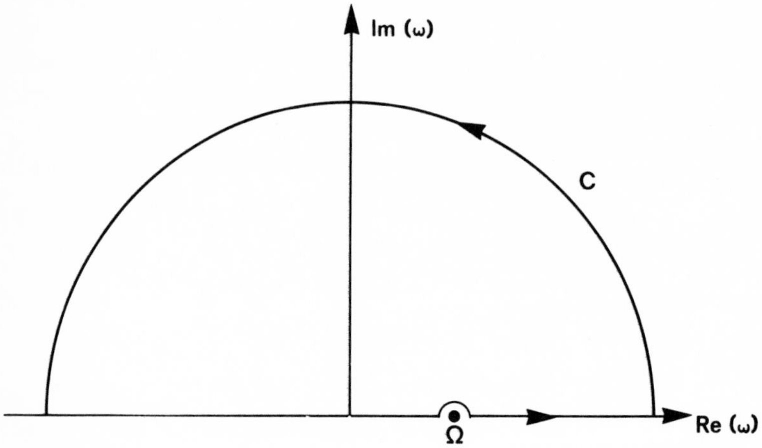

Figure 7.4

We know that $K(\omega)$ has no poles inside the contour, hence the integral has the value zero. Evaluating one-half the residue at $\omega = \Omega$ therefore gives

$$i\pi \; K(\Omega) + \mathrm{PV} \int_{-\infty}^{\infty} \frac{K(\omega)\; d\omega}{\omega - \Omega} = 0, \tag{49}$$

provided $K(\omega) \to 0$ as $|\omega| \to \infty$ in the upper half plane. Equating real and imaginary parts we have [with $K(\omega) = K_r(\omega) + iK_i(\omega)$]

$$\begin{aligned} K_r(\Omega) &= -\frac{1}{\pi}\,\mathrm{PV} \int_{-\infty}^{\infty} \frac{K_i(\omega)\; d\omega}{\omega - \Omega}, \\ K_i(\Omega) &= \frac{1}{\pi}\,\mathrm{PV} \int_{-\infty}^{\infty} \frac{K_r(\omega)\; d\omega}{\omega - \Omega}, \end{aligned} \tag{50}$$

which are the Kramers-Krönig relations. Thus the requirement of causality leads to a connection between the real and imaginary parts of $K(\omega)$ for a very general class of linear systems.

Another important relation of this type is obtained by considering the integral (see Figure 5)

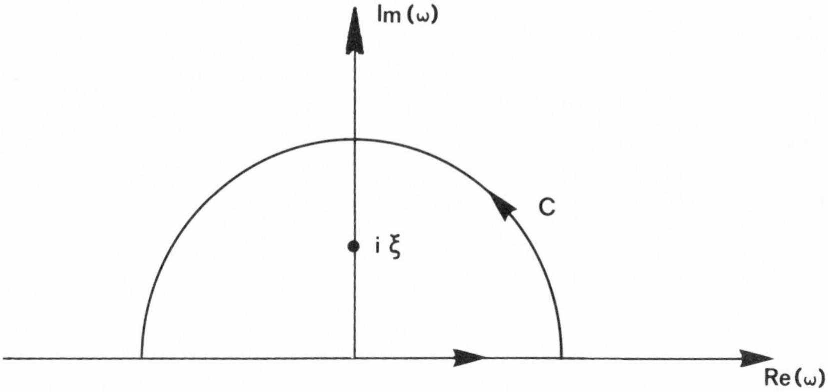

Figure 7.5

$$\int_C \frac{\omega\ K(\omega)}{\omega^2+\xi^2}\ d\omega = \pi i\ K(i\xi). \tag{51}$$

Exploiting the fact that $K_r(\omega)$ is an even function of (real) ω, this gives

$$K(i\xi) = \frac{2}{\pi}\int_0^\infty \frac{\omega\ K_i(\omega)\ d\omega}{\omega^2 + \xi^2}, \tag{52}$$

an important relation with many physical applications.

Problems

Prove the following general properties of the Fourier transform.

1. $\mathcal{F}[f^*(t)] = F^*(-\omega)$

2. If $f(t)$ is an even function, $F(\omega) = F_c(\omega)$.

3. If $f(t)$ is an odd function, $F(\omega) = -iF_s(\omega)$.

4. $\mathcal{F}[f(t/a) + b] = a\ e^{iab\omega}\ F(a\omega)$

5. $\mathcal{F}[e^{ibt}\ f(at)] = \frac{1}{a}\ F(\frac{\omega-b}{a})$

6. $\mathcal{F}_s[\cos(bt)\ f(at)] = \frac{1}{2a}\ [F_s(\frac{\omega+b}{a}) + F_s(\frac{\omega-b}{a})]$

7. $\mathcal{F}_s[\sin(bt)\ f(at)] = \frac{1}{2a}\ [F_c(\frac{\omega-b}{a}) - F_c(\frac{\omega+b}{a})]$

8. $\mathcal{F}_c[\cos(bt)\ f(at)] = \frac{1}{2a}\ [F_c(\frac{\omega+b}{a}) + F_c(\frac{\omega-b}{a})]$

9. $\mathcal{F}_c[\sin(bt)\ f(at)] = \frac{1}{2a}\ [F_s(\frac{\omega+b}{a}) - F_s(\frac{\omega-b}{a})]$

10. $\mathcal{F}_c[f^{(n)}(t)] = -2f^{(n-1)}(0) + \omega F_s[f^{(n-1)}(t)]$

11. $\mathcal{F}_s[f^{(n)}(t)] = -\omega F_c[f^{(n-1)}(t)]$

12. $\mathcal{F}_s[f''(t)] = 2\omega\ f(0) - \omega^2 F_s(\omega)$

13. $\mathcal{F}_c[f''(t)] = -2f'(0) - \omega^2 F_c(\omega)$

14. $\frac{1}{2\pi}\int_0^\infty F_c(\omega)G_c(\omega)\cos(\omega t)\ d\omega$

$$= \frac{1}{2}\int_0^\infty g(u)[f(t+u) + f(|t-u|)]\ du$$

15. $\frac{1}{2\pi}\int_0^\infty F_s(\omega)\ G_s(\omega)\cos(\omega t)\ d\omega$

$$= \frac{1}{2}\int_0^\infty g(u)\ [f(t+u) + f(|t-u|)]\ du$$

16. $\frac{1}{2\pi}\int_0^\infty F_s(\omega)\ G_c(\omega)\sin(\omega t)\ d\omega$

$$= \frac{1}{2}\int_0^\infty f(u)\ [g(|t-u|) - g(t+u)]\ du$$

$$= \frac{1}{2}\int_0^\infty g(u)\ [f(t+u) - f(|t-u|)]\ du$$

17. $\int_0^\infty f(t)\ g(t)\ dt = \frac{1}{2\pi}\int_0^\infty F_s(\omega)\ G_s(\omega)\ d\omega$

$$= \frac{1}{2\pi}\int_0^\infty F_c(\omega)\ G_c(\omega) d\omega$$

18. For a function $f(x)$ with Fourier transform $F(p)$, we define the quantities

$$\langle x^n\rangle = \int_{-\infty}^{\infty} x^n |f(x)|^2 dx,$$

$$(\Delta x)^2 = \langle x^2\rangle - \langle x\rangle^2,$$

$$\langle p^n\rangle = \frac{1}{2\pi}\int_{-\infty}^{\infty} p^n |F(p)|^2\, dp,$$

$$(\Delta p)^2 = \langle p^2\rangle - \langle p\rangle^2.$$

Show that[4]

$$(\Delta x)\,(\Delta p) \geq \frac{1}{2}$$

for any function $f(x)$.

[Consider the inequality

$$\int_{-\infty}^{\infty} |\{xf(x) - \langle x\rangle f(x)\} + \{f'(x)+i\langle p\rangle f(x)\}|^2\, dx \geq 0$$

where α is an arbitrary real number.]

Verify the following list of Fourier transforms:

19. $$f(t) = \begin{cases} e^{iat}, & p < t < q \\ 0, & t > q \quad \text{or} \quad t < p \end{cases}$$

$$F(\omega) = \frac{e^{ip(a+\omega)} - e^{iq(a+\omega)}}{\omega}$$

20. $$f(t) = \begin{cases} 1/t, & t > 1 \\ 0, & t < 1 \end{cases}$$

$$f(\omega) = -E_1(-i\omega)$$

21. $$f(t) = \cos(at^2)$$

$$F(\omega) = (\pi/a)^{1/2}\cos[(\omega^2/4a)-(\pi/4)]$$

22. $f(t) = \sin(at^2)$

$F(\omega) = (\pi/p)^{1/2} \sin[(\omega^2/4a)+(\pi/4)]$

23. $f(t) = |t|^{-\alpha}, \quad 0 < \mathrm{Re}(\alpha) < 1$

$$F(\omega) = \frac{2(-\alpha)!\ \sin(\pi\alpha)}{|\omega|^{1-\alpha}}$$

24. $$f(t) = \frac{e^{-a|t|}}{|t|^{1/2}}$$

$$F(\omega) = \left[\frac{\sqrt{a^2+\omega^2} + a}{a^2+\omega^2}\right]^{1/2}$$

25. $$f(t) = \begin{cases} (a^2-t^2)^{\nu}, & t < a \\ 0, & t > a \end{cases}$$

$$F(\omega) = 2^{\nu+1/2} \sqrt{\pi}\ \nu!\ a^{\nu+1/2}\ \omega^{-\nu-1/2}\ J_{\nu+1/2}\ (a\omega)$$

26. $$f(t) = \frac{\sinh\ (at)}{\sinh\ (\pi t)}, \quad -\pi < a < \pi$$

$$F(\omega) = \frac{\sin(a)}{\cosh(\omega) + \cos\ (a)}$$

27. $$f(t) = \frac{\cosh\ (at)}{\cosh\ (\pi t)}, \quad -\pi < a < \pi$$

$$F(\omega) = \frac{2\cos\ (a/2)\ \cosh(\omega/2)}{\cosh(\omega) + \cos(a)}$$

28. $$f(t) = \frac{\sin[b\sqrt{a^2 + t^2}]}{\sqrt{a^2 + t^2}}$$

$$F(\omega) = \begin{cases} 0, & |\omega| > b \\ \pi J_0\ (a\sqrt{b^2-\omega^2}), & |\omega| < b \end{cases}$$

29. $f(x) = \operatorname{erf}(ax)$

$$F(\omega) = \frac{e^{-\omega^2/4a^2}}{\omega}$$

30. $f(t) = e^{-t}$

$$F_s(\omega) = \frac{2\omega}{1+\omega^2}$$

31. $f(t) = e^{-t^2}$

$$F_s(\omega) = \sqrt{\pi}\, e^{-\omega^2/4}$$

32. $f(t) = \dfrac{\sin(t)}{t}$

$$F_s(\omega) = \ell n \left|\frac{1+\omega}{1-\omega}\right|$$

33. $$f(t) = \begin{cases} t(a^2-t^2)^{\nu}, & 0 < t < a \\ 0, & t > a \end{cases}$$

$$F_s(\omega) = 2^{\nu+1/2}\sqrt{\pi}\,\nu!\, a^{\nu+3/2}\,\omega^{-\nu-1/2}\, J_{\nu+3/2}(a\omega)$$

34. $f(t) = \dfrac{\cosh(at)}{\sinh(\pi t)}$

$$F_s(\omega) = \frac{\sinh(\omega)}{\cosh(\omega) + \cos(a)}$$

35. $F(t) = \dfrac{\sinh(at)}{\cosh(\pi t)}$

$$F_s(\omega) = \frac{2\sin(a/2)\sinh(\omega/2)}{\cosh(\omega) + \cos(a)}$$

Use the Parseval relations to evaluate the following integrals.

36. $$\int_0^\infty \frac{dt}{(a^2+t^2)(b^2+t^2)} = \frac{\pi}{2ab(a+b)}$$

37. $$\int_0^\infty \frac{t^2\,dt}{(a^2+t^2)(b^2+t^2)} = \frac{\pi}{2(a+b)}$$

38. $$\int_0^\infty \frac{\sin(at)\,\sin(bt)}{t^2}dt = \begin{cases} \pi a/2, & a < b \\ \pi b/2, & a > b \end{cases}$$

Poisson Summation Formula

39. Let $f(x)$ be an integrable function, and define

$$f_+(x) = f(x)h(x),$$

$$f_-(x) = f(x)h(-x),$$

$$F_+(\omega) = \mathscr{F}[f_+(x)],$$

$$F_-(\omega) = \mathscr{F}[f_-(x)].$$

Then show that

$$\frac{1}{2}f(0) + \sum_{n=1}^{\infty} f_+(\alpha n) = \frac{1}{\alpha}\sum_{m=-\infty}^{\infty} F_+(\frac{2\pi m}{\alpha}),$$

$$\frac{1}{2}f(0) + \sum_{n=-1}^{-\infty} f_-(\alpha n) = \frac{1}{\alpha}\sum_{m=-\infty}^{\infty} F_-(\frac{2\pi m}{\alpha}),$$

by representing $f_\pm(\alpha n)$ by its inverse Fourier transform and evaluating the integral (after performing the summation first) by residues. Adding these gives

$$\sum_{n=-\infty}^{\infty} f(\alpha n) = \frac{1}{\alpha}\sum_{m=-\infty}^{\infty} F(\frac{2\pi m}{\alpha}),$$

known as the Poisson summation formula.

40. Show that

$$\sum_{n=-\infty}^{\infty} \frac{1}{1+\alpha^2 n^2} = \frac{\pi}{\alpha}\coth\,(\frac{\pi}{\alpha})$$

41. $$\sum_{n=-\infty}^{\infty} e^{-n^2q^2} = \frac{\sqrt{\pi}}{q}\sum_{m=-\infty}^{\infty} e^{-m^2\pi^2/q^2}$$

42.
$$\sum_{n=-\infty}^{\infty} J_0(na) = \begin{cases} \frac{2}{a}, & 0 < a < 2\pi \\ \frac{2}{a} + 4\sum_{k=1}^{m} \frac{1}{\sqrt{a^2-4\pi^2k^2}}, & 2m\pi<a<2(m+1)\pi \end{cases}$$

43.
$$\sum_{n=-\infty}^{\infty} J_0(n\pi)\cos(n\pi a) = \frac{2}{\pi\sqrt{1-a^2}}, \quad -1 < a < 1$$

44.
$$\sum_{n=-\infty}^{\infty} \frac{(-1)^n}{\sqrt{a^2+\pi^2n^2}} = \frac{2}{\pi}\sum_{m=-\infty}^{\infty} K_0[(2m+1)a]$$

Footnotes

1. This method leaves the value of $f(0)$ undetermined, a matter of no practical consequence since inverse transforms are unique only to within a null function.

2. Alternatively, we could appeal to Bessel's integral (20.50) immediately to obtain the result (19).

3. We will see in Section 9 that (25) and (26) are valid for generalized functions with no additional assumptions.

4. In quantum mechanics, this is the uncertainty principle.

§8. APPLICATION TO PARTIAL DIFFERENTIAL EQUATIONS

The use of the Fourier transform to obtain a form of solution to a partial differential equation (together with associated boundary conditions) is a very general technique. For simple problems, the integral representation obtained as the solution will be amenable to exact analysis; more often the method converts the original problem to the technical matter of evaluating a difficult integral. Numerical methods may be necessary in general, although asymptotic and other useful information can often be obtained directly by appropriate methods. We illustrate some of the more simple problems in this section, leaving applications involving mixed boundary values, Green's functions, and transforms in several variables until later.

8.1. Potential Problems

Problems in electrostatics and steady-state heat conduction involve the solution of Laplace's equation

$$\nabla^2\phi = 0 \tag{1}$$

subject to prescribed boundary conditions on the function ϕ. We consider three examples.

Example 1:

$$\begin{aligned} &\nabla^2\phi(x,y) = 0, \quad -\infty < x < +\infty, \quad y \geq 0 \\ &\phi(x,0) = \psi(x) \end{aligned} \tag{2}$$

If we take the Fourier transform of (2) with respect to the variable x, we obtain the ordinary differential equation

$$\frac{d^2}{dy^2}\Phi(\omega,y) - \omega^2\Phi(\omega,y) = 0, \qquad \Phi(\omega,0) = \Psi(\omega), \tag{3}$$

with the solutions

$$\Phi(\omega,y) = A(\omega)\, e^{\omega y} + B(\omega)\, e^{-\omega y}. \tag{4}$$

To choose the functions $A(\omega)$ and $B(\omega)$--constants with respect to y--we need a further condition. This is because we have not specified the behavior of the solution for large y. Assuming that the solution is bounded for large y we obtain[1]

$$\Phi(\omega,y) = \begin{cases} \Psi(\omega)\, e^{-\omega y}, & Re(\omega) > 0 \\ \Psi(\omega)\, e^{+\omega y}, & Re(\omega) < 0. \end{cases} \tag{5}$$

The Fourier transform $\Phi(\omega,y)$ is a product, so we introduce a function $K(\omega,y)$ by $\Phi(\omega,y) = K(\omega,y)\Psi(\omega)$. From (7.11) we see that $K(\omega,y)$ is the transform of an elementary function, and on using the convolution theorem we obtain as the solution to (2) the general formula

$$\phi(x,y) = \frac{y}{\pi}\int_{-\infty}^{\infty} \frac{\psi(\xi)\, d\xi}{y^2+(x-\xi)^2}. \tag{6}$$

Example 2: Consider the electrostatic field produced by the arrangement shown in Figure 1, where a finite section of an infinite electrically conducting cylinder of radius a is held at potential V while the remainder is grounded. Using cylindrical polar coordinates, and the fact that the potential is axially symmetric, we obtain the equations

$$\frac{1}{r}\frac{\partial}{\partial r}\left(r\frac{\partial\phi}{\partial r}\right) + \frac{\partial^2\phi}{\partial z^2} = 0,$$

$$\phi(a,z) = \begin{cases} V, & |z| < \ell \\ 0, & |z| > \ell. \end{cases} \tag{7}$$

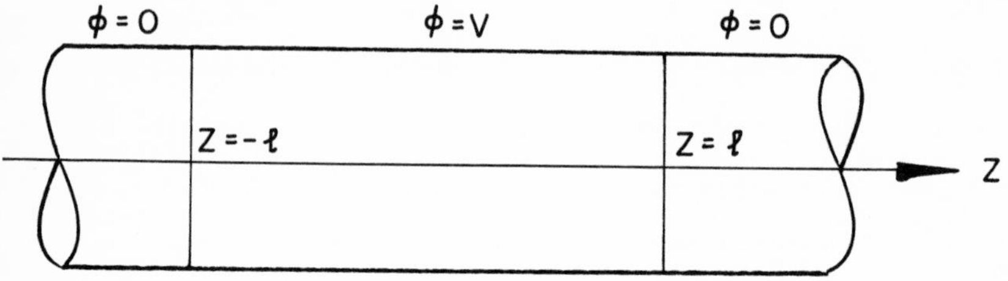

Figure 8.1

Now we take the Fourier transform with respect to z, so that equations (7) become

$$\frac{d^2}{dr^2}\Phi(r,\omega) + \frac{1}{r}\frac{d}{dr}\Phi(r,\omega) - \omega^2\Phi(r,\omega) = 0,$$
$$\Phi(a,\omega) = \frac{2V\sin(\omega\ell)}{\omega}. \tag{8}$$

The differential equation can be solved in terms of modified Bessel functions. The solution which is finite at $r = 0$ and satisfies the boundary condition at $r = a$ is

$$\Phi(r,\omega) = \frac{2V\sin(\omega\ell)}{\omega}\frac{I_0(\omega r)}{I_0(\omega a)}. \tag{9}$$

The expression for the potential follows immediately from the inversion integral, which may be evaluated over real values of ω. Noting that $\Phi(r,\omega)$ is an even function of ω, we can write the solution as a real integral, namely

$$\phi(r,z) = \frac{2V}{\pi}\int_0^\infty \frac{\cos(\omega z)\sin(\omega\ell)}{\omega}\frac{I_0(\omega r)}{I_0(\omega a)}\,d\omega. \tag{10}$$

This solution could also have been obtained by applying the Fourier cosine transform to the problem for $z \geq 0$, since the potential is obviously an even function of z, which implies that $\partial\phi/\partial z = 0$ at $z = 0$.

Example 3: As a final example of a potential problem, we find the electrostatic field inside an infinite conducting cylinder of radius a due to a point charge q on its axis.[2] If we take the origin of our (cylindrical) coordinates at the point charge, the potential has the form

$$\phi(r,z) = \frac{q}{\sqrt{r^2+z^2}} + u(r,z), \tag{11}$$

where the first term is the field due to a point charge in the absence of the boundary, and $u(r,z)$ is a harmonic function chosen to make $\phi = 0$ on the surface of the cylinder. Taking the Fourier transform with respect to z, we obtain from Laplace's equation $(\nabla^2 u = 0)$ and the boundary condition the equations

$$\begin{aligned} \frac{1}{r}\frac{d}{dr}\left(r\frac{dU(r,\omega)}{dr}\right) - \omega^2 U(r,\omega) &= 0, \\ U(a,\omega) = -q\int_{-\infty}^{\infty}\frac{e^{i\omega z}\,dz}{\sqrt{z^2+a^2}} & \\ = -2q\,K_0(\omega a). & \end{aligned} \tag{12}$$

The solution of (12) which is finite at $r = 0$ is

$$U(r,\omega) = -2q\,K_0(\omega a)\frac{I_0(\omega r)}{I_0(\omega a)}, \tag{13}$$

from which we have

$$\phi(r,z) = \frac{q}{\sqrt{r^2+z^2}} - \frac{2q}{\pi}\int_0^\infty \frac{K_0(\omega a)\,I_0(\omega r)}{I_0(\omega a)}\cos(\omega z)\,d\omega. \tag{14}$$

8.2. Water Waves: Basic Equations[3]

Water waves are perhaps the most easily observed oscillatory phenomenon in nature. We will discuss the application of integral transforms to the mathematical analysis of water waves in several places; in the present section we outline the basic equations to be solved and discuss some simple problems which are amenable to analysis via the Fourier transform.

Equations of Motion: We will briefly sketch those equations of hydrodynamics which are appropriate to the theory of water waves.[4] The basic assumption is that water may be regarded as an incompressible, inviscid fluid. We denote its density by ρ, and its velocity at a point $\underset{\sim}{r}$ and time t by $\underset{\sim}{v}(\underset{\sim}{r},t)$. A small element of the fluid is acted upon by two distinct forces: the pressure which acts across the boundary of the element, and external forces such as gravity (generally called body forces). Denoting the pressure by p and the body forces by $\underset{\sim}{F}$, Newton's second law gives the equation of motion

$$\rho \frac{D\underset{\sim}{v}}{Dt} = \underset{\sim}{F} - \underset{\sim}{\nabla} p, \tag{15}$$

where the operator D/Dt is the time rate of change for an observer moving with the element, at velocity $\underset{\sim}{v}$. In a stationary coordinate system, assuming that the body forces are simply gravitational, (15) becomes

$$\rho \frac{\partial \underset{\sim}{v}}{\partial t} + \rho(\underset{\sim}{v}\cdot\underset{\sim}{\nabla})\underset{\sim}{v} + \underset{\sim}{\nabla} p = \rho\underset{\sim}{g}, \tag{16}$$

where $\underset{\sim}{g}$ is a constant vector, the acceleration due to gravity. One consequence of (16) is that if $\underset{\sim}{\nabla} \times \underset{\sim}{v} = 0$ at one instant of time, then it is true for all time. We will make this assumption (irrotational flow). Then we can introduce a velocity potential ϕ from which $\underset{\sim}{v}$ is obtained by

$$\underset{\sim}{v} = \underset{\sim}{\nabla}\phi. \tag{17}$$

There is a fundamental conservation law of hydrodynamics which expresses the fact that fluid is neither created nor destroyed during its motion. For a fluid of arbitrary density $\rho(\underset{\sim}{r},t)$, it is

$$\frac{\partial\rho}{\partial t} + \underset{\sim}{\nabla}\cdot(\rho\underset{\sim}{v}) = 0. \tag{18}$$

In the present case, ρ is a constant, so (18) becomes $\underset{\sim}{\nabla}\cdot\underset{\sim}{v} = 0$. With (17), this implies

$$\nabla^2\phi = 0, \tag{19}$$

so that the velocity potential is a harmonic function.

Another equation can be obtained from (16) and (17). Writing $\underset{\sim}{v} = \underset{\sim}{\nabla}\phi$ in (16) and switching the order of differentiation in the first two terms, we have

$$\underset{\sim}{\nabla}\frac{\partial\phi}{\partial t} + \frac{1}{2}\underset{\sim}{\nabla}(v^2) + \underset{\sim}{\nabla}\frac{p}{\rho} - \underset{\sim}{g} = 0. \tag{20}$$

With the z-axis as the upward vertical direction, $\underset{\sim}{g} = -g\underset{\sim}{\nabla}z$, we can integrate (20) along any path in space (and within the fluid) to obtain Bernoulli's equation

$$\frac{\partial\phi}{\partial t} + \frac{1}{2}v^2 + \frac{p}{\rho} + gz = A(t), \tag{21}$$

where $A(t)$ is an arbitrary constant of the integration.

<u>Boundary Conditions</u>: We assume that the water has a boundary surface S with the property that any particle on the

surface remains on the surface. At a fixed boundary, such as the bottom of the sea, the normal velocity must be zero, so that the velocity potential has to satisfy the condition

$$\frac{\partial\phi}{\partial n} = 0. \tag{22}$$

At a free surface conditions are more complicated, since we will want to specify the pressure and let this determine where the surface is. If we assume that the equation of the surface is $z = \eta(x,y,t)$ then by differentiation with respect to t we have

$$\frac{\partial\phi}{\partial x}\frac{\partial\eta}{\partial x} + \frac{\partial\phi}{\partial y}\frac{\partial\eta}{\partial y} - \frac{\partial\phi}{\partial z} + \frac{\partial\eta}{\partial t} = 0 \quad \text{on} \quad z = \eta. \tag{23}$$

A second boundary condition follows from Bernoulli's equation. Assuming the pressure is equal to a prescribed function p_0 on S, we have

$$\frac{\partial\phi}{\partial t} + \frac{1}{2}|\underset{\sim}{\nabla}\phi|^2 + \frac{p_0}{\rho} + g\eta = A \quad \text{on} \quad z = \eta. \tag{24}$$

These boundary conditions are both nonlinear, and in addition must be used to determine η. Hence the general equations which govern the theory of water waves, even in this simple model, are extremely complicated and intractable to present analytic methods.

<u>Small Amplitude Waves</u>: We shall suppose that the elevation of the free surface η and the pressure p are small perturbations from equilibrium values $\eta = 0$ and $p = p_0$, and that the velocity of the flow is small. Then we can linearize (23) and (24) by neglecting products of perturbation quantities and applying the resulting conditions at $z = 0$ rather than $z = \eta$. Thus we have

$$g\eta + \frac{\partial\phi}{\partial t} + \frac{\delta p}{\rho} = 0 \quad \text{on} \quad z = 0, \tag{25}$$

$$\frac{\partial\eta}{\partial t} - \frac{\partial\phi}{\partial z} = 0 \quad \text{on} \quad z = 0, \tag{26}$$

where $\delta p = p - p_0$. We can eliminate η completely to get a boundary condition on ϕ, namely

$$\frac{\partial^2\phi}{\partial t^2} + g\frac{\partial\phi}{\partial z} + \frac{1}{\rho}\frac{\partial}{\partial t}(\delta p) = 0, \tag{27}$$

after which the free surface elevation η may be obtained from (25). Apart from the simplified form of these boundary conditions, we have the additional fact that the region in which ϕ must be determined and the boundary at which (25) and (26) must apply are fixed by the equilibrium solution.

8.3. Water Waves Generated by an Initial Surface Displacement

We consider first waves on water of infinite depth which are generated by an initial displacement of the surface elevation, the pressure at the surface being a given constant. In this section we confine our analysis to waves in two dimensions, that is, solutions of the relevant equations which are independent of y. For such functions, the Fourier transform (with respect to x) of (19) is

$$\frac{d^2}{dz^2}\Phi(\omega,z,t) - \omega^2\Phi(\omega,z,t) = 0, \tag{28}$$

and the solution for real ω which is bounded as $z \to -\infty$ is

$$\Phi(\omega,z,t) = A(\omega,t)\,e^{|\omega|z} \tag{29}$$

Now the boundary condition (27) with $p = 0$ yields for $A(\omega,t)$ the differential equation

$$\frac{\partial^2 A}{\partial t^2} + g|\omega|A = 0, \tag{30}$$

with the solution

$$A(\omega,t) = B(\omega)\sin(t\sqrt{|\omega|g}) + C(\omega)\cos(t\sqrt{|\omega|g}). \tag{31}$$

Suppose that the water is initially at rest, with surface elevation given by $\eta(x,0) = \eta_0(x)$. The condition $\phi(x,z,0) = 0$ gives $C(\omega) = 0$. Denoting the transform of η by H, (25) gives

$$g\,H_0(\omega) + \sqrt{|\omega|g}\;B(\omega) = 0. \tag{32}$$

Thus $\Phi(\omega,z,t)$ and $H(\omega,t)$ are uniquely determined. Explicitly,

$$\begin{aligned} \Phi(\omega,z,t) &= H_0(\omega)\left(\frac{g}{|\omega|}\right)^{1/2} \sin(t\sqrt{|\omega|g})e^{|\omega|z} \\ H(\omega,t) &= H_0(\omega)\cos(t\sqrt{|\omega|g}) \end{aligned} \tag{33}$$

The expressions for the velocity potential ϕ and surface elevation η follow immediately. The integrals for these expressions are intractable even for simple forms of $H_0(\omega)$; nevertheless they yield useful information, either through numerical studies or asymptotic analysis.

<u>An Asymptotic Form</u>: We examine in some detail the asymptotic form of the solution if we write

$$\eta_0(x) = \begin{cases} \frac{1}{2a}, & |x| < a \\ 0\;, & |x| > a. \end{cases} \tag{34}$$

Then the expression for $\phi(x,0,t)$ which we obtain from inverting (33a) is

$$\phi(x,0,t) = \frac{-\sqrt{g}}{2\pi}\int_0^\infty \frac{\sin(a\omega)}{a\omega}\{\sin(\omega x + t\sqrt{\omega g}) - \sin(\omega x - t\sqrt{\omega g})\}\frac{d\omega}{\sqrt{\omega}} \tag{35}$$

An important feature of this last expression is that it is unchanged if we replace x by $|x|$. Physically this must be so since the initial displacement $\eta_0(x)$ was symmetrical about $x = 0$, and hence the disturbance must propagate in both directions symmetrically. We now make the substitutions

$$\begin{aligned}\sqrt{\omega|x|} \pm \frac{t}{2}\sqrt{\frac{g}{|x|}} &= \alpha,\\ \frac{t}{2}\sqrt{\frac{g}{|x|}} &= \beta,\end{aligned} \tag{36}$$

after which the expression for ϕ becomes

$$\phi(x,0,t) = \frac{1}{\pi}\left(\frac{g}{|x|}\right)^{1/2}\left\{\int_\beta^\infty \frac{\sin[(a/|x|)(\alpha-\beta)^2]}{(a/|x|)(\alpha-\beta)^2}\sin(\alpha^2-\beta^2)\,d\alpha - \int_{-\beta}^\infty \frac{\sin[(a/|x|)(\alpha+\beta)^2]}{(a/|x|)(\alpha+\beta)^2}\sin(\alpha^2-\beta^2)\,d\alpha\right\}. \tag{37}$$

We now make some approximations. First, we note that the major contribution to each integral comes from the regions $|\alpha| \sim |\beta|$, so that if $a/|x| \ll 1$, we can write[5]

$$\phi(x,0,t) \simeq -\frac{2}{\pi}\left(\frac{g}{|x|}\right)^{1/2}\int_0^\beta \sin(\alpha^2-\beta^2)\,d\alpha, \qquad |x| \gg a. \tag{38}$$

The corresponding approximation to $\eta(x,t)$ is readily found to be

$$\begin{aligned}\eta(x,t) &= -\frac{1}{g}\frac{\partial\phi(x,0,t)}{\partial t}\\ &\simeq \frac{-t\,g^{1/2}}{\pi|x|^{3/2}}\int_0^\beta \cos(\alpha^2-\beta^2)\,d\alpha, \quad |x| \gg a.\end{aligned} \tag{39}$$

It is of interest to investigate these integrals for large β. The two integrals are the real and imaginary parts of

$$\int_0^\beta e^{i(\alpha^2-\beta^2)}\,d\alpha = e^{-i\beta^2+i\pi/4}\int_0^{\beta e^{-i\pi/4}} e^{-s^2}\,ds$$
$$= \frac{\sqrt{\pi}}{2}\,e^{-\beta^2+i\pi/4}\,\operatorname{erf}(\beta e^{-i\pi/4}). \tag{40}$$

For large β, the error function tends to unity, so that we have the approximation

$$\eta(x,t) \simeq \frac{-t\,g^{1/2}}{2\pi|x|^{3/2}}\cos\left[\frac{g\,t^2}{4|x|} - \frac{\pi}{4}\right], \quad gt^2 >> |x| >> a. \tag{41}$$

It is interesting to discuss the character of the motion furnished by this solution. The crests of the waves are given by the condition $gt^2/4|x| = (2n+\frac{1}{4})\pi$, hence they move at an increasing velocity as time progresses. Another feature is that the distance between two successive crests increases with time. Hence the waves furthest away from the initial disturbance move more rapidly and become longer as the pattern is drawn out. Simultaneously, new waves of shorter wave length continually appear and also propagate outwards. These conclusions remain unchanged if we consider the three dimensional case;[6] furthermore they may easily be observed by throwing a small stone into a calm pond.

8.4. Waves Due to a Periodic Disturbance: Radiation Condition

In this section we consider water waves generated by a pressure fluctuation which is periodic in time. We commence by investigating the solution of the hydrodynamic equations when the water is initially at rest with surface elevation $\eta = 0$, but subject to the pressure fluctuation

$$\delta p = \psi(x) \cos(\Omega t), \quad t \geq 0, \quad \Omega > 0. \tag{42}$$

For simplicity of the ensuing algebra, we replace the function $\cos(\Omega t)$ by $\exp(-i\Omega t)$ and afterwards take the real part of the solution. The Fourier transform $\Phi(\omega,z,t)$ is again given by (29), but now (30) is replaced by

$$\frac{\partial^2 A(\omega,t)}{\partial t^2} + g|\omega|\, A(\omega,t) = \frac{i\Omega}{\rho}\, \Psi(\omega)\, e^{-i\Omega t}, \tag{43}$$

which is obtained from the boundary condition (27). The solution is

$$A(\omega,t) = \frac{i\Omega\Psi(\omega)}{\rho(|\omega|g - \Omega^2)}\, e^{-i\omega t} + B(\omega)\, e^{it\sqrt{|\omega|g}} + C(\omega)\, e^{-it\sqrt{|\omega|g}}. \tag{44}$$

The initial condition $\Phi(\omega,z,0) = 0$ gives

$$B(\omega) + C(\omega) = \frac{-i\Omega\Psi(\omega)}{\rho(|\omega|g - \Omega^2)} \tag{45}$$

and from (25), together with the initial condition $\eta_0(x) = 0$ we obtain the further relation

$$B(\omega) - C(\omega) = \frac{i\sqrt{|\omega|g}\ \Psi(\omega)}{\rho(|\omega|g - \Omega^2)}. \tag{46}$$

It is now an easy matter to determine $\Phi(\omega,z,t)$; viz

$$\Phi(\omega,z,t) = \frac{i\Psi(\omega)\, e^{|\omega|z}}{\rho} \left[\frac{-\Omega\, e^{-i\Omega t}}{\Omega^2 - |\omega|g} + \frac{e^{it\sqrt{|\omega|g}}}{2(\Omega+\sqrt{|\omega|g})} + \frac{e^{-it\sqrt{|\omega|g}}}{2(\Omega-\sqrt{|\omega|g})} \right]. \tag{47}$$

Using the Fourier transform of (26) and noting that

$\partial\Phi/\partial z = |\omega|\Phi$, we obtain also the transform of the surface elevation as

$$H(\omega,t) = \frac{|\omega|\Psi(\omega)}{\rho(\Omega^2-|\omega|g)} [e^{-i\Omega t} - \cos(t\sqrt{|\omega|g}) + i(\frac{\Omega}{\sqrt{|\omega|g}}) \sin(t\sqrt{|\omega|g})]. \tag{48}$$

Steady-State and Transient Solutions: The inversion integral for $\eta(x,t)$ is over real values of ω, and it is evident from (48) that there is no singularity at $\Omega^2 = |\omega|g$, although each of the terms taken individually will lead to a singularity there. We will use this fact to split $\eta(x,t)$ into two terms: one term (coming from the factor $\exp(-i\Omega t)$ in the square brackets) is the steady periodic response to the periodic perturbation; the other, coming from the remaining two terms, dies away as t increases. To do this, we first write the inversion integral in the form

$$\eta(x,t) = \frac{1}{2\pi}\int_0^\infty [H(\omega,t)e^{-i\omega x} + H(-\omega,t)e^{i\omega x}]\, d\omega \tag{49}$$

and then deform the contour to the one shown in Figure 2 which avoids the point $\omega = \Omega^2/g$. This enables us to consider the various terms in (49) separately, and to write $\eta(x,t)$ as the sum of two functions $\eta^{(s)}(x,t)$ and $\eta^{(t)}(x,t)$, where

$$\eta^{(s)}(x,t) = \frac{e^{-i\Omega t}}{2\pi\rho}\int_C \frac{\omega}{\Omega^2-\omega g} [\Psi(\omega)e^{-i\omega x} + \Psi(-\omega)e^{i\omega x}]\, d\omega,$$

$$\eta^{(t)}(x,t) = \frac{1}{2\pi\rho}\int_C \frac{\sqrt{\omega}}{2\sqrt{g}} \left[\Psi(\omega)\left\{\frac{e^{-i\omega x+it\sqrt{\omega g}}}{\Omega+\sqrt{\omega g}} - \frac{e^{-i\omega x-it\sqrt{\omega g}}}{\Omega-\sqrt{\omega g}}\right\} + \Psi(-\omega)\left\{\frac{e^{i\omega x+it\sqrt{\omega g}}}{\Omega+\sqrt{\omega g}} - \frac{e^{i\omega x-it\sqrt{\omega g}}}{\Omega-\sqrt{\omega g}}\right\}\right] d\omega\,. \tag{50}$$

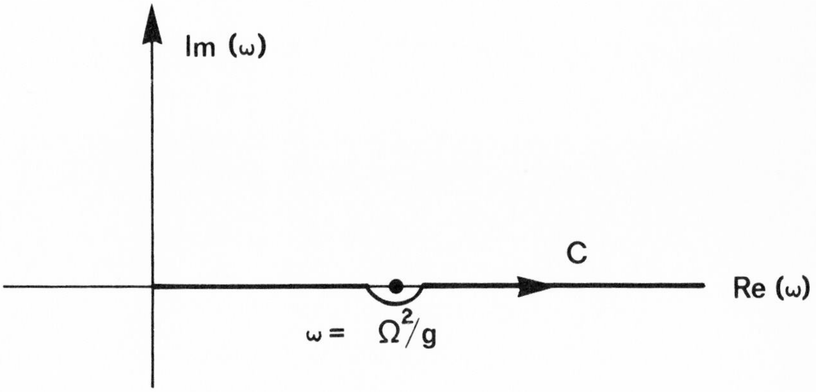

Figure 8.2

Now we may show that for any fixed x, $\eta^{(t)}$ is of order t^{-1} provided that $\Psi(\omega)$ falls off sufficiently fast as $\omega \to \infty$. To do this, we proceed as follows. First observe that the terms having the factor $\Omega + \sqrt{\omega g}$ as denominator may be evaluated as real integrals, and are of order t^{-1} for large t by Problem 2.1. Also, by the same consideration, the contribution from the other two terms which come from integrating along portions of the real axis are of order t^{-1}. If we denote the integral around the small semi-circle by I, the argument of $\omega - \Omega^2/g$ on the semi-circle by ϕ, and the minimum value of $|\Omega - \sqrt{\omega g}|$ by Δ, then we have the bound

$$
\begin{aligned}
|I| &\le A \int_{-\pi}^{0} e^{t\Delta \sin \phi} d\phi \\
&= \mathcal{O}(t^{-1})
\end{aligned}
\tag{51}
$$

so that $\eta^{(t)}$ is indeed a transient term, and the steady state is given by $\eta^{(s)}$.

Radiation Conditions: Suppose that $\psi(x) = 0$ for $|x| > a$; then $\Psi(\omega)$ is an analytic function which grows no faster than $\exp(a|\omega|)$ as $|\omega| \to \infty$. We may use this to estimate $\eta^{(s)}(x,t)$ for large x by deforming the integration contour to the imaginary ω axis. If $x > a$, we deform the contour to the positive imaginary axis for the term involving $\exp(i\omega x)$, and to the negative real axis for the term involving $\exp(-i\omega x)$. In the former case we must also pick up the residue at $\omega = \Omega^2/g$. Hence

$$\eta^{(s)}(x,t) = -\frac{\Omega^2 e^{-i\omega t}}{\pi\rho}\int_0^\infty \frac{\Psi(-i\xi)e^{-\xi x}}{\Omega^4+\xi^2 g^2}\,d\xi + \frac{i\Omega^2}{\rho g^2}\Psi(-\Omega^2/g)\,e^{-i\Omega t+i\Omega^2 x/g}. \tag{52}$$

For large x, the integral is of order x^{-1}, leaving the second term as the major contribution.[7] This describes traveling waves moving away from the disturbance at velocity g/Ω. Similar considerations show that for x large and negative, we again have outgoing traveling waves.

Our separation of $\eta(x,t)$ into transient and steady-state solutions has been quite tedious, and since we usually only want the steady state component, we now ask how this could be obtained directly. We therefore go back to the original equations, and assume that the pressure perturbation (42) is periodic for all t, $-\infty < t < \infty$. Consistent with this we must also assume that all other functions have the time dependence $\exp(-i\Omega t)$, which is equivalent to putting $B(\omega) = C(\omega) = 0$ in (44). This leads to functions $\Phi(\omega,z,t)$ and $H(\omega,t)$ which have poles on the real axis at $\omega = \pm\,\Omega^2/g$; consequently, the inversion contour must avoid these poles. All that our analysis of the transient terms has achieved has been to indicate the appropriate choice of contour. We could have done this more easily, however, by one of two methods.

(i) We could analyze the expression for $\eta^{(s)}(x,t)$ obtained for the various contour choices, and choose the contour which gives outgoing waves.

(ii) We could replace Ω by $\Omega + i\varepsilon$, so that the driving force is increasing exponentially. By taking the limit $\varepsilon \to 0$, we then recover the correct result. This procedure is known to physicists as "turning on the perturbation adiabatically" since the effect of the exponential growth is that there is no driving force for $t \to -\infty$.

Either of these procedures is simpler than the above analysis. The condition that the steady-state solution has outgoing waves only is known as a radiation condition.

Problems

1. Find the stationary temperature distribution $u(x,y)$ of a semi-infinite body $y \geq 0$, if the boundary is held at the temperature

$$u(x,0) = \begin{cases} T, & |x| < a \\ 0, & |x| < a. \end{cases}$$

2. Find the stationary temperature distribution $u(x,y)$ of a quadrant $x \geq 0$, $y \geq 0$, if the face $y = 0$ is held at zero temperature while the other, $x = 0$, is thermally insulated for $y \geq b$, while heat flows into the strip $0 \leq y < b$ at a constant density q. Find the distribution of heat flow through the face $y = 0$.

3. If a function $u(x,y)$ satisfies Laplace's equation in the quadrant $x \geq 0$, $y \geq 0$, and if it also satisfies the boundary conditions

$$u_x(0,y) = f(y), \quad y \geq 0$$
$$u(x,0) = 0 \quad , \quad x \geq 0$$

then show that

$$u_y(x,0) = -\frac{2}{\pi}\int_0^\infty \frac{t\ f(t)}{x^2+t^2}\,dt.$$

4. Show that the solution $\phi(x,y)$ of Laplace's equation in an infinite strip $-\infty < x < \infty$, $0 \leq y \leq a$, subject to the boundary conditions

$$\phi(x,0) = f(x),$$
$$\phi(x,a) = g(x),$$

is

$$\phi(x,y) = \frac{1}{2a}\sin\left(\frac{\pi y}{a}\right)\left\{\int_{-\infty}^\infty \frac{f(t)dt}{\cosh\pi(x-t)/a - \cos\pi y/a} + \int_{-\infty}^\infty \frac{g(t)dt}{\cosh\pi(x-t)/a + \cos\pi y/a}\right\}.$$

[Use Problem 7.26.]

5. Derive the solution to Problem 4 when the boundary conditions are

(i) $\phi(x,0) = f(x)$, $\phi_y(x,a) = g(x)$,

(ii) $\phi_y(x,0) = f(x)$, $\phi(x,a) = g(x)$,

(iii) $\phi_y(x,0) = f(x)$, $\phi_y(x,a) = g(x)$.

6. Solve Problem 4 in the special case

(i) $f(x) = g(x)\begin{cases} V_0, & |x| \geq 0 \\ 0\ , & |x| < 0 \end{cases}$

(ii) $f(x) = g(x)\begin{cases} V_0, & |x| \leq b \\ 0\ , & |x| > b \end{cases}$

7. Investigate the solutions of Laplace's equation in the semi-infinite strip $0 \leq x \leq \infty$, $0 \leq y \leq a$, using the Fourier sine or cosine transform as appropriate to the boundary condition at $x = 0$.

8. Show that the potential due to a point charge q placed on the axis of an infinite conducting cylinder of radius a is

$$\phi(r,z) = \frac{q}{\sqrt{r^2+z^2}} - \frac{2q}{\pi}\int_0^\infty \frac{K_0(\omega a)}{I_0(\omega a)} I_0(\omega r)\cos(\omega z)\, d\omega.$$

9. The end of a semi-infinite cylinder $0 \leq r \leq a$, $0 \leq z \leq \infty$ is held at constant temperature T_0, while the cylindrical surface is held at zero temperature. Show that the steady temperature distribution is given by

$$u(r,z) = T_0 \left[1 - \frac{2}{\pi}\int_0^\infty \frac{I_0(\omega r)}{I_0(\omega a)} \frac{\sin(\omega z)}{\omega}\, d\omega \right]$$

where $I_0(x)$ is a modified Bessel function.

10. By taking the Fourier sine transform in x, solve the one-dimensional diffusion equation

$$\frac{\partial^2 u}{\partial x^2} = \frac{1}{\kappa}\frac{\partial u}{\partial t}$$

on the line $x \geq 0$, subject to the boundary conditions

$$u(0,t) = f(t), \qquad t \geq 0,$$

$$u(x,0) = g(x), \qquad x \geq 0.$$

11. Solve Problem 9 if the first boundary condition is replaced by

$$u_x(0,t) = f(t), \qquad t > 0,$$

using the cosine transform.

12. Show that, if in considering (51) we took the contour around the pole on the other side, the conclusion regarding $\eta^{(t)}$ would be invalid.

13. Consider waves on water of finite depth h generated by an initial displacement $\eta_0(x)$ of the surface elevation. In particular, investigate the asymptotic form of the solution if

$$\eta_0(x) = \begin{cases} \frac{1}{2a}, & |x| < a \\ 0, & |x| > a. \end{cases}$$

14. Show that the radiation condition of Section 9.4 applies to the generation of water waves on water of finite depth by a periodic pressure fluctuation.

15. Consider (two-dimensional) waves on a stream of uniform depth h, whose unperturbed motion is a uniform velocity U in the positive x direction. Then the velocity potential may be written as $Ux + \phi(x,z,t)$, and the free surface conditions, after linearizing, are

$$\frac{p}{\rho} + g\eta + \frac{\partial\phi}{\partial t} + U\frac{\partial\phi}{\partial x} = 0,$$

$$\frac{\partial\eta}{\partial t} + U\frac{\partial\eta}{\partial x} - \frac{\partial\phi}{\partial z} = 0.$$

Show that if $p(x,t) = p(x)\cos(\Omega t)$, $t > 0$ with the motion undisturbed initially, then the following behaviour is predicted:[8]

(i) If $U^2 > gh$, the disturbance dies out both upstream and downstream of the region where $p(x)$ is nonzero.

(ii) If $U^2 < gh$, the disturbance dies out upstream, but at any downstream point there is, after sufficient time has elapsed, a steady periodic disturbance.

Footnotes

1. Note that Φ is not a meromorphic function even if Ψ is.

2. This problem anticipates some of the discussions of Section 11.

3. The standard reference on water waves is STOKER (1957).

4. A lucid exposition may be found in CURLE & DAVIES (1968), Ch. 21.

5. This follows because in this case

$$\frac{\sin(a/|x|)(\alpha \pm \beta)^2}{(a/|x|)(\alpha\pm\beta)^2} \simeq 1$$

when $(\alpha\pm\beta) \simeq 0$.

6. STOKER (1957), Ch. 4.

7. The result follows from Watson's lemma.

8. This problem is considered by K. K. Puri, J. Eng. Math. (1970), 4, 283.

§9. GENERALIZED FUNCTIONS

The subject of generalized functions is an enormous one, and we refer the reader to one of the excellent modern books[1] for a full account of the theory. We will sketch in this section some of the more elementary aspects of the theory, because the use of generalized functions adds considerably to the power of the Fourier transform as a tool.

9.1. The Delta Function

Generalized functions have their origin in Dirac's delta function, denoted $\delta(x-x_0)$, which is typically defined in books on quantum mechanics by: "$\delta(x-x_0)$ is zero everywhere except at $x = x_0$, where it is infinite; moreover it has the property that

$$\int_{-\infty}^{\infty} f(x)\ \delta(x-x_0)\ dx = f(x_0) \tag{1}$$

for any function which is sufficiently well behaved." Now it is evident that this definition is inconsistent, since if $\delta(x-x_0)$ is a function in the ordinary sense, then $\int\delta(x-x_0)f(x)dx = 0$ regardless of whether or not $\delta(x-x_0)$ is infinite at $x = x_0$. Thus if we wish to use (1), we must generalize the concept of a function so as to give the required formula (1) a precise meaning.

In the applications of mathematics to physical problems, functions are used to represent variables. For example, $E(t)$ might represent a voltage at time t. Now it is impossible to observe the instantaneous value of a voltage; we can only measure the effect of the voltage acting during a finite time interval. To consider a concrete example, suppose that the measuring process is linear, so that the measured value $\tilde{E}(t)$ is

$$\tilde{E}(t) = \int_0^\tau k(t')\, E(t-t')\, dt'. \tag{2}$$

In this situation, $E(t)$ can not be measured for any choice of function k, and so a theory which dealt with values of $\tilde{E}(t)$ directly would be sufficient. The essential difference between using $\tilde{E}(t)$ and $E(t)$ is seen more clearly by setting $t = 0$ in (2), for then we see that $\tilde{E}(0)$ is a function of the function $k(t')$. Functions which act on functions rather than numbers are usually called functionals; $\tilde{E}(t)$ is a functional, assigning a value to each pair of functions $k(t')$ and $E(t)$.

Other examples of functionals are the Fourier and Laplace transforms. The Fourier transform, for instance, assigns the value $F(\omega)$ to the function pair $f(x)$ and $\exp(i\omega x)$. The essential difference between these examples and the generalized functions which we are about to define is that we may evaluate the functionals by classical methods involving the use of functions. We have already seen that there is no such interpretation possible for (1); rather we may define the delta functional by

$$L[f] = f(x_0) \tag{3}$$

instead of (1). This we will now proceed to do, using a more convenient notation than (3).

9.2. Test Functions and Generalized Functions

We begin by defining the range of our generalized functions; that is, we define the functions, called "test functions", on which the functionals act. We choose[2] the set of all complex-valued functions of a real variable having the properties:

(i) Each function $\phi(x)$ has derivatives of every order for all x,

(ii) Each function $\phi(x)$ is zero outside some finite interval $a < x < b$. This interval is arbitrary, depending on the particular test function.

An example of a test function is

$$\phi(x) = \begin{cases} e^{-1/(1-x^2)}, & |x| < 1 \\ 0 & , |x| > 1. \end{cases} \tag{4}$$

The conditions imposed on the test functions are very restrictive,[3] so it is reassuring to note that for any continuous function $f(x)$ which is absolutely integrable there are test functions which are arbitrarily close, i.e., for any $\varepsilon > 0$ we may find a test function $\phi(x)$ such that

$$|f(x) - \phi(x)| < \varepsilon, \quad -\infty < x < \infty. \tag{5}$$

Such a function may be constructed as follows:

Choose a so that $|f(x)| < \varepsilon$ for $|x| > a$, and then let the functions $\phi_\alpha(x)$ be constructed from (4) by

$$\phi_\alpha(x) = \frac{\phi(x/\alpha)}{\int_{-\infty}^{\infty} \phi(x/\alpha)\, dx} . \tag{6}$$

Then it is not difficult to show that the functions $\psi_\alpha(x)$ defined by

$$\psi_\alpha(x) = \int_{-a}^{a} \phi_\alpha(x-x')\, f(x')\, dx' \tag{7}$$

are test functions, and that we may choose α_0 so that for all $\alpha > \alpha_0$, $\psi_\alpha(x)$ satisfies (5).

Properties of Test Functions: Some of the simplest and most useful properties of our test functions are as follows:

(i) They form a linear space. In particular this means that a finite linear combination of test functions is again a test function.

(ii) If $\phi(x)$ is a test function and $f(x)$ an infinitely differentiable function, then $f(x)\,\phi(x)$ is again a test function.

(iii) The Fourier transforms of test functions have a particularly simple form. Suppose that $\phi(x) = 0$ for $|x| > a$; then

$$\Phi(\omega) = \int_{-a}^{a} \phi(x)\, e^{i\omega x}\, dx. \tag{8}$$

Now this integral may be differentiated with respect to ω, so that it is an entire function.[4] Moreover, if we write $\omega = \sigma + i\tau$, then

$$\begin{aligned} |\Phi(\omega)| &< \int_{-a}^{a} |\phi(x)|\, e^{-\tau x}\, dx \\ &< A\, e^{a|\tau|} \end{aligned} \tag{9}$$

so that $\Phi(\omega)$ is an entire analytic function whose growth for large $|\omega|$ is bounded by an exponential function. Conversely, given any function $\Phi(\omega)$ with these properties it is easy to show that it is the Fourier transform of a test function.

Linear Functionals: A complex-valued function f of a real variable x may be defined as a rule which assigns a complex number [the value of $f(x)$] to each real x. The key to the theory of generalized functions is that this concept be relinquished in favor of a less restrictive one so that (1) can be given a precise meaning. This is afforded by the concept of a linear functional, that is, a rule, denoted $\langle f, \rangle$, which associates with every test function ϕ some complex number $\langle f,\phi\rangle$, such that

$$\langle f,\alpha\phi + \beta\psi\rangle = \alpha\langle f,\phi\rangle + \beta\langle f,\psi\rangle , \tag{10}$$

where ϕ and ψ are test functions and α and β are arbitrary constants. The important thing to note is that the linear functional assigns a value to each test function, not to each value of x.

An important class of linear functionals is the following: if $f(x)$ is any function which is integrable, then we define $\langle f,\phi\rangle$ by

$$\langle f,\phi\rangle = \int_{-\infty}^{\infty} f(x)\ \phi(x)\ dx. \tag{11}$$

Generalized Functions: Since the concept of continuity is of prime importance in the theory of ordinary functions, we define a similar concept for linear functionals. We will say that the sequence ϕ_n of test functions converges to zero if there is some interval $|x| \le a$ outside which all the $\phi_n(x)$ vanish, and inside which they converge to zero uniformly. Further, we will say that a linear functional $\langle f, \rangle$ is continuous if the sequence of numbers $\langle f,\phi_n\rangle$ tends to zero whenever the sequence of test functions ϕ_n converges to zero. Finally, we define generalized functions as the set of all continuous linear functionals acting on a set of test functions. In particular (11) defines a generalized function for any integrable function $f(x)$, since it is easy to show that it is continuous.[5] Functionals of this type, which correspond to ordinary integrable functions, are said to be regular generalized functions.[6] All others, (e.g., the delta function) are said to be singular.

We will often denote the generalized function $\langle f, \ \rangle$ simply by f. Thus we refer to the delta function as $\delta(x-x_0)$, although it is not a function of x, and what is actually defined is the operation on test functions

$$\langle\delta(x-x_0),\phi(x)\rangle = \phi(x_0). \tag{12}$$

It may readily be seen that this defines a continuous linear functional, i.e., a generalized function.

<u>Addition and Multiplication</u>: Suppose that f and g are generalized functions and $\psi(x)$ is infinitely differentiable (but not necessarily a test function). Then we define the generalized functions $f+g$ and ψf by

$$\langle f+g,\phi\rangle = \langle f,\phi\rangle + \langle g,\phi\rangle, \tag{13}$$

$$\langle\psi f,\phi\rangle = \langle f,\psi\phi\rangle. \tag{14}$$

Equation (14) is possible because $\psi\phi$ is a test function. As an example we define the generalized function $x\,\delta(x)$:

$$\begin{aligned}\langle x\delta(x),\phi(x)\rangle &= \langle\delta(x),x\phi(x)\rangle \\ &= 0\cdot\phi(0) \\ &= 0,\end{aligned} \tag{15}$$

which we write symbolically as

$$x\,\delta(x) = 0, \tag{16}$$

since the zero generalized function maps every test function to zero. Equation (16) shows that the equation $x\, f(x) = 0$ has as one possible solution $f(x) = A\,\delta(x)$, where A is an arbitrary constant. This is quite different from the situation with ordinary functions, for which the solution would be $f(x) = 0$ for $x \neq 0$, $f(0) = A$, giving the zero generalized function when this is substituted in (11) regardless of the value of A.

Generalized Functions over Finite Intervals: The test functions considered above are defined in the infinite interval $-\infty < x < \infty$. It is sometimes useful to use a finite interval $a \leq x \leq b$. In this case the additional restrictions $\phi^{(n)}(a) = \phi^{(n)}(b) = 0$, $n = 0,1,2,\ldots$ are applied. This ensures that the generalized functions have the same properties as for an infinite interval.

9.3. Elementary Properties

The manipulations of functions which are commonly useful in applied mathematics, such as differentiation, infinite summation, and changes of order of limiting processes, are accompanied by various restrictions on their validity. Nevertheless these conditions are often ignored in applications; for example, the delta function arose from the changing of orders of integration in quantum mechanics. We shall show in this section that the natural extensions of everyday concepts from functions to generalized functions leads to the lifting of many restrictions, and this is why generalized functions find so many applications.

Transformation of Variables: Given an integrable function

$f(x)$, we may define a regular generalized function by (11). Suppose, however, that we wish to use the function $f(g(x))$ in (11), where $g(x)$ is an infinitely differentiable monotone function satisfying $g(x) \to \pm\infty$ as $x \to \pm\infty$. We may relate the two generalized functions by

$$\begin{aligned} \langle f(g(x)),\ \phi(x)\rangle &= \int_{-\infty}^{\infty} f(g(x))\ \phi(x)\ dx \\ &= \int_{-\infty}^{\infty} [f(g)\phi(x(g))/g'(x(g))]\ dg \\ &= \langle f(g),\psi(g)\rangle, \end{aligned} \tag{17}$$

$$\psi(g) = \phi(x(g))/g'(x(g)).$$

Because of the properties of $g(x)$, $x(g)$ exists and is infinitely differentiable, and thus $\psi(g) = \phi(x(g))/g'(x(g))$ is a test function of g, so that the last line gives the meaning of $f(g(x))$ in terms of $f(x)$. We may use this relation to define the meaning of $f(g(x))$ when $f(x)$ is a singular generalized function; if $f(x)$ symbolically represents such a generalized function, then $f(g(x))$ is defined by (17) also. In this case, the intermediate calculations have no significance as the steps in an argument, rather they lead to a final result which is true by definition. As an example, $\delta(g(x))$ is the functional

$$\langle\delta(g(x)),\phi(x)\rangle = \phi(a)/g'(a), \tag{18}$$

where a is the unique solution of $g(a) = 0$.

<u>Differentiation</u>: We again commence with a regular generalized function (11) where the function $f(x)$ has a first derivative which is also differentiable. Then we may define

the generalized function $f'(x)$ by (11), and relate it to the original generalized function by integrating by parts, viz.

$$\begin{aligned} \langle f',\phi\rangle &= \int_{-\infty}^{\infty} f'(x)\ \phi(x)\ dx \\ &= -\int_{-\infty}^{\infty} f(x)\ \phi'(x)\ dx \qquad (19) \\ &= -\ \langle f,\phi'\rangle. \end{aligned}$$

Now we define the derivative of an arbitrary generalized function by $\langle f',\phi\rangle = -\langle f,\phi'\rangle$, and also denote this definition by

$$\int_{-\infty}^{\infty} f'(x)\ \phi(x)\ dx = -\int_{-\infty}^{\infty} f(x)\ \phi'(x)\ dx. \qquad (20)$$

If $f(x)$ is not a function satisfying the conditions for validity of equation (19), then (20) is a symbolic way of writing the definition. It emphasizes the important principle that we define the properties of generalized functions so that desirable manipulations are always valid.

Example 1: Consider the Heaviside step function, $h(x)$, which has the value unity for $x > 0$ and zero for $x < 0$. In the ordinary sense it may not be differentiated, but as a generalized function this causes no trouble. From (20) we have

$$\begin{aligned} \int_{-\infty}^{\infty} h'(x)\ \phi(x)\ dx &= -\int_{-\infty}^{\infty} h(x)\ \phi'(x)\ dx \\ &= \phi(0), \end{aligned} \qquad (21)$$

which is the defining relation for the delta function, hence

$$h'(x) = \delta(x). \qquad (22)$$

Example 2: Using (22) we may differentiate--in the

generalized sense--a function whose only problem is a finite number of finite discontinuities. We have represented such a

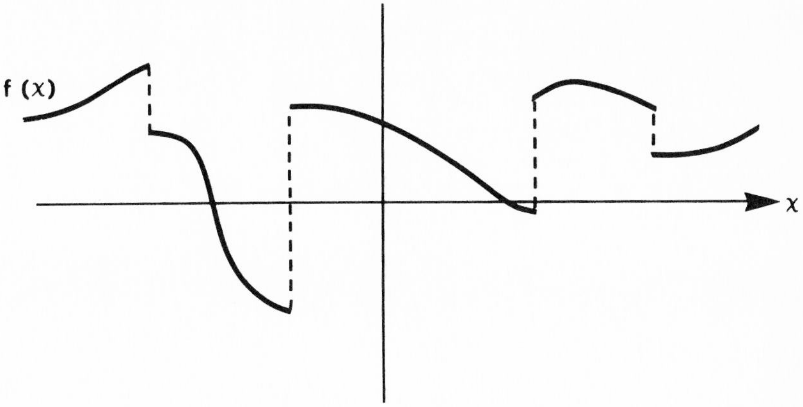

Figure 9.1

function schematically in Figure 1. In precise terms we assume that we may write

$$f(x) = f_1(x) + \sum_{k=1}^{n} A_k\ h(x-x_k), \tag{23}$$

where $f_1(x)$ is differentiable in the ordinary sense. Then we have

$$f'(x) = f_1'(x) + \sum_{k=1}^{n} A_k\ \delta(x-x_k) \tag{24}$$

for the generalized function.

<u>Convergence Properties</u>: A sequence of ordinary functions f_ν would be said to converge to the function f if for each x the numbers $f_\nu(x)$ converged to the number $f(x)$. Similarly, we shall say that the generalized functions f_ν

converge to the generalized function f if for each test function $\phi(x)$ the numbers $<f_\nu,\phi>$ converge to the number $<f,\phi>$. This definition is the same for integer values of ν or continuous values of ν. We give two examples:

(i) Let

$$f_\alpha(x) = e^{-\alpha x} h(x), \quad \alpha > 0. \tag{25}$$

Then for any test function ϕ

$$\lim_{\alpha\to 0} \int_{-\infty}^{\infty} f_\alpha(x)\, \phi(x)\, dx = \int_0^{\infty} \phi(x)\, dx; \tag{26}$$

hence we write

$$\lim_{\alpha\to 0} f_\alpha(x) = h(x). \tag{27}$$

(ii) Let

$$f_\nu(x) = \begin{cases} \nu, & |x| \leq 1/2\nu \\ 0, & |x| > 1/2\nu. \end{cases} \tag{28}$$

Then we have, for any test function ϕ,

$$\begin{aligned} \lim_{\nu\to\infty} \int_{-\infty}^{\infty} f_\nu(x)\, \phi(x)\, dx &= \lim_{\nu\to\infty} \nu \int_{-1/2\nu}^{1/2\nu} \phi(x)\, dx \\ &= \phi(0), \end{aligned} \tag{29}$$

which may be written as

$$\lim_{\nu\to\infty} f_\nu(x) = \delta(x). \tag{30}$$

This example shows that a sequence of regular generalized functions may converge to a singular generalized function.

Regular and Singular Generalized Functions: We will state here two important facts without proof.[7] First, if a sequence of generalized functions converges, then it converges to a generalized function. Second, every generalized function is the limit of a sequence of regular generalized functions. As an example of the first property, we mention the generalized function x^{-1}. This has no meaning as a regular generalized function, since the integral

$$\int_{-\infty}^{\infty} \frac{\phi(x)}{x}\,dx \tag{31}$$

diverges. However, the functions $f_\varepsilon(x) = 1/x$ for $|x| > \varepsilon$ and 0 for $|x| \leq \varepsilon$ define regular generalized functions, and the limit as $\varepsilon \to 0$ is also defined for all test functions. Hence we may define the singular generalized function x^{-1} by

$$\int_{-\infty}^{\infty} \frac{\phi(x)}{x}\,dx = \lim_{\varepsilon\to 0} \int_{|x|>\varepsilon} \frac{\phi(x)}{x}\,dx. \tag{32}$$

This is well known as the principal value of the integral.

<u>Differentiation of Sequences:</u> For any convergent sequence of generalized functions and any test function ϕ, we can write

$$\begin{aligned} \lim \langle f'_\nu,\phi\rangle &= -\lim\langle f_\nu,\phi'\rangle \\ &= -\langle f,\phi'\rangle \\ &= \langle f',\phi\rangle. \end{aligned} \tag{33}$$

Hence we can reverse the order of taking a limit and differentiating. In particular, if f_ν is the partial sum of an infinite series whose sum is f, then (33) shows that we may differentiate term by term.

As an example, consider the series $\sin x + \frac{1}{2}\sin 2x + \frac{1}{3}\sin 3x + \ldots$ which converges uniformly to a function

$f(x)$ equal to $(\pi-x)/2$ for $0 < x < 2\pi$ and periodic with period 2π for other real x. For any test function, we can multiply this series by $\phi(x)$ and integrate term by term, so that we can write

$$f(x) = \sin x + \frac{1}{2}\sin 2x + \frac{1}{3}\sin 3x + \ldots \qquad (34)$$

in the sense of generalized functions. Differentiating term by term we obtain the further relation

$$\frac{1}{2} + \cos x + \cos 2x + \cos 3x + \ldots = \pi \sum_{n=-\infty}^{\infty} \delta(x-2\pi n). \quad (35)$$

If we apply this to a test function $\phi(x)$ with Fourier transform $\Phi(\omega)$, then on writing the cosine as a sum of complex exponentials we obtain

$$\sum_{n=-\infty}^{\infty} \Phi(n) = 2\pi \sum_{n=-\infty}^{\infty} \phi(2\pi n), \qquad (36)$$

which is known as the Poisson summation formula.[8]

Consider the functions $g_\varepsilon(x) = \ln|x|\, h(|x| - \varepsilon)$, which converge to the integrable function $\ln|x|$ as $\varepsilon \to 0$. Now $g_\varepsilon'(x) = h(|x| - \varepsilon)/x + \ln\varepsilon\, [\delta(x-\varepsilon) - \delta(x+\varepsilon)]$, and on taking the limit $\varepsilon \to 0$ we readily show that

$$\lim_{\varepsilon\to 0} \langle g_\varepsilon'(x),\phi(x)\rangle = \langle \frac{1}{x}, \phi(x)\rangle. \qquad (37)$$

Thus $d(\ln|x|)/dx = 1/x$. Furthermore, we can apply this result to $\ln(x+iy)$, $y > 0$, where the branch of the logarithm is defined by

$$\ln(x+iy) = \frac{1}{2}\ln(x^2+y^2) + i\arctan(y/x). \qquad (38)$$

As $y \to 0$, the function $\ln(x+iy)$ converges to a function $f(x)$ defined by

$$f(x) = \ln |x| + i\pi h(-x). \tag{39}$$

Differentiating and using (37), we obtain the useful result[9]

$$\lim_{y \to 0} \frac{1}{x+iy} = \frac{1}{x} - i\pi \delta(x), \tag{40}$$

where the generalized function x^{-1} is the principal value integral when applied to a test function.

9.4. Analytic Functionals

We have already shown that the Fourier transform of a test function is an entire analytic function of ω which grows at most exponentially for large $|\omega|$. Let $\Phi(\omega)$ be such a function; then it is the Fourier transform of some test function $\phi(x)$. Hence for real ω we can write

$$\begin{aligned} |\Phi(\omega)| &= \left|\int_{-\infty}^{\infty} \phi(x)\, e^{i\omega x}\, dx\right| \\ &\leq \int_{-\infty}^{\infty} |\phi(x)|\, dx = A \end{aligned} \tag{41}$$

so that $\Phi(\omega)$ is bounded as $|\omega| \to \infty$. Moreover, the nth derivative of $\phi(x)$ is also a test function, so its Fourier transform is bounded for real ω. Applying (7.27), this means that

$$|\Phi(\omega)| < A_n |\omega|^{-n}, \qquad n = 0,1,2,\ldots \tag{42}$$

i.e., the functions $\Phi(\omega)$ fall off faster than any finite power of ω. Also, by using (7.27) we see that $\Phi(\omega)$ is infinitely differentiable. Thus we may use this set of functions[10] to set up generalized functions exactly as for the original test functions, and all the properties we have proved above will again apply.[11] In particular, regular generalized functions corresponding to (11) may be constructed; for integrable functions $F(\omega)$ we write

$$\langle F,\Psi\rangle = \int_{-\infty}^{\infty} F(\omega)\ \Psi(\omega)\ d\omega. \tag{43}$$

A particularly useful class of generalized functions over the test functions $\Psi(\omega)$ makes use of the fact that the latter are entire analytic functions. Then we define an analytic functional as

$$\langle G,\Psi\rangle = \int_{\Gamma} G(\omega)\ \Psi(\omega)\ d\omega, \tag{44}$$

where Γ is a given contour whose specification is an integral part of the definition of G.

Examples

(i) Consider the function $G(\omega) = (\omega-\omega_0)^{-1}$, where ω_0 is real. We may use it to define two different analytic functionals, namely

$$\begin{aligned} \langle G_+,\Psi\rangle &= \int_{ia-\infty}^{ia+\infty} \frac{\Psi(\omega)}{\omega-\omega_0}\, d\omega \\ \langle G_-,\Psi\rangle &= \int_{-ia-\infty}^{-ia+\infty} \frac{\Psi(\omega)}{\omega-\omega_0}\, d\omega \end{aligned} \tag{45}$$

where $a > 0$. From the property of residues, we see that

$$\langle G_+ - G_-,\Psi\rangle = -2\pi i\ \Psi(\omega_0), \tag{46}$$

which is written, in the notation of generalized functions, as

$$G_+(\omega) - G_-(\omega) = -\ 2\pi i\ \delta(\omega-\omega_0). \tag{47}$$

(ii) Motivated by the last example, we consider the functional

$$\frac{1}{2\pi i} \int_C \frac{\Phi(\omega)}{\omega - a}\, d\omega \tag{48}$$

where C encircles the point $\omega = a$ in a positive direction. From residue theory, the value of the integral is $\Psi(a)$, hence (48) is the generalized function $\delta(\omega - a)$. Analytic functionals are thus seen to encompass a wider class of generalized functions than regular functionals, at least in some respects.

(iii) The function $\exp(\omega^2)$ may be used to construct the analytic functional

$$\langle F, \Phi \rangle = \int_{-i\infty}^{i\infty} e^{\omega^2}\, \Phi(\omega)\, d\omega. \tag{49}$$

9.5. Fourier Transforms of Generalized Functions

The concept of the Fourier transform of a generalized function is a very powerful one and has many practical applications. To motivate the definition, we first consider those regular generalized functions which are constructed from ordinary functions having Fourier transforms for real ω. Then Parseval's relation (7.34) may be applied in either of two ways, viz.

$$\begin{aligned} \int_{-\infty}^{\infty} f(x)\, \phi(x)\, dx &= \frac{1}{2\pi} \int_{-\infty}^{\infty} F(\omega)\, \Phi(-\omega)\, d\omega \\ \int_{-\infty}^{\infty} f(x)\, \phi(-x)\, dx &= \frac{1}{2\pi} \int_{-\infty}^{\infty} F(\omega)\, \Phi(\omega)\, d\omega \end{aligned} \tag{50}$$

The integrals on the right hand side are particular types of analytic functionals, so we introduce the notation

$$\langle F, \Phi \rangle = \int_{-\infty}^{\infty} F(\omega)\, \Phi(\omega)\, d\omega. \tag{51}$$

Now (50) is a correspondence between generalized functions

over the test functions $\phi(x)$ and generalized functions over their Fourier transforms $\Phi(\omega)$. In the appropriate notation, we have

$$\langle F,\Phi(\omega)\rangle = 2\pi \langle f,\phi(-x)\rangle \tag{52}$$

and

$$\langle F,\Phi(-\omega)\rangle = 2\pi\langle f,\phi(x)\rangle. \tag{53}$$

Either of these relations will define a generalized function F corresponding to the generalized function f, even if f is not regular. Therefore we use (52) and (53) to define the Fourier transform of any generalized function. Thus by definition every generalized function f has a Fourier transform in this sense. We consider some simple examples.

(i)

$$f(x) = \delta(x-x_0). \tag{54}$$

Here (52) gives $\langle F,\Phi(\omega)\rangle = 2\pi\phi(-x_0)$. Using the Fourier inversion theorem to represent $\phi(-x_0)$, we obtain the analytic functional[12]

$$\langle F,\Phi(\omega)\rangle = \int_{-\infty}^{\infty} e^{+i\omega x_0}\,\Phi(\omega)\,d\omega. \tag{55}$$

(ii)

$$f(x) = e^{-i\omega_0 x} \tag{56}$$

Here (52) gives

$$\begin{aligned}\langle F,\Phi(\omega)\rangle &= 2\pi \int_{-\infty}^{\infty} e^{-i\omega_0 x}\,\phi(-x)\,dx \\ &= 2\pi\,\Phi(\omega_0)\end{aligned} \tag{57}$$

which is equivalent to $F = 2\pi\,\delta(\omega-\omega_0)$. From (48) above, we may also write F as an analytic functional, even though it is a singular generalized function.

Elementary Properties: Because of the restrictions on the test functions, we may apply (7.27) repeatedly to obtain

$$\mathscr{F}[\phi^{(n)}(x)] = (-i\omega)^n\ \Phi(\omega)$$
$$\mathscr{F}[(ix)^n\ \phi(x)] = \Phi^{(n)}(\omega) \tag{58}$$

for arbitrary n. We use these results to show that they also apply to arbitrary generalized functions. Considering first the derivatives of the generalized function $f(x)$, (58) yields

$$\begin{aligned} 2\pi\ \langle f^{(n)}(x),\phi(x)\rangle &= 2\pi\ (-1)^n\ \langle f(x),\phi^{(n)}(x)\rangle \\ &= (-1)^n\ \langle F(\omega),(i\omega)^n\ \Phi(-\omega)\rangle \\ &= \langle(-i\omega)^n F(\omega),\Phi(-\omega)\rangle \end{aligned} \tag{59}$$

and comparison with (53) shows that $(-i\omega)^n\ F(\omega)$ is the Fourier transform of $f^{(n)}(x)$. A similar argument shows that $F^{(n)}(\omega)$ is the Fourier transform of $(ix)^n\ f(x)$.

Another important property of Fourier transforms follows directly from their definition. Suppose the sequence of generalized functions f_ν converges; then the sequence of Fourier transforms F_ν also converges. This is frequently useful in finding the Fourier transform of a singular generalized function as we shall show.

Examples:

(i)

$$f(x) = \delta'(x)$$
$$\langle F,\Phi\rangle = \int_{-\infty}^{\infty} (-i\omega)\ \Phi(\omega)\ d\omega \tag{60}$$

(ii)

$$f(x) = x$$
$$F(\omega) = -i\omega\ \delta'(\omega) \tag{61}$$

(iii)
$$f(x) = e^{-i\omega_0 x}\, h(x), \tag{62}$$

where $h(x)$ is the unit step function. The simplest approach is to define the regular generalized functions $f_\varepsilon(x) = \exp(-\varepsilon x)\, f(x)$, $\varepsilon > 0$, and then take the limit $\varepsilon \to 0$. In this way we obtain

$$\begin{aligned} \langle F,\Phi\rangle &= \lim_{\varepsilon\to 0} \int_{-\infty}^{\infty} \frac{i\Phi(\omega)}{\omega - \omega_0 + i\varepsilon}\, d\omega \\ &= \int_C \frac{i\,\Phi(\omega)}{\omega - \omega_0}\, d\omega \end{aligned} \tag{63}$$

where the contour C is parallel to the real axis, and passes above the singularity at $\omega = \omega_0$.

Problems

1. Show that

$$\phi(x) = \int_a^b \phi_\alpha(x-y)\, dy,$$

with ϕ_α defined by (4), is a test function, and that

$$\phi(x) = 1, \quad a + \alpha < x < b - \alpha.$$

2. Show that if f is infinitely differentiable, there is a test function ϕ with the property

$$\phi(t) = f(t), \quad a < t < b.$$

3. Show that if two regular generalized functions are equal ($\langle f,\phi\rangle = \langle g,\phi\rangle$ for all ϕ), then at any point where $f(x)$ and $g(x)$ are continuous functions,

$$f(x) = g(x).$$

4. Show that

(i) $$\frac{\partial}{\partial\alpha}\,\delta(\alpha x) = -\frac{1}{\alpha^2}\,\delta(x)$$

(ii) $$\frac{\partial}{\partial\alpha}\,\delta'(\alpha x) = -\frac{2}{\alpha^3}\,\delta'(x)$$

(iii) $$f(x)\,\delta(x) = f(0)\,\delta(x)$$

(iv) $$f(x)\,\delta'(x) = -f'(0)\,\delta(x) + f(0)\delta'(x)$$

(v) $$\delta(e^{\alpha x}-\beta) = \frac{1}{\alpha\beta}\,\delta\left(x - \frac{\ell n\beta}{\alpha}\right)$$

5. $$\lim_{a\to\infty}\int_0^a \cos(\omega x)\,d\omega = \pi\delta(x)$$

6. $$\lim_{\omega\to\infty}\omega^{\alpha}\sin(\omega x) = 0$$

7. $$\lim_{\nu\to\infty}\frac{\nu}{\sqrt{\pi}}\,e^{-\nu^2x^2} = \delta(x)$$

8. $$\lim_{\alpha\to\infty}\frac{\alpha}{2}\,e^{-\alpha|x|} = \delta(x)$$

9. Consider the Fourier series

$$\sum_{n=-\infty}^{\infty} a_n e^{inx},$$

where

$$a_n < A\,n^k,\quad n \neq 0,\quad k > 0.$$

Show that if $p \geq k + 2$, then

$$\sum_{n=-\infty}^{\infty} a_n e^{inx} = a_0 + f^{(p)}(x)$$

where

$$f(x) = \sum_{\substack{n\neq 0\\ n=-\infty}}^{\infty} \frac{a_n}{(in)^p}\,e^{inx}.$$

10. Evaluate

$$\sum_{n=-\infty}^{\infty} n^m e^{inx}, \quad m > 0.$$

11. $$x^n \delta^{(m)}(x) = \begin{cases} 0, & m < n \\ \dfrac{(-1)^n m!}{(m-n)!} \delta^{(m-n)}(x), & m \geq n \end{cases}$$

12. $$\delta^{(m)}(ax+b) = \frac{1}{a^m |a|} \delta^{(m)}(x + b/a)$$

13. $$\lim_{\Delta \to 0} \frac{\delta(x+\Delta) - \delta(x)}{\Delta} = \delta'(x)$$

14. Consider the regular functional x_+^λ defined by

$$\langle x_+^\lambda, \phi \rangle = \int_0^\infty x^\lambda \phi(x)\, dx, \quad \mathrm{Re}(\lambda) > -1.$$

Show that for $\mathrm{Re}(\lambda) > -n-1$, analytic continuation in λ yields

$$\langle x_+^\lambda, \phi \rangle = \int_0^1 x^\lambda [\phi(x) - \phi(0) - x\phi'(0) - \dots - \frac{x^{n-1}}{(n-1)!}\phi^{(n-1)}(0)]dx$$

$$+ \int_1^\infty x^\lambda \phi(x)\, dx + \sum_{k=1}^{n} \frac{\phi^{(k-1)}(0)}{(k-1)!(\lambda+k)}.$$

15. Show that

$$\frac{d}{dx} x_+^\lambda = \lambda x_+^{\lambda-1}, \quad \lambda \neq 0,-1,-2,\dots .$$

16. The generalized function x_-^λ is defined by

$$\langle x_-^\lambda, \phi \rangle = e^{i\pi\lambda} \int_{-\infty}^0 (-x)^\lambda \phi(x) dx, \quad \mathrm{Re}(\lambda) > -1,$$

and by analytic continuation for $\mathrm{Re}(\lambda) < -1$. Show that the generalized functions $(x \pm i0)^\lambda$ are related to $x_\pm^\lambda$ by

$$(x \pm i0)^\lambda = \lim_{\varepsilon\to 0} (x \pm i\varepsilon)^\lambda$$
$$= x_+^\lambda + e^{\pm i\pi\lambda} x_-^\lambda$$

and that they are entire functions of λ.

17. The generalized function x^{-1} was defined in Section 9.3 as the principal value integral, and we showed that

$$(x \pm i0)^{-1} = x^{-1} \mp i\pi\delta(x).$$

We could define x^{-n} by differentiating this expression, viz.

$$(x \pm i0)^{-n} = x^{-n} \pm (-1)^n \frac{i\pi}{(n-1)!} \delta^{(n-1)}(x)$$

Derive an explicit formula for x^{-n}, and in particular show that

(i) $\langle x^{-n},\phi\rangle = \dfrac{1}{(n-1)!} \langle x^{-1},\phi^{(n-1)}\rangle,$

(ii) $\langle x^{-2},\phi\rangle = \displaystyle\int_0^\infty \frac{\phi(x)+\phi(-x)-2\phi(0)}{x^2}\, dx.$

18. Show that the general solution of

$$x^m f(x) = 0$$

is

$$f(x) = \sum_{k=0}^{m-1} \alpha_k \delta^{(k)}(x)$$

where α_k are arbitrary constants.
[First show that, if $\phi(x)$ is an arbitrary test function and $\psi(x)$ is a test function satisfying $\psi(0) = 1$ and $\psi'(0) = \psi''(0) = \ldots = \psi^{(m-1)}(0) = 0$, then

$$\chi(x) = \phi(x) - \psi(x) \sum_{k=0}^{m-1} \frac{x^k \phi^{(k)}(0)}{k!}$$

is also a test function, for which $\langle f,\chi\rangle = 0.$]

19. By using the Fourier transform, show that the general solution of

$$f^{(n)}(x) = 0$$

is

$$f(x) = \sum_{k=0}^{n-1} \alpha_k x^k,$$

where α_k are arbitrary constants.

20. Show that

$$\mathscr{F}[e^{-a|x|}] = \frac{2a}{a^2+\omega^2}, \quad a < \mathrm{Im}(\omega) < a .$$

21. $$\mathscr{F}[h(x)] = \pi\,\delta(\omega) + \frac{i}{\omega}$$

22. $$\mathscr{F}[e^{ax}] = 2\pi\,\delta(\omega-ia)$$

23. $$\mathscr{F}[\sin ax] = -i\pi[\delta(\omega+a)-\delta(\omega-a)]$$

24. $$\mathscr{F}[x_\pm^\lambda] = \pm ie^{\pm i\pi\lambda/2}\,\lambda!\,(\omega \pm i0)^{-\lambda-1}$$

25. $$\mathscr{F}[(x \pm i0)^\lambda] = \frac{2\pi e^{\pm i\pi\lambda/2}}{(-\lambda-1)!}\,\omega_\mp^{-\lambda-1}$$

26. $$\mathscr{F}[x^{-1}] = i\pi\,\mathrm{sgn}(\omega)$$

27. $$\mathscr{F}[x^{-2}] = -\pi|\omega|$$

28. $$\mathscr{F}[x^{-m}] = \frac{i^m\pi}{(m-1)!}\,\omega^{m-1}\,\mathrm{sgn}(\omega)$$

29. Prove that every ultradistribution[11] has the convergent Taylor series

$$F(\omega+\alpha) = \sum_{n=0}^{\infty} \frac{\alpha^n}{n!} F^{(n)}(\omega)$$

for all α.

30. Using the Taylor series, show that

$$\mathcal{F}[e^{ax}] = 2\pi\ \delta(\omega - ia).$$

Footnotes

1. ZEMANIAN (1965), GELFAND & SHILOV (1964).

2. This set of test functions is usually referred to as D. Another important possibility is the set S of infinitely differentiable functions which fall to zero faster than any power of $1/x$ as $|x| \to \infty$. The reason for our choice of D will become apparent later.

3. In particular, every trial function is identically zero for $|x| > a$ for some a; any function of a complex variable with this property must have essential singularities at points on the real axis.

4. It may readily be shown that the Fourier transforms of functions in S (see footnote 2) are again in S. This is why we choose to use D; the Fourier transforms of functions in D are all entire functions satisfying (9). This set of functions is usually denoted Z.

5. By some standard theorems of classical analysis, we may write the series of inequalities

$$\left|\int_{-a}^{a} f(x)\phi_n(x)\ dx\right| < \int_{-a}^{a} |f(x)||\phi_n(x)|dx$$

$$< \max_{-a<x<a} |\phi_n(x)| \int_{-a}^{a} |f(x)|dx,$$

and if the sequence $\phi_n(x)$ converges uniformly to zero, so must the right-hand side of this inequality.

6. Thus the functional $\tilde{E}(t)$ which we discussed in Section 9.1 is regular.

7. See GELFAND & SHILOV (1964), p. 15.

8. See problem (7.39).

9. The proof is easy because (37) and (38) are both regular, whereas to prove (40) directly is more difficult.

10. It is interesting to note that D and Z (see footnote 4) are both subspaces of S, although D and Z have no functions in common except zero.

11. Generalized functions defined over Z (see footnote 4) are sometimes called ultradistributions, in distinction from generalized functions over D, which are called distributions.

12. If we ignore the fact that exp $(i\omega t)$ is not a test function and write

$$\int_{-\infty}^{\infty} e^{i\omega x}\, \delta(x-x_0)\, dx = e^{i\omega x_0}$$

we lose essential information regarding the contour of the analytic functional.

§10. GREEN'S FUNCTIONS

One approach to the solution of non-homogeneous boundary value problems is by means of the construction of functions known as Green's functions. Historically, the concept originated with work on potential theory published by Green in 1828. Green's work has provided the germs of a much wider formulation for solving a variety of eigenvalue, boundary value, and inhomogeneous problems, particularly since the advent of generalized functions. We shall not attempt a systematic treatment in this book; rather we will discuss problems and methods where integral transform techniques are useful.[1] In particular, we will discuss in this section problems where the Fourier transform in one variable is applicable.

10.1. One-dimensional Green's Functions

We consider the linear second-order equation

$$L[y] = p(x)\, y''(x) + q(x)\, y'(x) + r(x)\, y(x) = f(x) \tag{1}$$

where the coefficients $p(x)$, $q(x)$, $r(x)$ and the function $f(x)$ are given. If we impose on the solution $y(x)$ the boundary conditions

$$\begin{aligned} a_1 y(a) + a_2 y'(a) &= 0 \\ b_1 y(b) + b_2 y'(b) &= 0 \qquad (a < b) \end{aligned} \tag{2}$$

then we have a two-point boundary value problem of a type quite common in applications. Any two solutions of this problem differ by a solution of the homogeneous problem

$$L[y] = 0 \tag{3}$$

together with the boundary conditions. If this latter problem has no solution other than $y(x) \equiv 0$, then it follows that the original problem can have only one solution, which we will now construct.

Let $u_L(x)$ be the (unique) solution of the homogeneous problem (3) satisfying the boundary condition at the left-hand boundary $x = a$. Similarly, let $u_R(x)$ be the solution satisfying the boundary condition at the right-hand boundary. The functions $u_L(x)$ and $u_R(x)$ cannot be linearly dependent, since there are no solutions of (3) satisfying both boundary conditions. We may therefore use these functions to construct a solution of the original boundary value problem by the method of variation of parameters. We write

$$y(x) = v_L(x)\ u_L(x) + v_R(x)\ u_R(x), \tag{4}$$

where we must determine suitable functions $v_L(x)$ and $v_R(x)$. We have a considerable amount of freedom in choosing these functions; following the usual procedure we require that

$$v_L'(x)\ u_L(x) + v_R'(x)\ u_R(x) = 0. \tag{5}$$

Substitution of (4) and (5) into (1) yields

$$\begin{aligned} v_L'(x) &= -[u_R(x)/p(x)\ W(x)]\ f(x), \\ v_R'(x) &= [u_L(x)/p(x)\ W(x)]\ f(x), \\ W(x) &= u_L(x)\ u_R'(x) - u_R(x)\ u_L'(x). \end{aligned} \tag{6}$$

We must also satisfy the boundary conditions; from (4) and (5) we can write these as $v_L(b) = 0$ and $v_R(a) = 0$. Hence we can integrate the expressions given in (6) to obtain

$$v_L(x) = \int_x^b \frac{u_R(\xi)}{p(\xi)\ W(\xi)} f(\xi)\ d\xi,$$
$$v_R(x) = \int_a^x \frac{u_L(\xi)}{p(\xi)\ W(\xi)} f(\xi)\ d\xi. \tag{7}$$

Equations (4) and (7) may now be combined to read

$$y(x) = \int_a^b g(x,\xi)\ f(\xi)\ d\xi,$$
$$g(x,\xi) = \begin{cases} \dfrac{u_L(x)\ u_R(\xi)}{p(\xi)\ W(\xi)}, & x < \xi \\ \dfrac{u_R(x)\ u_L(\xi)}{p(\xi)\ W(\xi)}, & x > \xi. \end{cases} \tag{8}$$

The function $g(x,\xi)$ thus defined is known as the Green's function for the given boundary value problem.

Adjoint Problem: Suppose $u(x)$ and $v(x)$ are two functions, twice differentiable but otherwise arbitrary. Using integration by parts, we can derive the identity

$$\int_a^b v(x)\ L[u(x)]dx = \int_a^b v[pu'' + qu' + r]\ dx$$
$$= \int_a^b u[(pv)'' - (qv)' + rv]\ dx \tag{9}$$
$$+ [p(vu'-uv') + uv(q-p')]_a^b$$

This process has introduced a new differential operator, which we denote by $L^\dagger$; it is defined by

$$L^\dagger[v] = (pv)'' - (qv)' - rv. \tag{10}$$

We also define adjoint boundary conditions as follows: $v(x)$ satisfies the adjoint boundary condition to (2) if

$$\int_a^b vL[u]\,dx = \int_a^b u\,L^\dagger[v]\,dx \tag{11}$$

for every $u(x)$ satisfying (2). From (2) and (9), we see this is equivalent to the boundary conditions

$$\begin{aligned}
&[a_1p(a) + a_2p'(a) - a_2q(a)]\,v(a) + a_2p(a)\,v'(a) = 0,\\
&[b_1p(b) + b_2p'(b) - b_2q(b)]\,v(b) + b_2p(b)\,v'(b) = 0.
\end{aligned} \tag{12}$$

Finally, we say that the boundary value problem $L^\dagger[v] = f$ subject to the condition (12) is the adjoint problem to (1) and (2).

We will denote the Green's function for the adjoint problem by $g^\dagger(x,\xi)$. It may be constructed by the same procedure as was followed for $g(x,\xi)$. We now proceed to do this. First (see Problem 1) note that the left-hand and right-hand solutions to the homogeneous adjoint problem are related to $u_L(x)$ and $u_R(x)$ by

$$\begin{aligned}
u_L^\dagger(x) &= \frac{u_L(x)}{p(x)\;W(x)}\,,\\
u_R^\dagger(x) &= \frac{u_R(x)}{p(x)\;W(x)}\,.
\end{aligned} \tag{13}$$

Also, the Wronskians are related by $p(x)\;W^\dagger(x) = 1/p(x)W(x)$. Hence, from (8) we have

$$\begin{aligned}
g^\dagger(x,\xi) &= \begin{cases} \dfrac{u_L^\dagger(x)\;u_R^\dagger(\xi)}{p(\xi)\;W^\dagger(\xi)}\,, & x < \xi\\[2ex] \dfrac{u_R^\dagger(x)\;u_L^\dagger(\xi)}{p(\xi)\;W^\dagger(\xi)}\,, & x > \xi \end{cases}\\
&= g(\xi,x).
\end{aligned} \tag{14}$$

Thus the two Green functions are related quite simply by

$g^{\dagger}(x,\xi) = g(\xi,x)$.

Self-adjoint Systems: If the boundary value problem is identical with its adjoint, it is said to be self-adjoint. For boundary conditions of the type given in (2), a self adjoint boundary value problem is one for which $q = p'$, so that we may write

$$L[u] = (pu')' + ru. \tag{15}$$

The Green's function for such a problem has the important property of symmetry: $g(x,\xi) = g(\xi,x)$. Furthermore, since we may easily show that $p(x)\ W(x)$ is a constant, (8) simplifies to

$$g(x,\xi) = \begin{cases} \Delta\ u_L(x)\ u_R(\xi), & x < \xi \\ \Delta\ u_R(x)\ u_L(\xi), & x > \xi \end{cases}$$

$$\Delta = \frac{1}{p(\xi)\ W(\xi)} \tag{16}$$

for a self-adjoint problem. These results are often of practical importance as a step in the process of constructing Green's functions for partial differential equations.

10.2. Green's Functions as Generalized Functions

We restrict our comments in this section to Green's functions for problems over an infinite range, that is, the boundary conditions (2) are taken in the limit $a \to -\infty$, $b \to \infty$. Obviously, $g(x,\xi)$ defines a generalized function which depends on a parameter x, viz.

$$\langle g(x,\xi),\ \phi(\xi)\rangle = \int_{-\infty}^{\infty} g(x,\xi)\ \phi(\xi)\ d\xi. \tag{17}$$

Let us consider the differential equation for $g(x,\xi)$. We operate on the function $y(x) = \langle g,\phi\rangle$ with L; then by (8) we have

$$\int_{-\infty}^{\infty} L[g(x,\xi)]\ \phi(\xi)\ d\xi = \phi(x), \tag{18}$$

which may be written as

$$L[g(x,\xi)] = \delta(x-\xi). \tag{19}$$

We have derived (19) from the original definitions for L and g; however, we may consider the reverse process. Instead of beginning with $g(x,\xi)$ as a given function, we ask for solutions of (19) subject to the boundary conditions

$$\begin{aligned} &\lim_{x\to-\infty} [a_1 g(x,\xi) + a_2 g_x(x,\xi)] = 0 \\ &\lim_{x\to+\infty} [b_1 g(x,\xi) + b_2 g_x(x,\xi)] = 0. \end{aligned} \tag{20}$$

It may be shown that if $p(x) \neq 0$ for $-\infty < x < \infty$, then the only solutions of the homogeneous problem $L[g] = 0$, where g is a generalized function, are the solutions in the ordinary sense.[2] (That is, they are the regular functionals constructed from the usual solutions.) Since we have assumed that there are no such solutions, the Green's function, if it exists, will be unique. Using (8), we therefore write

$$\begin{aligned} g(x,\xi) = &\frac{u_L(x)\ u_R(\xi)}{p(\xi)\ W(\xi)}\ h(\xi-x) \\ &+ \frac{u_R(x)\ u_L(\xi)}{p(\xi)\ W(\xi)}\ h(x-\xi) \end{aligned} \tag{21}$$

We may apply the operator L to this generalized function directly using the results of Section 9.3, and show that it satisfies (19).

<u>Fourier Transform</u>: This viewpoint is particularly useful if

the operator L is sufficiently simple for the Fourier transform to be of use. As an example, we consider the equation

$$\left(\frac{\partial^2}{\partial x^2} - a^2\right) g(x,\xi) = \delta(x-\xi), \qquad (22)$$

leaving aside the question of boundary conditions temporarily. On taking the Fourier transform, we have

$$(\omega^2+a^2)\ G(\omega,\xi) = -e^{i\omega\xi}. \qquad (23)$$

In addition, there is a restriction on the contour of inversion, which is brought about by the fact that the Fourier transform of $\delta(x-\xi)$ is not the function $\exp(i\omega\xi)$ but an analytic functional whose contour begins at $\omega = -\infty$ and ends at $\omega = +\infty$. This information is insufficient to specify the Fourier inversion integral for $G(\omega,\xi)$, since there are poles at $\omega = \pm ia$. The choice of contour with respect to these poles is equivalent to a choice of boundary conditions, and is in no way restricted or determined by (22) and (23), which have in fact yielded all their available information. We illustrate by constructing three examples from (23).

(i) If we write

$$g(x,\xi) = \frac{-1}{2\pi}\int_{-\infty}^{\infty} \frac{e^{i\omega(\xi-x)}}{\omega^2+a^2}\, d\omega, \qquad (24)$$

then by the application of standard techniques for contour integrals we obtain

$$g(x,\xi) = \frac{-1}{2a}\, e^{-a|x-\xi|}. \qquad (25)$$

(ii) Choosing a different contour which is consistent with the restrictions on the end points, we may write

$$g(x,\xi) = -\frac{1}{2\pi}\int_{i\gamma-\infty}^{i\gamma+\infty} \frac{e^{i\omega(\xi-x)}}{\omega^2+a^2}\,d\omega,\ \gamma > a, \tag{26}$$

and we obtain a different function, viz.

$$g(x,\xi) = \begin{cases} 0, & x < \xi \\ \frac{1}{a}\sinh[a(x-\xi)], & x > \xi. \end{cases} \tag{27}$$

(iii) We may choose the contour of Figure 1, which gives

$$g(x,\xi) = \frac{1}{2a} e^{a|x-\xi|}. \tag{28}$$

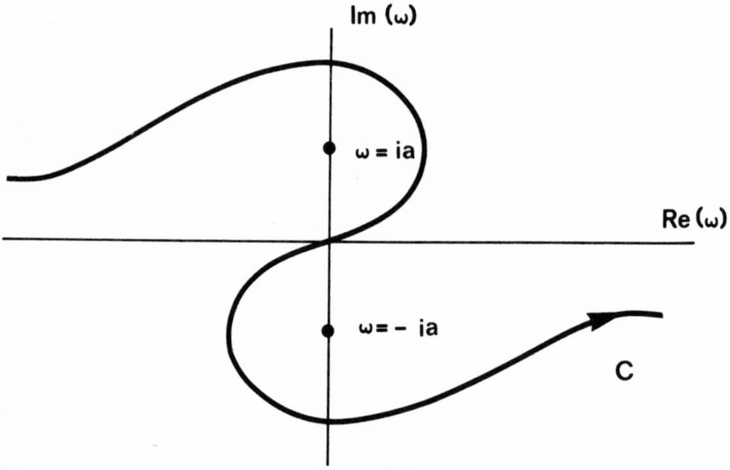

Figure 10.1

10.3. Poisson's Equation in Two Dimensions

We consider first the equation

$$\nabla^2 g(\underset{\sim}{r},\underset{\sim}{r}') = \delta(\underset{\sim}{r}-\underset{\sim}{r}') = \delta(x-x')\,\delta(y-y') \tag{29}$$

where r = (x,y) is unrestricted. The problem does not

have a unique solution, although any two solutions differ by a harmonic function, i.e., a solution of the homogeneous equation. It can be shown quite readily that a solution is $g(\underset{\sim}{r},\underset{\sim}{r}') = (1/2\pi)\ \ell n\ |\underset{\sim}{r}-\underset{\sim}{r}'|$. We shall show how to construct this function from the defining equation via the Fourier transform. On taking the transform with respect to x we have

$$\left(\frac{\partial^2}{\partial y^2} - \omega^2\right) G(\omega,y,x',y') = e^{i\omega x'}\ \delta(y-y'). \tag{30}$$

Now we know that for any ω this equation has more than one solution; the corresponding Green's functions differ only by harmonic functions. In order to obtain for G a form which has an inverse Fourier transform in the usual sense, we choose the solution of (30) as

$$G(\omega,y,x',y') = \begin{cases} \dfrac{-e^{i\omega x'}}{2\omega}\ e^{-\omega|y-y'|}, & \mathrm{Re}(\omega) > 0 \\[2ex] \dfrac{e^{i\omega x'}}{2\omega}\ e^{\omega|y-y'|}, & \mathrm{Re}(\omega) < 0. \end{cases} \tag{31}$$

It is important to note that this function is not analytic across the imaginary axis and is also singular at $\omega = 0$. For the inversion contour we may use any curve which begins at $-\infty$ and ends at $+\infty$. [This requirement comes from taking the Fourier transform of $\delta(x-x')$.] We choose a contour which crosses the imaginary axis only once, at $\omega = i\alpha \neq 0$. The resulting inversion integral is

$$\frac{1}{2\pi}\left\{\int_{-\infty}^{i\alpha} \frac{e^{-i\omega(x-x')+\omega|y-y'|}}{2\omega}\,d\omega - \int_{i\alpha}^{\infty} \frac{e^{-i\omega(x-x')-\omega|y-y'|}}{2\omega}\,d\omega\right\} \tag{32}$$

$$= -\frac{1}{4\pi}\{E_1\,(\alpha(x-x')+i\alpha|y-y'|) + E_1(\alpha(x-x')-i\alpha|y-y'|)\}.$$

It is shown in Appendix C that the exponential integral can be written in the form $E_1(z) = \ell n(z) + \phi(z)$, where $\phi(z)$ is an entire function. Hence (32) may be written as

$$g(\underset{\sim}{r},\underset{\sim}{r}') = \frac{1}{2\pi}[\ell n\ \alpha + \ell n|r-r'| + \mathrm{Re}\{\phi(-\alpha(x-x')-i\alpha(y-y'))\}] \quad (33)$$

The first and third terms are harmonic functions (depending on the choice of contour through α), and we discard them. This leaves us with $(1/2\pi)\ \ell n|\underset{\sim}{r}-\underset{\sim}{r}'|$, and because of its fundamental significance in the solution of problems in regions with boundaries, we call it the elementary solution of (29). Denoting this elementary solution by $e(\underset{\sim}{r},\underset{\sim}{r}')$ we have

$$e(\underset{\sim}{r},\underset{\sim}{r}') = \frac{1}{2\pi}\ \ell n|\underset{\sim}{r}-\underset{\sim}{r}'|. \quad (34)$$

The significance of this function is that if we define the function $u(\underset{\sim}{r})$ by

$$u(\underset{\sim}{r}) = \int e(\underset{\sim}{r},\underset{\sim}{r}')\ f(\underset{\sim}{r}')\ d\underset{\sim}{r}', \quad (35)$$

then it will satisfy Poisson's equation

$$\nabla^2 u(\underset{\sim}{r}) = f(\underset{\sim}{r}).$$

This relationship holds for a wide class of functions and generalized functions $f(\underset{\sim}{r})$; in particular it holds for any function $f(\underset{\sim}{r})$ which is absolutely integrable.

Poisson's Equation in a Bounded Region: Suppose we consider a region R of the plane whose boundary we denote S. We can solve the equations

$$\begin{aligned} \nabla^2 g(\underset{\sim}{r},\underset{\sim}{r}') &= \delta(\underset{\sim}{r}-\underset{\sim}{r}'), \\ g(\underset{\sim}{r},\underset{\sim}{r}') &= 0, \quad \underset{\sim}{r} \text{ on } S_1, \\ \underset{\sim}{n}\cdot\underset{\sim}{\nabla} g(\underset{\sim}{r},\underset{\sim}{r}') &= 0, \quad \underset{\sim}{r} \text{ on } S_2. \end{aligned} \tag{36}$$

where S_1 and S_2 together constitute the whole boundary S. We will show that this function can be used to construct a solution to the boundary value problem

$$\begin{aligned} \nabla^2 u(\underset{\sim}{r}) &= f(\underset{\sim}{r}), \\ u(\underset{\sim}{r}) &= \phi_1(\underset{\sim}{r}), \quad \underset{\sim}{r} \text{ on } S_1, \\ \underset{\sim}{n}\cdot\underset{\sim}{\nabla} u(\underset{\sim}{r}) &= \phi_2(\underset{\sim}{r}), \quad \underset{\sim}{r} \text{ on } S_2, \end{aligned} \tag{37}$$

and so it is called the Green's function for this boundary value problem. To construct the solution, we first define the region R_ε to be the region enclosed by the boundaries, with a circle of radius ε centered at $\underset{\sim}{r} = \underset{\sim}{r}'$ deleted. R is the region in which we desire the solution to (37). In R_ε we have $\nabla^2 g(\underset{\sim}{r},\underset{\sim}{r}') = 0$, since the function $g(\underset{\sim}{r},\underset{\sim}{r}') - e(\underset{\sim}{r},\underset{\sim}{r}')$ is harmonic in R_ε, and $e(\underset{\sim}{r},\underset{\sim}{r}')$ is harmonic in R_ε. Now we multiply (37) by g, and integrate over R_ε using Green's theorem to change an area integral to a line integral around the boundary. Explicitly, using polar coordinates for the small circle of radius ε, this gives

$$\begin{aligned} &\int_{R_\varepsilon} g(\underset{\sim}{r},\underset{\sim}{r}')\, f(\underset{\sim}{r})\, d\underset{\sim}{r} \\ &= \int_{R_\varepsilon} [g(\underset{\sim}{r},\underset{\sim}{r}')\, \nabla^2 u(\underset{\sim}{r}) - u(\underset{\sim}{r})\, \nabla^2 g(\underset{\sim}{r},\underset{\sim}{r}')]\, d\underset{\sim}{r} \\ &= \int_S [g(\underset{\sim}{r},\underset{\sim}{r}')\, \phi_2(\underset{\sim}{r}) - \phi_1(\underset{\sim}{r})\, \underset{\sim}{n}\cdot\underset{\sim}{\nabla}\, g(\underset{\sim}{r},\underset{\sim}{r}')]\, d\ell \\ &\quad + \frac{\varepsilon}{2\pi} \int_0^{2\pi} [\ell n\, \varepsilon\, \frac{\partial u}{\partial r} + \frac{u}{\varepsilon}]\, d\theta. \end{aligned} \tag{38}$$

On taking the limit $\varepsilon \to 0$ and using the boundary conditions for $g(\underset{\sim}{r},\underset{\sim}{r}')$, this gives the representation

$$u(\underset{\sim}{r}') = \int_R g(\underset{\sim}{r},\underset{\sim}{r}')\, f(\underset{\sim}{r})\, d\underset{\sim}{r}$$
$$+ \int_{S_1} \phi_1(\underset{\sim}{r})\, \underset{\sim}{n}\cdot\underset{\sim}{\nabla}\, g(\underset{\sim}{r},\underset{\sim}{r}')\, d\ell \tag{39}$$
$$- \int_{S_2} \phi_2(\underset{\sim}{r})\, g(\underset{\sim}{r},\underset{\sim}{r}')\, d\ell.$$

Symmetry of the Green's Function: By a similar trick we may show a most important result, namely $g(\underset{\sim}{r},\underset{\sim}{r}') = g(\underset{\sim}{r}',\underset{\sim}{r})$. To do so, we use the fact that $\nabla^2 g(\underset{\sim}{r},\underset{\sim}{r}') = 0$ and $\nabla^2 g(\underset{\sim}{r},\underset{\sim}{r}'') = 0$ in the region obtained by deleting circles of radius ε with centers at $\underset{\sim}{r}'$ and $\underset{\sim}{r}''$ from R. Application of Green's theorem then gives

$$0 = \int [g(\underset{\sim}{r},\underset{\sim}{r}')\, \nabla^2 g(\underset{\sim}{r},\underset{\sim}{r}'') - g(\underset{\sim}{r},\underset{\sim}{r}'')\, \nabla^2 g(\underset{\sim}{r},\underset{\sim}{r}')]\, d\underset{\sim}{r}$$
$$= \int_S [g(\underset{\sim}{r},\underset{\sim}{r}')\, \underset{\sim}{n}\cdot\underset{\sim}{\nabla}\, g(\underset{\sim}{r},\underset{\sim}{r}'') - g(\underset{\sim}{r},\underset{\sim}{r}'')\, \underset{\sim}{n}\cdot\underset{\sim}{\nabla}\, g(\underset{\sim}{r},\underset{\sim}{r}')]\, d\ell \tag{40}$$
$$+ \frac{\varepsilon}{2\pi} \int_0^{2\pi} \left[\frac{g(\underset{\sim}{r},\underset{\sim}{r}'')}{\varepsilon} - \frac{g(\underset{\sim}{r},\underset{\sim}{r}')}{\varepsilon}\right] d\theta.$$

The products which occur in the integral over S are zero because of the boundary conditions, and on letting $\varepsilon \to 0$, the last integral yields $g(\underset{\sim}{r}',\underset{\sim}{r}'') = g(\underset{\sim}{r}'',\underset{\sim}{r}')$.

An Example: We will find the Green's function determined by the equations

$$\nabla^2 g(\underset{\sim}{r},\underset{\sim}{r}') = \delta(\underset{\sim}{r}-\underset{\sim}{r}'), \quad -\infty < x < \infty, \quad 0 \le y \le \ell,$$
$$g(\underset{\sim}{r},\underset{\sim}{r}') = 0 \quad , \quad \underset{\sim}{r} \text{ on the boundaries.} \tag{41}$$

Taking the Fourier transform in x, we have

$$\left(\frac{\partial^2}{\partial y^2} - \omega^2\right) G(\omega,y,x',y') = e^{i\omega x'}\, \delta(y-y'), \tag{42}$$

which is the same equation as (30). This time, however, there are two boundary conditions, and on using (8) we have

$$G(\omega,y,x',y') = -e^{i\omega x'} \frac{\sinh(\omega y^<) \sinh[\omega(\ell-y^>)]}{\omega \sinh \omega\ell} , \tag{43}$$

where $y^<$ is the smaller of y and y', and $y^>$ is the larger of y and y'. The Fourier inversion of G gives a unique result, provided we require $g(\underset{\sim}{r},\underset{\sim}{r}') \to 0$ as $|\underset{\sim}{r}-\underset{\sim}{r}'| \to 0$. If this is the case, the Fourier transform must exist on the real axis, hence we choose the real axis for the inversion integral. This point is not irrelevant, since G has poles on the imaginary axis at $\omega\ell = 2\pi i n$, $n = \pm1,\pm2, \pm3,\ldots$. Choosing a contour which does not pass between the poles at $\pm 2\pi i$ will lead to a Green's function which grows exponentially as $|\underset{\sim}{r}-\underset{\sim}{r}'| \to \infty$. We conclude, then, that the solution of (41) is

$$g(x,y,x',y') \\ = -\frac{1}{2\pi}\int_{-\infty}^{\infty} \frac{e^{-i\omega(x-x')} \sinh(\omega y^<) \sinh[\omega(\ell-y^>)]}{\omega \sinh(\omega\ell)} d\omega . \tag{44}$$

<u>Relation to Images</u>: In applications, it may be useful to leave g as an integral, since it cannot be evaluated in finite closed form. An infinite expansion may, however, be obtained, as we now show. We first write (44) as twice the integral over the range $0 < \omega < \infty$, and then apply the expansion

$$\frac{\sinh(\omega y^<) \sinh[\omega(\ell-y^>)]}{\sinh(\omega\ell)} = \frac{1}{2}\sum_{n=0}^{\infty} e^{-(2n+1)\omega\ell} \\ \times \{e^{\omega(\ell+y^<-y^>)} + e^{-\omega(\ell+y^<-y^>)} - e^{\omega(\ell-y^<-y^>)} - e^{-\omega(\ell-y^<-y^>)}\} . \tag{45}$$

If we replace the lower limit of integration by ε, we can express each term in the integral as an exponential integral. Using the result $E_1(-\varepsilon\alpha) = \ln(\varepsilon\alpha) - \gamma + O(\varepsilon\alpha)$, we can then take the limit $\varepsilon \to 0$ to get

$$g(x,y,x',y') = \frac{1}{2\pi} \operatorname{Re} \sum_{n=-\infty}^{\infty} \ln\left[\frac{y-y'+2n\ell +i(x-x')}{y+y'+2n\ell +i(x-x')}\right] \tag{46}$$

which is the form of solution which results from applying the method of images to the problem.

Normal Derivative: It is of interest to calculate the normal derivative, $\underset{\sim}{n}\cdot\nabla\underset{\sim}{g}$, on the boundary. For $y = 0$, we obtain from (44)

$$-g_y(x,0,x',y') = +\frac{1}{2\pi}\int_{-\infty}^{\infty} \frac{e^{-i\omega(x-x')}\sinh[\omega(\ell-y')]}{\sinh(\omega\ell)}\,d\omega. \tag{47}$$

If $y' \to 0$, the integral tends to $\delta(x-x')$; if $y' = \ell$, it is identically zero. Both of these properties are necessary for the validity of (39) for the present system.

10.4. Helmholtz's Equation in Two Dimensions

It is frequently of interest to consider solutions of the inhomogeneous wave equation

$$\nabla^2\phi - \frac{1}{c^2}\frac{\partial^2\phi}{\partial t^2} = f \tag{48}$$

which have the time dependence $\exp(-i\Omega t)$. This converts the wave equation to Helmholtz's equation

$$(\nabla^2+k^2)\phi = f, \quad k = \Omega/c \tag{49}$$

which is generally solved subject to boundary conditions. We will consider Green's functions for this latter equation, commencing with the elementary solution in two dimensions, that is, the solution of

$$(\nabla^2+k^2)\; e(\underset{\sim}{r},\underset{\sim}{r}') = \delta(\underset{\sim}{r}-\underset{\sim}{r}') \tag{50}$$

with no boundary conditions required on a finite boundary. The most straightforward method of solving (50) is to look for solutions of $(\nabla^2+k^2)\; e(\underset{\sim}{r},\underset{\sim}{r}') = 0$ which have a singularity at $\underset{\sim}{r} = \underset{\sim}{r}'$; these turn out to be the Hankel functions of argument $k|\underset{\sim}{r}-\underset{\sim}{r}'|$. The arbitrary constants may then be chosen by applying Green's theorem to the integral

$$\int e(\underset{\sim}{r},\underset{\sim}{r}')[\nabla^2+k^2]\; \phi(\underset{\sim}{r})\; d\underset{\sim}{r} \tag{51}$$

where ϕ is a test function, exactly as was done in connection with (37-40) above. We leave this approach to the problems, choosing rather to attack (50) directly. On taking the Fourier transform with respect to x, we obtain

$$\left[\frac{\partial^2}{\partial y^2} - \omega^2 + k^2\right] E(\omega,y,x',y') = e^{i\omega x'}\; \delta(y-y'). \tag{52}$$

Suppose that we choose the integration contour for inversion so that $\mathrm{Re}(\sqrt{\omega^2-k^2}) \geq 0$ on it; then the solution which is bounded for large $|y-y'|$ is

$$E(\omega,y,x',y') = \frac{-e^{i\omega x'}\; e^{-\sqrt{\omega^2-k^2}|y-y'|}}{2\sqrt{\omega^2-k^2}}, \tag{53}$$

and substitution into the Fourier inversion theorem gives explicit forms for the elementary solution. Figure 2 shows the only two contours consistent with the condition $\mathrm{Re}(\sqrt{\omega^2-k^2}) \geq 0$; they lead to two different elementary solutions. We will evaluate the inversion integral for contour C_1. First, we introduce the polar coordinates r, θ by

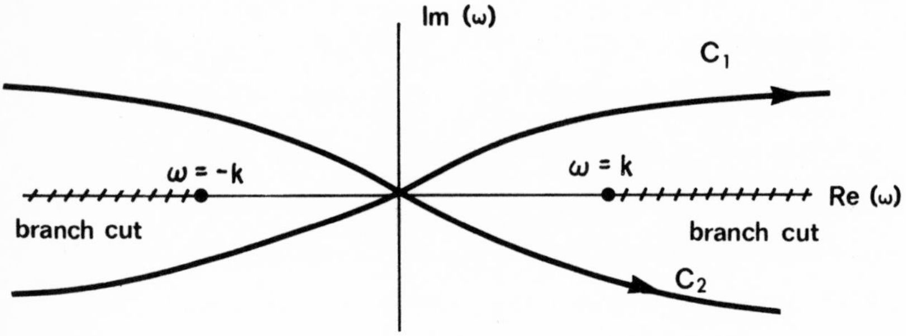

Figure 10.2.

$$x - x' = \rho \cos \theta$$
$$|y-y'| = \rho \sin \theta \qquad (54)$$

and then carry out the change of variables

$$\omega = -k \cos (\phi+it),$$
$$= - k \cos \phi \cosh t + ik \sin \phi \sinh t, \qquad (55)$$
$$-\infty < t < \infty .$$

The path described by this new variable is a hyperbola, passing between two branch points at $\omega = \pm k$, and is shown in Figure 3.

In terms of t, we may write

$$-i\omega(x-x') - \sqrt{\omega^2-k^2}\ |y-y'| = ik\rho \cos(\phi-\theta+it),$$

$$(56)$$

$$\frac{d\omega}{\sqrt{\omega^2-k^2}} = - dt,$$

where we have chosen the branch of $\sqrt{\omega^2-k^2}$ which makes it

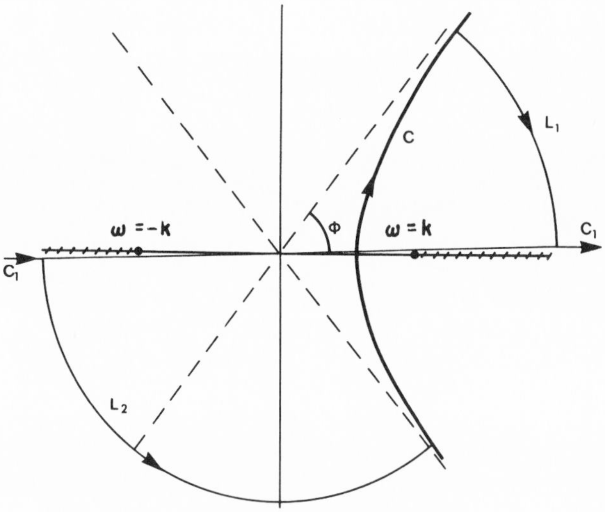

Figure 10.3

agree with the choice appropriate for C_1. Now we may deform the contour C_1 to the hyperbola provided the contributions from the arcs L_1 L_2 are of order R^{-1} (see Figure 3). On L_1, this imposes the requirement $\phi \leq \theta$; on L_2, $\phi \geq \theta$. The only consistent choice is $\phi = \theta$, whereupon the Fourier inversion integral of (53) becomes

$$\begin{aligned} e(\underset{\sim}{r},\underset{\sim}{r}') &= \frac{-1}{4\pi}\int_{-\infty}^{\infty} e^{ik\rho\cosh t}\,dt \\ &= \frac{-i}{4} H_0^{(1)}(k\rho). \end{aligned} \tag{57}$$

The identification of the integral as a Hankel function follows from (20.66). Using the results of Section 20 we may write down asymptotic forms for $e(\underset{\sim}{r},\underset{\sim}{r}')$ for small and large $|\underset{\sim}{r}-\underset{\sim}{r}'|$; they are

$$e(\underset{\sim}{r},\underset{\sim}{r}') \sim \frac{1}{2\pi} \ln \rho, \quad k\rho \ll 1$$
$$\sim \frac{e^{-3i\pi/4}}{4} \left(\frac{2}{\pi k\rho}\right)^{1/2} e^{ik\rho}, \quad k\rho \gg 1. \tag{58}$$

The logarithmic singularity at $\rho = 0$ is characteristic of problems involving the Laplacian operator in two dimensions; the asymptotic form for large ρ shows that we have outgoing waves as the boundary condition at infinity. It is easy to show that the inversion of (53) using the contour C_2 leads to

$$e(\underset{\sim}{r},\underset{\sim}{r}') = \frac{i}{4} H_0^{(2)}(kr)$$
$$\sim \frac{e^{3i\pi/4}}{4} \left(\frac{2}{\pi k\rho}\right)^{1/2} e^{-ik\rho}, \quad k\rho \gg 1. \tag{59}$$

This corresponds to incoming waves at infinity. Thus we have two distinct elementary solutions; the decision on whether to use either one, or a linear combination of the two as the solution of (50) depends on the physical content of the problem at hand.

<u>Helmholtz's Equation in a Bounded Region</u>: The developments of the previous section may now be replaced. If we define a Green's function by

$$(\nabla^2 + k^2)\, g(\underset{\sim}{r},\underset{\sim}{r}') = \delta(\underset{\sim}{r}-\underset{\sim}{r}'),$$
$$g(\underset{\sim}{r},\underset{\sim}{r}') = 0, \qquad \underset{\sim}{r} \text{ on } S_1, \tag{60}$$
$$\underset{\sim}{n}\cdot\underset{\sim}{\nabla} g(\underset{\sim}{r},\underset{\sim}{r}') = 0, \qquad \underset{\sim}{r} \text{ on } S_2,$$

it leads to the same representation (39) for the solution to the inhomogeneous problem

$$(\nabla^2 + k^2)\, u(\underset{\sim}{r}) = f(\underset{\sim}{r}),$$
$$u(\underset{\sim}{r}) = \phi_1(\underset{\sim}{r}), \quad \underset{\sim}{r} \text{ on } S_1, \qquad (61)$$
$$\underset{\sim}{n}\cdot\underset{\sim}{\nabla}\, u(\underset{\sim}{r}) = \phi_2(\underset{\sim}{r}), \quad \underset{\sim}{r} \text{ on } S_2.$$

Furthermore, we can again show that

$$g(\underset{\sim}{r},\underset{\sim}{r}') = g(\underset{\sim}{r}',\underset{\sim}{r}). \qquad (62)$$

Specific examples are left to the reader (Problems 5-8).

Problems

1. Show that the functions defined in (13) do satisfy the stated differential equation and the adjoint boundary condition, and that the Wronskian is given by

$$p(x)\, W[u_L^\dagger,\, u_R^\dagger] = \frac{1}{p(x)W[u_L, u_R]}.$$

2. Show that, for a self-adjoint second-order differential equation,

$$\frac{d}{dx}[p(x)\; W(x)] = 0.$$

3. Find elementary solutions of

$$(\nabla^2 + k^2)\, e(\underset{\sim}{r}) = \delta(\underset{\sim}{r})$$

in two dimensions by using Hankel functions directly.

4. Show that the Green's function for Poisson's equation in a three-dimensional half-space $-\infty < x < \infty$, $-\infty < y < \infty$, $z > 0$, subject to $g = 0$ on $z = 0$ is

$$-\frac{1}{4\pi R} + \frac{1}{4\pi R'}$$

where

$$R = \sqrt{(x-x')^2 + (y-y')^2 + (z-z')^2} ,$$

$$R' = \sqrt{(x-x')^2 + (y-y')^2 + (z+z')^2} .$$

5. Show that the Green's function for Helmholtz's equation in the three-dimensional half-space $z > 0$, satisfying $\partial g/\partial n = 0$ on $z = 0$ and having the form of outgoing waves for large R, is

$$-e^{ikR}/R - e^{ikR'}/R'$$

with R and R' defined in Problem 4.

6. A metal disc of radius a is set into an infinite metal wall, separated by a thin insulator. If the potential of the disc oscillates at frequency Ω, show that the potential far from the disc has the approximate form

$$\phi \simeq \frac{Va^2}{R} e^{ikR-i\Omega t} \frac{J_1(ka \sin \theta)}{ka \sin \theta} ,$$

$$\Omega = ck, \quad \tan \theta = \frac{\sqrt{x^2+y^2}}{2} .$$

7. Show that the Green's function for Helmholtz's equation in the strip $-\infty < x < \infty$, $0 \leq y \leq a$, satisfying $g = 0$ when $y = 0$ or a, is

$$g(x,x',y,y') = -\frac{1}{2\pi} \int_0^\infty \frac{\sinh(sy^<)\sinh s(a-y^>)}{s \sinh (s\ell)} e^{-i\omega|x-x'|} d\omega$$

where

$$s^2 = \omega^2 - k^2, \quad y^< = \min(y,y'), \quad y^> = \max(y,y').$$

Using (45) express the solution as an infinite series of Hankel functions.

8. Consider the boundary value problem[5]

$$(\nabla^2 + k^2)\,\phi(x,y) = 0\,,$$
$$\phi(0,y) = 0, \qquad y < 0,$$
$$\phi_x(0,y) = f(y), \quad y > 0.$$

Show that, in polar coordinates R,θ ($x = R\sin\theta$, $y = R\cos\theta$) the function

$$x^{1/2}\int_{-\cos\theta}^{1} \frac{v(R,\cos\alpha)}{\sqrt{\cos\alpha + \cos\theta}}\, d(\cos\alpha)$$

is the solution, provided that v satisfies

$$\frac{\partial^2 v}{\partial\rho^2} + \frac{1}{\rho}\frac{\partial v}{\partial\rho} + \frac{\partial^2 v}{\partial\zeta^2} + k^2 v = 0,$$

$$\rho = R\sin\theta,$$

$$\zeta = R\cos\theta,$$

and

$$\lim_{\rho\to 0} \rho\frac{\partial v}{\partial\rho} = \begin{cases} \dfrac{\sqrt{2}}{\pi\sqrt{|\zeta|}}\, f(-\zeta), & \zeta < 0 \\ 0 \qquad\qquad\quad , & \zeta > 0. \end{cases}$$

Using the Green's function for Helmholtz's equation in three dimensions, deduce that

$$v(R,\cos\theta) = -\frac{1}{\pi\sqrt{2}}\int_0^\infty \rho^{1/2} f(\rho)\, \frac{\exp(ik\sqrt{\rho^2+R^2+2\rho R\cos\theta})}{\sqrt{\rho^2 + R^2 + 2\rho R\cos\theta}}\, d\rho.$$

9. Consider the boundary value problem[6]

$$(\nabla^2+k_1^2)(\nabla^2+k_2^2)\,\psi(\underset{\sim}{r}) = 0,$$

$$\left.\begin{aligned} \psi(\underset{\sim}{r}) &= f(\underset{\sim}{r}) \\ \nabla^2\psi(\underset{\sim}{r}) &= g(\underset{\sim}{r}) \end{aligned}\right\}\ \underset{\sim}{r} \text{ on } S,$$

with $k_1^2 \neq k_2^2$. Show that, if we define ψ_1 and ψ_2 by

$$(\nabla^2+k_{1,2}^{\;2})\psi = \psi_{2,1},$$

then the solution may be written

$$\psi = \frac{\psi_2 - \psi_1}{k_1^{\;2} - k_2^{\;2}},$$

with the functions $\psi_{1,2}$ determined by

$$(\nabla^2 + k_{1,2}^2)\psi_{1,2} = 0,$$

$$\psi_{1,2}(\underset{\sim}{r}) = k_{2,1}^2\ f(\underset{\sim}{r}) + g(\underset{\sim}{r}),\quad \underset{\sim}{r}\ \text{on}\ S.$$

Hence represent the solution in terms of the Green's function for Helmholtz's equation with satisfies $g = 0$ on S.

10. Consider the boundary value problem

$$(\nabla^2+k^2)^2\psi = 0,$$

$$\left.\begin{aligned}\psi(\underset{\sim}{r}) &= f(\underset{\sim}{r})\\ \nabla^2\psi(\underset{\sim}{r}) &= g(\underset{\sim}{r})\end{aligned}\right\}\ \underset{\sim}{r}\ \text{on}\ S.$$

By applying the limiting procedure $k_1 \to k_2$ to Problem 9, show that

$$\psi(\underset{\sim}{r}) = \int f(\underset{\sim}{r}')\ \frac{\partial}{\partial n}\ G(\underset{\sim}{r},\underset{\sim}{r}')\ dS$$

$$- \frac{1}{2k}\int_S \{k^2 f(\underset{\sim}{r}') + g(\underset{\sim}{r}')\}\ \frac{\partial^2}{\partial k\partial n}\ G(\underset{\sim}{r},\underset{\sim}{r}')\ dS'\ .$$

11. Find an integral representation for the solution of

$$\nabla^4\phi(x,y,z) = 0,\quad -\infty < x < \infty,\quad -\infty < y < \infty,\quad z > 0,$$

subject to

$$\phi(x,y,0) = f(x,y),$$
$$[\nabla^2\phi]_{z=0} = g(x,y).$$

Footnotes

1. Excellent accounts are given in STAKGOLD (1968) and MORSE & FESHBACH (1953), Ch. 7.

2. See GELFAND & SHILOV (1964), pp. 39ff.

3. The difference between any two solutions of $\nabla^2 g = \delta(\underset{\sim}{r}-\underset{\sim}{r}')$ satisfies $\nabla^2\phi = 0$; therefore we may write $g(\underset{\sim}{r},\underset{\sim}{r}') = e(\underset{\sim}{r}-\underset{\sim}{r}') + \phi(\underset{\sim}{r},\underset{\sim}{r}')$.

4. The normal derivative of (47) is a Fourier transform which is given in Problem 7.26.

5. This problem is adapted from a paper by W. E. Williams, Q. J. Mech. Appl. Math., (1973), 26, 397, where some more general results may be found.

6. Problems 9-11 are based on results given by G. S. Argawal, A. J. Devaney and D. N. Pattenayak, J. Math. Phys. (1973), 14, 906.

§11. FOURIER TRANSFORMS IN TWO OR MORE VARIABLES

11.1. Basic Notation and Results

The theory of Fourier transforms of a single variable may be extended to functions of several variables. Thus, if $f(x,y)$ is a function of two variables, the function $F(\xi,\eta)$ defined by

$$F(\xi,\eta) = \int_{-\infty}^{\infty} dx \int_{-\infty}^{\infty} dy\ e^{i(\xi x + \eta y)}\ f(x,y) \tag{1}$$

is the two-dimensional Fourier transform of $f(x,y)$, and, provided that the inversion formula (7.6) may be applied twice, we have

$$f(x,y) = \frac{1}{(2\pi)^2} \int_{i\gamma_1-\infty}^{i\gamma_1+\infty} d\xi \int_{i\gamma_2-\infty}^{i\gamma_2+\infty} d\eta\ e^{-i(\xi x+\eta y)} F(\xi,\eta). \tag{2}$$

An important point to note about this formula is that it involves functions of more than one complex variable. The theory of such functions is exceedingly complicated, and there are no well developed techniques of the same generality and power as for functions of one complex variable. Usually it is necessary to treat each variable in turn, temporarily regarding the others as constant. Some of the subtleties which emerge will become evident in this and later sections, through concrete examples.

An elegant notation may be used if the variables are components of a vector; thus for a function $f(\underset{\sim}{r})$ in n dimensions we write

$$\begin{aligned} F(\underset{\sim}{k}) &= \int f(\underset{\sim}{r})\ e^{i\underset{\sim}{k}\cdot\underset{\sim}{r}}\ d^n\underset{\sim}{r} \\ f(\underset{\sim}{r}) &= \frac{1}{(2\pi)^n} \int F(\underset{\sim}{k}) e^{-i\underset{\sim}{k}\cdot\underset{\sim}{r}}\ d^n\underset{\sim}{k}. \end{aligned} \tag{3}$$

Elementary Properties: Formal manipulations, which we leave to the reader as an exercise, lead to the following parallels to the properties derived in Section 7.2.

(i) Derivatives:[1]

$$\mathscr{F}[\nabla f(\mathbf{r})] = -i\mathbf{k}\, F(\mathbf{k}) \tag{4}$$

$$\mathscr{F}[\mathbf{r} f(\mathbf{r})] = -i\nabla_{\mathbf{k}} F(\mathbf{k}) \tag{5}$$

or

$$\begin{aligned} \nabla_{\mathbf{r}} &\leftrightarrow -i\mathbf{k} \\ \nabla_{\mathbf{k}} &\leftrightarrow +i\mathbf{r} \end{aligned} \tag{6}$$

(ii) Translations:

$$\mathscr{F}[f(\mathbf{r}-\mathbf{a})] = e^{i\mathbf{k}\cdot\mathbf{a}}\, F(\mathbf{k}) \tag{7}$$

$$\mathscr{F}[e^{i\mathbf{q}\cdot\mathbf{r}} f(\mathbf{r})] = F(\mathbf{k}+\mathbf{q}) \tag{8}$$

(iii) Convolutions: If

$$h(\mathbf{r}) = \int g(\mathbf{r}-\mathbf{r}')\, f(\mathbf{r}')\, d^n\mathbf{r}', \tag{9}$$

then

$$H(\mathbf{k}) = G(\mathbf{k})\, F(\mathbf{k}). \tag{10}$$

Also

$$\mathscr{F}[f(\mathbf{r})\, g(\mathbf{r})] = \frac{1}{(2\pi)^n} \int F(\mathbf{k}-\mathbf{k}')\, G(\mathbf{k}')\, d^n\mathbf{k}'. \tag{11}$$

(iv) Parseval relation:

$$\int f(\mathbf{r})\, g^*(\mathbf{r})\, d^n\mathbf{r} = \frac{1}{(2\pi)^n} \int F(\mathbf{k})\, G^*(\mathbf{k})\, d^n\mathbf{k}. \tag{12}$$

Illustrative Example: As a simple application, we will re-derive the results of Section 10.4 using a two variable transform. The algebraic manipulations involved are trivial, but the analysis of the inversion integral already exhibits some interesting and illuminating subtleties. We want to solve the equation[2]

$$(\nabla^2 + k^2)\, e(\underset{\sim}{r}) = \delta(\underset{\sim}{r}). \tag{13}$$

Taking the two dimensional Fourier transform, we have

$$(k^2 - q^2)\, E(\underset{\sim}{q}) = 1, \tag{14}$$

where

$$E(\underset{\sim}{q}) = \int e^{i\underset{\sim}{q}\cdot\underset{\sim}{r}}\, e(\underset{\sim}{r})\, d^2\underset{\sim}{r}. \tag{15}$$

If we solve (14) for $E(\underset{\sim}{q})$ and substitute into (3b), we obtain

$$e(\underset{\sim}{r}) = \frac{1}{(2\pi)^2} \int \frac{e^{-i\underset{\sim}{q}\cdot\underset{\sim}{r}}}{k^2-q^2}\, d^2\underset{\sim}{q}. \tag{16}$$

This solution is not unique, since the Fourier transform of a delta function does not specify the inversion contour.[3] We denote the components of $\underset{\sim}{q}$ by ξ and η; it is our intention first to evaluate the η integral for each value of ξ which is needed, and then to integrate over ξ. Great care is needed at this point, since the η integral depends critically on the position of the η contour relative to each value of ξ. Hence, we choose the ξ contour first; our choice is indicated in Figure 1. On this contour, $0 < \arg(\xi^2-k^2) < \pi$, so we may define the function $s(\xi,k)$ by

$$s(k,\xi) = \sqrt{\xi^2-k^2}\,, \qquad 0 < \arg(s) < \pi/2. \tag{17}$$

Turning to the η integral, we need to evaluate

$$I(\xi,y) = -\int_{-\infty}^{\infty} \frac{e^{-i\eta y}\, d\eta}{(\eta-is)(\eta+is)}\,. \tag{18}$$

We integrate along the real axis, since the poles lie off it

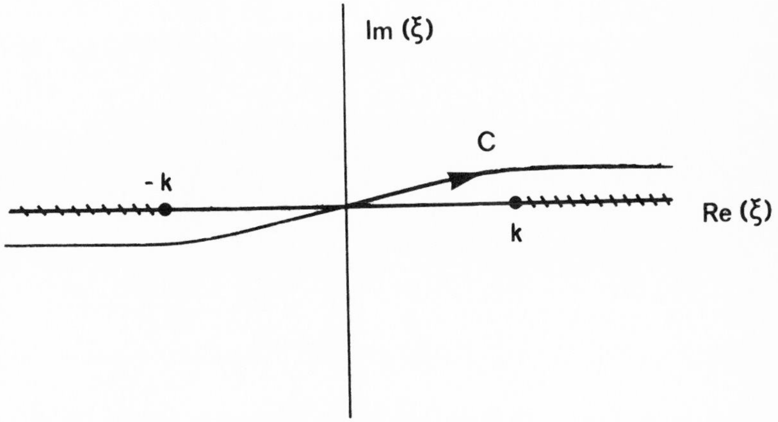

Figure 11.1

by virtue of (17). By residues, we have

$$I(\xi,y) = -\frac{\pi e^{-s|y|}}{s}, \tag{19}$$

and on using this in (16), we obtain

$$e(\underset{\sim}{r}) = -\frac{1}{4\pi}\int \frac{e^{-i\eta x}\, e^{-\sqrt{\eta^2-k^2}\,|y|}}{\sqrt{\eta^2-k^2}}\, d\eta, \tag{20}$$

a result already obtained in Section 10.4.

<u>Use of Radiation Condition</u>: The difficulty in the above treatment arises because k is a real quantity, so that we must choose an inversion contour which goes off the real axis. One method of treating such problems, discussed in detail in Section 9.4, is the application of radiation conditions. Thus we replace k by $k + i\varepsilon$; then both components of $\underset{\sim}{q}$ may be confined to real values.

Changing to polar coordinates, with $\underset{\sim}{r}$ defining the polar direction, (16) becomes

$$e(\underset{\sim}{r}) = \frac{1}{(2\pi)^2}\int_0^\infty q\,dq \int_0^{2\pi} d\theta\, \frac{e^{iqr\cos\theta}}{k^2 - q^2}. \tag{21}$$

The θ-integral is one form of Bessel's integral for $J_0(qr)$, hence

$$e(\underset{\sim}{r}) = \frac{1}{2\pi}\int_0^\infty q\,dq\, \frac{J_0(qr)}{k^2 - q^2}. \tag{22}$$

This integral defines an analytic function of k for $0 < \arg(k) < \pi$, and in particular, if $k = ia$, Problem 20.23 yields

$$\begin{aligned} e(\underset{\sim}{r}) &= -\frac{1}{2\pi} K_0(ar) \\ &= -\frac{1}{2\pi} K_0(-ikr). \end{aligned} \tag{23}$$

Having evaluated the integrals, we must set $\varepsilon = 0$. Equation (20.79) then shows that

$$e(\underset{\sim}{r}) = -\frac{i}{4} H_0^{(1)}(kr) \tag{24}$$

in agreement with (10.57).

11.2. Diffraction of Scalar Waves

The mathematical solution of diffraction problems is generally very difficult, and explicit exact formulas are known for only a small number of relatively simple cases. Fortunately, a large number of problems of interest in optics (and other fields) may be usefully approximated by a method which is due to Kirchhoff. We discuss here the diffraction of a scalar wave, satisfying the equation

$$\nabla^2\phi - \frac{1}{c^2}\frac{\partial^2\phi}{\partial t^2} = 0, \tag{25}$$

by an aperture in a plane screen at $z = 0$. The basic idea is to express the solution at an arbitrary point in terms of the values in the aperture; these aperture values are then set equal to the strength of the incident wave, calculated in the absence of the screen. (If we can calculate the aperture function in the presence of the screen, the problem is exactly solved.) In the following we denote the aperture plane by A, and the remainder of the $z = 0$ plane by B. Also, corresponding to any three-dimensional position vector $\underset{\sim}{r}$ having components x,y, and z, we introduce the two-dimensional vector $\underset{\sim}{s}$ having components x and y.

We consider monochromatic waves, i.e. we set

$$\phi(\underset{\sim}{r},t) = \phi(\underset{\sim}{r})e^{-i\Omega t},$$

$$\mathrm{Re}(\Omega) > 0, \quad \mathrm{Im}(\Omega) > 0, \tag{26}$$

where we have taken $\mathrm{Im}(\Omega) > 0$ as a radiation condition. After determining the necessary solutions, we may set $\mathrm{Im}(\Omega) = 0$. We introduce the two-dimensional Fourier transforms

$$\phi(\underset{\sim}{r}) = \int e^{i\underset{\sim}{q}\cdot\underset{\sim}{s}}\,\Phi(\underset{\sim}{q},z)d^2\underset{\sim}{q}, \tag{27}$$

$$\Phi(\underset{\sim}{q},z) = \frac{1}{(2\pi)^2}\int e^{-i\underset{\sim}{q}\cdot\underset{\sim}{s}}\,\phi(\underset{\sim}{s},z)d\underset{\sim}{s}. \tag{28}$$

Substitution into (25) yields the following ordinary differential equation for $\Phi(\underset{\sim}{q},z)$:

$$\frac{d^2\Phi}{dz^2} - (q^2-k^2)\Phi = 0, \qquad k = \Omega/c, \tag{29}$$

with the general solution

$$\Phi(\underset{\sim}{q},z) = F(\underset{\sim}{q})e^{-z\sqrt{q^2-k^2}} + G(\underset{\sim}{q})e^{z\sqrt{q^2-k^2}}. \tag{30}$$

From (26) we have $-\pi < \arg(q^2-k^2) < 0$ for all real $\underset{\sim}{q}$, and thus we may choose the square root by $-\pi/2 < \arg(\sqrt{q^2-k^2}) < 0$. Hence the second term in (30) is an incoming wave, contrary to our physical requirement that the diffracted wave be outgoing, so that $G(\underset{\sim}{q}) = 0$. Thus $F(\underset{\sim}{q}) = \Phi(\underset{\sim}{q},0)$, which may be obtained from $\Phi(\underset{\sim}{s},0)$ by putting $z = 0$ in (28). Equation (27) may be written as

$$\begin{aligned}
\phi(\underset{\sim}{s},z) &= \int d^2\underset{\sim}{q}\; e^{i\underset{\sim}{q}\cdot\underset{\sim}{s}}\; e^{-z\sqrt{q^2-k^2}} \\
&\quad \times \left\{ \frac{1}{(2\pi)^2} \int d^2\underset{\sim}{s}'\; e^{-i\underset{\sim}{q}\cdot\underset{\sim}{s}'}\; \phi(\underset{\sim}{s}',0) \right\} \\
&= -\frac{1}{4\pi^2} \int d^2\underset{\sim}{s}'\; \phi(\underset{\sim}{s}',0) \\
&\quad \times \left\{ \frac{d}{dz} \int d^2\underset{\sim}{q}\; \frac{e^{i\underset{\sim}{q}\cdot(\underset{\sim}{s}-\underset{\sim}{s}') - z\sqrt{q^2-k^2}}}{\sqrt{q^2-k^2}} \right\} \qquad (31) \\
&= \frac{1}{2\pi} \int d^2\underset{\sim}{s}'\; \phi(\underset{\sim}{s}',0)\; \frac{d}{dz}\left(\frac{e^{ikR}}{R}\right), \\
& \qquad R^2 = z^2 + |s-s'|^2.
\end{aligned}$$

The evaluation of the $\underset{\sim}{q}$ integral, which leads to the last step, is dealt with in Problem 15. This formula is particularly appropriate if ϕ satisfies the boundary condition $\phi = 0$ on B, for then the integral in (31) extends only over the aperture A. Thus, in this case we have expressed ϕ in terms of its value in A.

<u>Fraunhofer and Fresnel Diffraction</u>: In practice, these formulas have their most important applications in an asymptotic limit which we now derive. Suppose that the aperture is small compared to z; then if we choose the origin to lie in A, and we have $|\underset{\sim}{s}'| \ll z$ in (31), we may write

$$R = |\underset{\sim}{r} - \underset{\sim}{s}'|$$
$$\simeq r - \hat{\underset{\sim}{r}}\cdot\underset{\sim}{s}' + \frac{\underset{\sim}{s}'\cdot\underset{\sim}{s}' - (\hat{\underset{\sim}{r}},\underset{\sim}{s}')^2}{2r}. \tag{32}$$

This replacement is made in the exponential function; elsewhere we simply write $R \simeq r$. The case where it is necessary to retain the quadratic terms in (32) is called "Fresnel diffraction," the simpler case, $R \simeq r - \hat{\underset{\sim}{r}}\cdot\underset{\sim}{s}'$, is called "Fraunhofer diffraction." For Fraunhofer diffraction, (31) becomes

$$\phi(\underset{\sim}{s},z) \simeq -\frac{ze^{ikr}}{2\pi r^2}\int d^2\underset{\sim}{s}'\, e^{-ik\hat{\underset{\sim}{r}}\cdot\underset{\sim}{s}'}\,\phi(\underset{\sim}{s}',0)$$
$$= -\frac{2\pi z}{r^2}\, e^{ikr}\, F(k\underset{\sim}{s}/r). \tag{33}$$

Thus the angular distribution of the wave diffracted by an aperture in this limit is determined principally by the Fourier transform of the illumination of the aperture. If the aperture is narrow compared with the wavelength, the diffraction pattern will be broad; conversely, a wide aperture produces little diffraction.

11.3. Retarded Potentials of Electromagnetism

In the classical theory of electromagnetism, the electric and magnetic fields $\underset{\sim}{E}$ and $\underset{\sim}{H}$ satisfy Maxwell's equations, which in c.g.s. units are

$$\begin{aligned}
\underset{\sim}{\nabla}\cdot\underset{\sim}{E} &= 4\pi\rho,\\
\underset{\sim}{\nabla}\times\underset{\sim}{E} &= -\frac{1}{c}\frac{\partial H}{\partial t},\\
\underset{\sim}{\nabla}\cdot\underset{\sim}{H} &= 0,\\
\underset{\sim}{\nabla}\times\underset{\sim}{H} &= \frac{1}{c}\frac{\partial E}{\partial t} + \frac{4\pi}{c}\underset{\sim}{j},
\end{aligned} \tag{34}$$

where ρ and $\underset{\sim}{j}$ are the sources, i.e., the charges and currents which generate the fields. It is usual to express $\underset{\sim}{E}$ and $\underset{\sim}{H}$ in terms of a vector potential $\underset{\sim}{A}$ and a scalar potential ϕ via the relations

$$\underset{\sim}{H} = \underset{\sim}{\nabla} \times \underset{\sim}{A}, \qquad E = -\frac{1}{c}\frac{\partial \underset{\sim}{A}}{\partial t} - \underset{\sim}{\nabla}\phi, \tag{35}$$

together with the subsidiary condition (called Coulomb gauge)

$$\underset{\sim}{\nabla}\cdot\underset{\sim}{A} + \frac{1}{c}\frac{\partial\phi}{\partial t} = 0. \tag{36}$$

After some simple algebra, it is readily shown that $\underset{\sim}{A}$ and ϕ are determined from $\underset{\sim}{j}$ and ρ by

$$\left[\nabla^2 - \frac{1}{c^2}\frac{\partial^2}{\partial t^2}\right]\underset{\sim}{A} = -\frac{4\pi}{c}\underset{\sim}{j}, \qquad \left[\nabla^2 - \frac{1}{c^2}\frac{\partial^2}{\partial t^2}\right]\phi = -4\pi\rho. \tag{37}$$

The problem is to express the potential in terms of the sources by an explicit formula. Such a formula involves the Green's function. We consider here the equation for ϕ, and introduce the four-dimensional Fourier transform

$$\Phi(\underset{\sim}{k},\omega) = \int \phi(\underset{\sim}{r},t)\, e^{i(\underset{\sim}{k}\cdot\underset{\sim}{r}+\omega t)}\, d^3\underset{\sim}{r}\, dt, \tag{38}$$

with a similar definition for $\overline{\rho}(\underset{\sim}{k},\omega)$. Equation (37b) now reduces to simple algebra, with the solution

$$\Phi(\underset{\sim}{k},\omega) = \frac{4\pi}{k^2 - \omega^2/c^2}\,\overline{\rho}(\underset{\sim}{k},\omega). \tag{39}$$

From the product form of (39), we deduce that the Fourier transform of the Green's function is $4\pi/(k^2-\omega^2/c^2)$. This is only defined once we apply boundary conditions. We take the radiation condition $\mathrm{Im}(\omega) > 0$, so that

$$\frac{1}{2\pi^2}\int \frac{e^{-i\underset{\sim}{k}\cdot\underset{\sim}{r}}}{k^2-\omega^2/c^2}\, d^3\underset{\sim}{k} = \frac{e^{i\omega r/c}}{r}\,. \tag{40}$$

The ω is not defined in the usual sense; but since we are dealing with generalized functions we use the result $\mathscr{F}[\delta(t-t')] = \exp(i\omega t')$ to get

$$G(\underset{\sim}{r}-\underset{\sim}{r}',\ t-t') = \frac{\delta(t-t' - |\underset{\sim}{r}-\underset{\sim}{r}'|/c)}{|\underset{\sim}{r}-\underset{\sim}{r}'|} \tag{41}$$

and

$$\phi(\underset{\sim}{r},t) = \int \frac{\rho(\underset{\sim}{r}',t-|\underset{\sim}{r}-\underset{\sim}{r}'|/c)}{|\underset{\sim}{r}-\underset{\sim}{r}'|}\, d^3\underset{\sim}{r}',$$

$$A(r,t) = \frac{1}{c}\ \frac{\underset{\sim}{j}(\underset{\sim}{r}',t-|\underset{\sim}{r}-\underset{\sim}{r}'|/c)}{|\underset{\sim}{r}-\underset{\sim}{r}'|}\, d^3\underset{\sim}{r}'. \tag{42}$$

This is called a retarded solution because the source at $(\underset{\sim}{r}',t')$ only influences the potential at $\underset{\sim}{r}$ at a later time t related by $t' = t - |\underset{\sim}{r}-\underset{\sim}{r}'|/c$. Advanced potentials are obtained by taking $\mathrm{Im}(\omega) < 0$.

<u>Lienard-Wiechert Potentials</u>: For a point charge q at position $\underset{\sim}{r}_0(t)$ with velocity $\underset{\sim}{v}(t) = \dot{\underset{\sim}{r}}_0(t)$, we have

$$\begin{aligned}\rho(\underset{\sim}{r},t) &= q\delta(\underset{\sim}{r}-\underset{\sim}{r}_0),\\ \underset{\sim}{j}(\underset{\sim}{r},t) &= q\delta(\underset{\sim}{r}-\underset{\sim}{r}_0)\ \underset{\sim}{v}(\underset{\sim}{t}),\end{aligned} \tag{43}$$

$$\phi(\underset{\sim}{r},t) = \int \frac{\delta(\underset{\sim}{r}'-\underset{\sim}{r}_0)}{|\underset{\sim}{r}'-\underset{\sim}{r}_0|}\, d^3\underset{\sim}{r}', \tag{44}$$

$$\underset{\sim}{A}(\underset{\sim}{r},t) = \frac{1}{c}\int \frac{\delta(\underset{\sim}{r}'-\underset{\sim}{r}_0)\ \underset{\sim}{v}(t-|\underset{\sim}{r}'-\underset{\sim}{r}_0|/c)}{|\underset{\sim}{r}'-\underset{\sim}{r}_0|}\, d^3\underset{\sim}{r}'. \tag{45}$$

The integrals must be evaluated carefully, since

$\underset{\sim}{r}_0 = \underset{\sim}{r}(t-|\underset{\sim}{r}'-\underset{\sim}{r}_0|/c)$ is a function of $\underset{\sim}{r}'$. The simplest procedure is to introduce a new variable $\underset{\sim}{s} = \underset{\sim}{r}'-\underset{\sim}{r}_0$; then it is readily shown that the Jacobian of this transformation is $1 + \underset{\sim}{R}\cdot\underset{\sim}{v}(t')/R$, where $\underset{\sim}{R} = \underset{\sim}{r}_0-\underset{\sim}{r}$ and $t' = t-R/C$. Now the delta function is simply $\delta(\underset{\sim}{s})$, so we have

$$\phi(\underset{\sim}{r},t) = \frac{ec}{cR + \underset{\sim}{R}\cdot\underset{\sim}{v}(t')} ,$$
$$\underset{\sim}{A}(\underset{\sim}{r},t) = \frac{e\underset{\sim}{v}(t')}{cR+\underset{\sim}{R}\cdot\underset{\sim}{v}(t')} . \tag{46}$$

These are the Lienard-Wiechert potentials.

Problems

Verify the following Fourier transform pairs.

1. $f(x,y) = e^{-\alpha r}/r, \quad r^2 = x^2 + y^2$

   $F(\xi,\eta) = 2\pi/\sqrt{k^2+\alpha^2}, \quad k^2 = \xi^2 + \eta^2$

2. $f(x,y) = xe^{-\alpha r}/r^2, \quad r^2 = x^2 + y^2$

   $F(\xi,\eta) = 2\pi\xi\ [1 - k/\sqrt{k^2+\alpha^2}]/k^2, \quad k^2 = \xi^2 + \eta^2$

3. $f(\underset{\sim}{r}) = e^{-a^2r^2}, \quad \underset{\sim}{r} = (x,y,z)$

   $F(\underset{\sim}{k}) = (\pi^{3/2}/a^3)e^{-k^2/4a^2}$

4. $f(\underset{\sim}{r}) = e^{-\alpha r}/r, \quad \underset{\sim}{r} = (x,y,z)$

   $F(\underset{\sim}{k}) = \dfrac{4\pi}{k^2+\alpha^2}$

5. $f(\underset{\sim}{r}) = \begin{cases} 1, & r < a \\ 0, & r > a \end{cases}$

   $F(\underset{\sim}{k}) = \dfrac{4\pi}{k^3}\{\sin(ka) - ka\cos(ka)\}$

6. Show that the vector operators (in three dimensions) transform as

$$\nabla\phi(\underset{\sim}{r}) \rightarrow -\, i\underset{\sim}{k}\ \Phi(\underset{\sim}{k}),$$

$$\underset{\sim}{\nabla}\cdot\underset{\sim}{u}(\underset{\sim}{r}) \rightarrow -\, i\underset{\sim}{k}\cdot\underset{\sim}{U}(\underset{\sim}{k}),$$

$$\underset{\sim}{\nabla}\times\underset{\sim}{u}(\underset{\sim}{r}) \rightarrow -\, i\underset{\sim}{k}\times\underset{\sim}{U}(\underset{\sim}{k}).$$

7. Apply a two-variable Fourier transform to the problem

$$\frac{\partial u}{\partial t} = \kappa\nabla^2 u,$$

$$u(r,\theta,0) = f(r),$$

to show that

$$u(r,\theta,t) = \frac{-1}{4\pi\kappa t}\int_0^\infty \rho f(\rho)e^{-\rho^2/4\kappa t}\, I_0(\rho r/2\kappa t)\, d\rho.$$

8. At a plane boundary $z = a$ between two dielectrics, the electrostatic potential satisfies the boundary conditions

$$[\phi]_{z=a^-} = [\phi]_{z=a^+}$$

$$[\varepsilon\partial\phi/\partial z]_{z=a^-} = [\varepsilon\partial\phi/\partial z]_{z=a^+}$$

Using a two variable Fourier transform in x and y, find the Green's function for Laplace's equation in an infinite region which consists of a plane slab of material of thickness ℓ and dielectric constant ε in vacuum (dielectric constant 1).

9. Solve the initial value problem

$$\nabla^2 u(x,y,t) = \frac{1}{c^2}\frac{\partial^2 u(x,y,t)}{\partial t^2},$$

$$u(x,y,0) = f(x,y),$$

$$u_t(x,y,0) = 0,$$

using a two-variable Fourier transform.

10. Solve the boundary value problem

$$\nabla^4 u(x,y) = f(x,y), \quad -\infty \le x \le \infty, \quad y \ge 0,$$
$$u_x(x,0) = 0,$$
$$u_y(x,0) = 0,$$

by using a Fourier transform in x and a Fourier cosine transform in y.

11. Solve the heat diffusion equation

$$\frac{\partial u}{\partial t} = \kappa\nabla^2 u + \phi$$

where $\phi(\underset{\sim}{r},t)$ is the rate of production of heat, subject to the initial condition

$$u(\underset{\sim}{r},0) = f(\underset{\sim}{r}).$$

In particular, show that

$$u(\underset{\sim}{r},t) = \pi^{-3/2} \int f(\underset{\sim}{r}+2\underset{\sim}{s}\sqrt{\kappa t})\, e^{-s^2}\, d\underset{\sim}{s}$$
$$+ \pi^{-3/2} \int_0^t dt' \int \phi(\underset{\sim}{r}+2\underset{\sim}{s}\sqrt{\kappa t'},\, t')e^{-s^2}\, d\underset{\sim}{s}.$$

12. The solution of the Dirichlet problem

$$(\nabla^2-\lambda^2)u(\underset{\sim}{x},z) = 0,$$
$$u(\underset{\sim}{x},0) = g(\underset{\sim}{x}),$$
$$\underset{\sim}{x} = (x_1\ x_2,\dots,\ x_{n-1}),$$

may be written

$$u(\underset{\sim}{x},z) = \int g(\underset{\sim}{x}')\, L(\underset{\sim}{x}-\underset{\sim}{x}',z)\, d\underset{\sim}{x}'.$$

Show that[4]

$$L(\underset{\sim}{x},z) = 2(\lambda/2\pi)^{n/2}\, K_{n/2}(\lambda\rho),$$

$$\rho^2 = |\underset{\sim}{x}|^2 + z^2.$$

13. Initially, the half-space $x \geq 0$ is at a constant temperature. From time $t = 0$, the plane boundary $x = 0$ is held at the temperature $f(y)$. By applying the Laplace transform in x and the Fourier transform in y, derive the formula

$$u(x,y,t) = \frac{x}{\pi} \int_{-\infty}^{\infty} \frac{f(s)}{\rho^2} e^{-\rho^2/4\kappa t} ds,$$

$$\rho^2 = x^2 + (y-s)^2,$$

for the subsequent temperature distribution.

14. By using the Laplace transform in x and the Fourier transform in y, show that the solution of the wave equation

$$\nabla^2 u = \frac{1}{c^2} \frac{\partial^2 u}{\partial t^2}$$

in the region $x \geq 0$, $-\infty \leq y \leq \infty$, subject to the initial conditions

$$u(x,y,0) = 0$$

$$u_t(x,y,0) = 0$$

and the boundary condition

$$u(0,y,t) = f(y)$$

is

$$u(x,y,t) = \begin{cases} 0, & t < x/c \\ \dfrac{xct}{\pi} \displaystyle\int_{y-y'}^{y+y'} \frac{f(s)}{\rho^2} \frac{ds}{\sqrt{c^2t^2-\rho^2}}, & t > x/c \end{cases}$$

$$y' = \sqrt{v^2t^2-x^2},$$

$$\rho^2 = x^2 + (y-s)^2.$$

15. Evaluate the $\underset{\sim}{q}$ integral in (31) using the methods of Section 10.4.

16. Show that the Green's function defined by the equations[5]

$$\{1 + \underset{\sim}{v}\cdot\underset{\sim}{\nabla}\}G(\underset{\sim}{r}-\underset{\sim}{r}',\underset{\sim}{v}) = \delta(\underset{\sim}{r}-\underset{\sim}{r}'), \quad z > 0, \quad z' > 0,$$

$$G(\underset{\sim}{s}-\underset{\sim}{r},\underset{\sim}{v}) = 0, \quad \underset{\sim}{s} = (s_x,s_y,0), \quad z > 0, \quad v_z < 0,$$

is given by

$$G(\underset{\sim}{r}-\underset{\sim}{r}',\underset{\sim}{v}) = \frac{1}{(2\pi)^3}\int \frac{e^{-i\underset{\sim}{k}\cdot(\underset{\sim}{r}-\underset{\sim}{r}')}}{1 + i\underset{\sim}{k}\cdot\underset{\sim}{v}}\, d^3\underset{\sim}{k}.$$

Double Laplace Transforms:[6] If $f(x,y)$ is defined in the quadrant $x \geq 0$, $y \geq 0$, we define the double Laplace transform $F(p,q)$ of $f(x,y)$ by

$$F(p,q) = \int_0^\infty \int_0^\infty f(x,y)e^{-px-qy}\, dx\, dy = \mathscr{L}_2[f(x,y)].$$

Prove the following general properties of double transforms [under suitable restrictions on $f(x,y)$].

17. $\mathscr{L}_2[f(x+y)] = \dfrac{\bar{F}(q) - \bar{F}(p)}{p-q}$

where

$$\bar{F}(p) = \mathscr{L}[f(x)].$$

18. $$\mathscr{L}_2[f(x-y)] = \frac{\bar{F}(p) + \bar{F}(q)}{p+q}, \quad f \text{ even}$$

$$= \frac{\bar{F}(p) - \bar{F}(q)}{p+q}, \quad f \text{ odd}.$$

19. $\mathscr{L}_2[\partial u/\partial x] = pU(p,q) - \bar{U}_0(q)$

where[7]

$$\bar{U}_0(q) = \mathscr{L}[u(0,y);\ y \to q].$$

20. $\mathscr{L}_2[\partial^2 u/\partial x^2] = p^2U(p,q) - p\,\bar{U}_0(q) - \bar{U}_1(q)$

where

$$\bar{U}_1(q) = \mathscr{L}[u_x(0,y);\ y \to q].$$

21. Solve the partial differential equation[8]

$$\frac{\partial u}{\partial x} = \frac{\partial u}{\partial y}, \quad x \geq 0, \quad y \geq 0$$

subject to

$$u(x,0) = a(x)$$

using the double Laplace transform.

[Hint: $U(p,q)$ must be analytic for $\text{Re}(p) > \alpha$, $\text{Re}(q) > \beta$, for some fixed α,β. This imposes a restriction on the possible value of $u(0,y)$, and thus determines the solution uniquely.]

22. Solve the heat conduction problem

$$\frac{\partial u}{\partial t} = \kappa \frac{\partial^2 u}{\partial x^2}, \quad x \geq 0,$$

$$u(x,0) = 0, \quad x \geq 0,$$

$$u(0,t) = T_0, \quad t \geq 0,$$

using the double Laplace transform.

23. Consider the wave equation

$$\frac{\partial^2 u}{\partial x^2} = \frac{1}{c^2}\frac{\partial^2 u}{\partial t^2}, \quad x \geq 0, \quad t \geq 0,$$

$$u(x,0) = f(x),$$

$$u_t(x,0) = g(x),$$

$$u(0,t) = 0.$$

Show how the solution, which may be constructed by D'Alembert's method, can be recovered using the double Laplace transform.[9]

Footnotes

1. These results apply either to functions having the necessary behavior at infinity to allow integration by parts, or to generalized functions with no restrictions.

2. $\delta(\underset{\sim}{r}) = \delta(x)\delta(y)$. The theory of generalized functions may be extended quite simply to several variables, but we do not need to concern ourselves with the details here.

3. See Section 9.5.

4. This result is given in I. N. Sneddon, J. Eng. Math. (1974), 8, 177, together with a discussion of the connection with the half-space Dirichlet problem for Laplace's equation.

5. This is an example of the collisionless linear transport equation. See Section 19.6 for an example of the use of this Green's function in the solution of the linear transport equation with collisions.

6. See DITKIN & PRUDNIKOV (1970) for more information on double Laplace transforms.

7. We use the notation $\mathscr{L}[f(x,y); y \to p]$ so as to indicate which variable is transformed. Thus $\mathscr{L}[f(x,y); y \to p]$ is a function of x and p.

8. See J. C. Jaeger, Bull. Am. Math. Soc. (1940), 46, 687.

9. The application of the double Laplace transform to a more general second-order partial differential equation in the quadrant $x \geq 0$, $y \geq 0$ is discussed in K. Evans and E. A. Jackson, J. Math. Phys. (1971), 12, 2012.

Part III: Other Important Transforms

§12. MELLIN TRANSFORMS

12.1. Definitions

In this and the next two sections we study the Mellin transform, which, while closely related to the Fourier transform, has its own peculiar uses. In particular, it turns out to be a most convenient tool for deriving expansions, although it has many other applications. We recall first that the Fourier transform pair can be written in the form

$$A(\omega) = \int_{-\infty}^{\infty} a(t)e^{i\omega t}\, dt, \qquad \alpha < \text{Im}(\omega) < \beta , \tag{1}$$

and

$$a(t) = \frac{1}{2\pi}\int_{i\gamma-\infty}^{i\gamma+\infty} A(\omega)e^{-i\omega t}\, d\omega, \qquad \alpha < \gamma < \beta. \tag{2}$$

The Mellin transform and its inverse follow if we introduce the variable changes

$$\begin{aligned} p &= i\omega, \\ x &= e^{t}, \\ f(x) &= A(\ell n\ x), \end{aligned} \tag{3}$$

so that (1) and (2) become

$$F(p) = \int_0^\infty x^{p-1} f(x)\, dx, \qquad \alpha < \mathrm{Re}(p) < \beta, \tag{4}$$

$$f(x) = \frac{1}{2\pi i} \int_{c-i\infty}^{c+i\infty} x^{-p} F(p)\, dp. \tag{5}$$

Equation (4) is the Mellin transform, and (5) is the Mellin inversion formula. The transform normally exists only in the strip $\alpha < \mathrm{Re}(p) < \beta$, and the inversion contour must lie in this strip.

12.2. Simple Examples

We now study three simple examples which illustrate the most important and peculiarly useful features of the Mellin transform.

(i)
$$f(x) = e^{-\alpha x}, \qquad \alpha > 0, \tag{6}$$

$$F(p) = \int_0^\infty e^{-\alpha x} x^{p-1}\, dx$$

$$= \frac{(p-1)!}{\alpha^p}, \qquad \mathrm{Re}(p) > 0. \tag{7}$$

By the inversion formula we thus have the integral representation

$$f(x) = \frac{1}{2\pi i} \int_{c-i\infty}^{c+i\infty} (p-1)!\, (\alpha x)^{-p} dp, \qquad c = \mathrm{Re}(p) > 0. \tag{8}$$

From the asymptotic behavior of $(p-1)!$ for large p, we readily conclude that the contour of the inversion integral can be closed in the left-hand half-plane for any value of x, leading to the expansion

$$e^{-\alpha x} = \sum_{r=0}^{\infty} \frac{(-1)^r}{r!} (\alpha x)^r \tag{9}$$

corresponding to the poles and residues of the integrand.

(ii) $$f(x) = (1+\beta x)^{-\gamma}, \quad \gamma > 0, \quad |\arg\beta| \neq \pi \tag{10}$$

$$F(p) = \int_0^\infty \frac{x^{p-1}dx}{(1+\beta x)^\gamma} = \beta^{-p}\int_0^\infty \frac{y^{p-1}dy}{(1+y)^\gamma} \tag{11}$$

The substitution $y = z/(1-z)$ reduces the integral to the standard form

$$F(p) = \beta^{-p}\int_0^1 z^{p-1}(1-z)^{\gamma-p-1}dz = \beta^{-p}\,\frac{(p-1)!(\gamma-p-1)!}{(\gamma-1)!}, \tag{12}$$

where for the integral to converge, we must have

$$0 < \mathrm{Re}(p) < \gamma. \tag{13}$$

The inversion formula then gives us

$$(\gamma-1)!\ f(x) = \frac{1}{2\pi i}\int_{c-i\infty}^{c+i\infty} (p-1)!\ (\gamma-p-1)!\ (\beta x)^{-p}dp, \tag{14}$$

where the contour separates the two sets of poles as indicated in Figure 1.

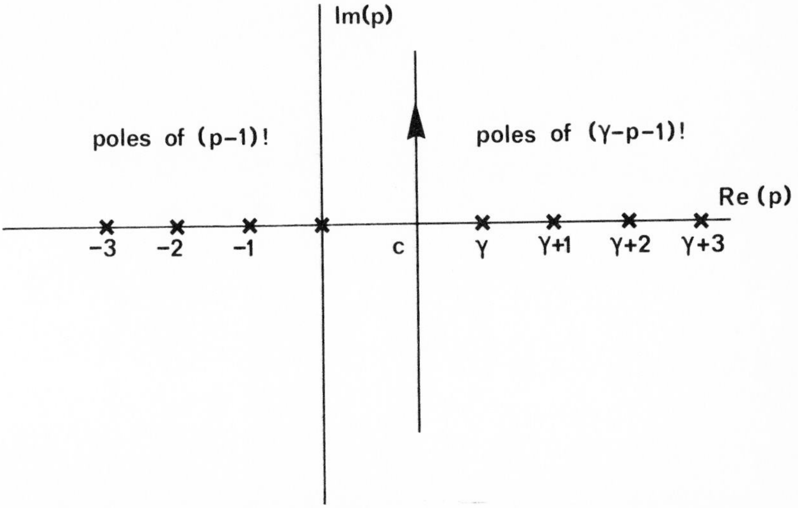

Figure 12.1

In order to close the contour so as to utilize the poles and residues of the integrand, we must first consider the asymptotic form of the integrand for large $|p|$. From Appendix A, we see that

$$|(p-1)!\ (\gamma-p-1)!\ (\beta x)^{-p}| \sim A|\beta x|^{-\mathrm{Re}(p)}, \quad |p| \to \infty, \tag{15}$$

and thus we can close in the left-hand half-plane if $|\beta x| < 1$ and in the right-hand half-plane if $|\beta x| > 1$. This leads immediately to both ascending and descending expansions, a common feature of the Mellin transform inversion.

<u>Ascending Expansion</u>: If we close the contour to the left the poles are those of $(p-1)!$ Evaluating the residues at these poles we have

$$\begin{aligned} f(x) &= \frac{1}{(\gamma-1)!} \sum_{r=0}^{\infty} \frac{(-1)^r}{r!} (\gamma+r-1)!\ (\beta x)^r, \\ &= 1 - \gamma(\beta x) + \frac{\gamma(\gamma+1)}{2!} (\beta x)^2 - \dots . \end{aligned} \tag{16}$$

This last expression is just the binomial expansion of $f(x)$.

<u>Descending Expansion</u>: The poles of $(\gamma-p-1)!$ are at $p = r+\gamma$, $r = 0,1,2,\dots$, with residues $(-1)^{r+1}/r!$ Therefore, closing the contour to the right we have the expansion

$$(\gamma-1)!\ f(x) = (-1) \sum_{r=0}^{\infty} \frac{(-1)^{r+1}}{r!} (r+\gamma-1)!\ (\beta x)^{-r-\gamma}, \tag{17}$$

where the additional factor (-1) arises since we are closing the contour in the negative (clockwise) direction. Written out explicitly the expansion is

$$f(x) = (\beta x)^{-\gamma}\left[1 - \frac{\gamma}{\beta x} + \frac{\gamma(\gamma+1)}{2!(\beta x)^2} - \frac{\gamma(\gamma+1)(\gamma+2)}{3!(\beta x)^2} + \dots\right], \tag{18}$$

which is the binomial expansion valid for large values of $|\beta x|$.

(iii) The exponential integral is

$$E_1(x) = \int_x^\infty \frac{e^{-u}}{u}\,du. \tag{19}$$

Denoting by $\Omega(p)$ the Mellin transform of $E_1(x)$, we have

$$\begin{aligned}\Omega(p) &= \int_0^\infty x^{p-1}\,dx \int_1^\infty \frac{e^{-wx}}{w}\,dw \\ &= \int_1^\infty \frac{dw}{w} \int_0^\infty x^{p-1}\, e^{-wx}\,dx \qquad (20)\\ &= \frac{(p-1)!}{p}, \quad \mathrm{Re}(p) > 0.\end{aligned}$$

Thus,

$$E_1(x) = \frac{1}{2\pi i}\int_{c-i\infty}^{c+i\infty} (p-1)!\; x^{-p}\,\frac{dp}{p}, \quad c > 0. \tag{21}$$

Closing the contour to the left, which is permissible because of the asymptotic form of $\Omega(p)$, we recover a sum of residues at $p = 0,-1,-2,\ldots$. The pole at $p = 0$ is a double pole with residue $-\ln x - \gamma$, where γ is Euler's constant (equal to the value of $d\,\ln(\alpha!)/d\alpha$ at $\alpha = 0$); the other poles are simple poles. Our ascending expansion appears as

$$E_1(x) = -\ln x - \gamma - \sum_{k=1}^\infty \frac{(-1)^k\, x^k}{k!\;k}. \tag{22}$$

In this case, as with the exponential function [example (i)], we do not recover a descending expansion, because there are no singularities in the right-hand half-plane. The real reason is deeper than this; the exponential function has an essential singularity at infinity, and hence no expansion in powers of $1/x$. It is easy to see, either from (19) or (20), that $E_1(x) \sim \exp(-x)$ for large x; consequently we consider the function

$$f(x) = e^{x} E_1(x). \tag{23}$$

Taking the Mellin transform we obtain

$$\begin{aligned} F(p) &= \int_0^\infty x^{p-1}\, dx \int_1^\infty \frac{e^{-x(w-1)}}{w}\, dw \\ &= \int_1^\infty \frac{dw}{w} \int_0^\infty e^{-x(w-1)}\, x^{p-1}\, dx \\ &= (p-1)!\int_1^\infty w^{-1}\, (w-1)^{-p}\, dw \\ &= (p-1)!\,(-p)!\,(p-1)!, \quad 0 < \mathrm{Re}(p) < 1. \end{aligned} \tag{24}$$

There are now poles in both half-planes, but we cannot close the contour to the right and throw away the integral around the large semi-circle, because $F(p)$ grows exponentially as $p \to +\infty$.

<u>Ascending Expansion</u>: Closing the contour to the left, we must evaluate the residues at the double poles of $[(p-1)!]^2$. This can be done by writing

$$F(p)\, x^{-p} = \frac{\pi^2}{\sin^2(\pi p)}\, \frac{x^{-p}}{(-p)!}, \tag{25}$$

leading to the expansion

$$\begin{aligned} E_1(x) &= e^{-x} \sum_{k=0}^{\infty} \frac{x^k}{k!}\, [\psi(k) - \ell n\, x], \\ \psi(\alpha+1) &= \frac{d}{d\alpha}\, \ell n(\alpha!). \end{aligned} \tag{26}$$

<u>Descending Expansion</u>: The inversion integral gives

$$E_1(x) = \frac{e^{-x}}{2\pi i} \int_{c-i\infty}^{c+i\infty} F(p)\, x^{-p}\, dp, \quad 0 < c < 1. \tag{27}$$

The contour may not be closed to the right; however we can shift it a finite distance, since $F(p)$ goes to zero exponentially as $p \to c \pm i\infty$. Thus we can write

$$E_1(x) = e^{-x}\left\{\sum_{k=1}^{n} \frac{(-1)^{k+1}k!}{x^{k+1}} + \frac{1}{2\pi i}\int_{c+n-i\infty}^{c+n+i\infty} F(p)x^{-p}dp\right\}. \quad (28)$$

We leave it to the reader to verify that the remainder term is of order x^{-n-2} as $x \to \infty$, showing that we have recovered the well-known asymptotic expansion

$$E_1(x) \sim e^{-x} \sum_{k=1}^{\infty} \frac{(-1)^{k+1}k!}{x^{k+1}} . \quad (29)$$

12.3. Elementary Properties

Mellin transforms have a number of important elementary properties, which we now investigate. We use the notation

$$\mathcal{M}[f;p] = \int_0^\infty f(x)\, x^{p-1}\, dx \quad (30)$$

where this simplifies the appearance of the results.

Derivatives:

$$\mathcal{M}[f';p] = \int_0^\infty f'(x)\, x^{p-1}\, dx$$

$$= \left[f(x)\, x^{p-1}\right]_0^\infty - (p-1)\int_0^\infty f(x)\, x^{p-2}\, dx. \quad (31)$$

We assume that $F(p)$ exists for $\alpha < \mathrm{Im}(p) < \beta$; consequently we must have

$$\lim_{x\to 0} x^p f(x) = 0, \qquad \mathrm{Re}(p) > \alpha,$$

$$\lim_{x\to\infty} x^p f(x) = 0, \qquad \mathrm{Re}(p) < \beta, \quad (32)$$

and thus

$$\mathcal{M}[f';p] = -(p-1)\, F(p-1), \quad \alpha < \mathrm{Im}(p-1) < \beta. \quad (33)$$

Powers:

$$\mathcal{M}[x^{\mu}f(x);p] = \int_0^{\infty} x^{\mu}f(x)\ x^{p-1}\ dx = F(p+\mu). \tag{34}$$

Laplacian in Plane-Polar Coordinates: In two dimensions, the Laplace operator ∇^2 is

$$\nabla^2 f(r,\phi) = f_{rr} + \frac{1}{r} f_r + \frac{1}{r^2} f_{\phi\phi}. \tag{35}$$

If we take the Mellin transform $F(p,\phi)$ of f with respect to the radial variable, we obtain the simple relation

$$\mathcal{M}[\nabla^2 f;p] = \left\{\frac{d^2}{d\phi^2} + (p-2)^2\right\} F(p-2,\phi) \tag{36}$$

by the application of (33) and (34). Thus problems involving this operator may be simplified by use of the Mellin transform.

Convolutions: If

$$h(x) = \int_0^{\infty} y^{\mu}f(xy)\ g(y)\ dy, \tag{37}$$

then

$$\begin{aligned} H(p) &= \int_0^{\infty} x^{p-1}\ dx \int_0^{\infty} y^{\mu}\ f(xy)\ g(y)\ dy \\ &= \int_0^{\infty} y^{\mu}g(y)dy \int_0^{\infty} x^{p-1}\ f(xy)\ dx \\ &= \int_0^{\infty} y^{\mu-p}g(y)dy \int_0^{\infty} t^{p-1}\ f(t)\ dt \\ &= F(p)\ G(\mu-p+1). \end{aligned} \tag{38}$$

Similarly, if

$$k(x) = \int_0^{\infty} y^{\mu}\ f(x/y)\ g(y)\ dy, \tag{39}$$

then

$$K(p) = F(p)\ G(\mu+p+1). \tag{40}$$

A further relation which is sometimes useful concerns $\mathscr{M}[fg;p]$; it is

$$\mathscr{M}[fg;p] = \int_0^\infty f(x)\, g(x)\, x^{p-1}\, dx$$

$$= \int_0^\infty g(x)\, x^{p-1}\, dx\, \frac{1}{2\pi i}\int_{c-i\infty}^{c+i\infty} F(s)x^{-s}\, ds \tag{41}$$

$$= \frac{1}{2\pi i}\int_{c-i\infty}^{c+i\infty} F(s)\, G(p-s)\, ds.$$

12.4. Potential Problems in Wedge-Shaped Regions

Consider the boundary-value problem

$$\nabla^2 u = 0, \quad 0 \le r < \infty, \quad -\alpha \le \theta \le \alpha, \tag{42}$$

$$u(r,\pm\alpha) = \begin{cases} 1, & 0 \le r < a \\ 0, & r > a, \end{cases} \tag{43}$$

which determines the solution of a potential problem in an infinite wedge of angle 2α. If we assume that $u(r,\theta)$ is bounded as $r \to 0$, and that as $r \to \infty$, $u(r,\theta) \sim r^{-\beta}$ for some $\beta > 0$, then the Mellin transform of u with respect to r exists. Before applying (36) to the partial differential equation, we multiply by r^2 so as to obtain an equation for $U(p,\theta)$ rather than $U(p-2,\theta)$. Then (42) and (43) transform to

$$\left[\frac{d^2}{d\theta^2} + p^2\right] U(p,\theta) = 0,$$
$$U(p,\alpha) = U(p,-\alpha) = a^p/p, \tag{44}$$

wihh the solution

$$U(p,\alpha) = \frac{a^p \cos p\theta}{p \cos p\alpha}\ . \tag{45}$$

We shall consider the inversion of Mellin transforms of this type in the next section; anticipating these results we obtain the solution of our potential problem as

$$u(r,\theta) = \begin{cases} 1 - \frac{1}{\pi} \arctan \left[\frac{2(ar)^{\beta} \cos \beta\theta}{a^{2\beta} - r^{2\beta}} \right], & 0 \leq r < a \\ \frac{1}{\pi} \arctan \left[\frac{2(ar)^{\beta} \cos \beta\theta}{r^{2\beta} - a^{2\beta}} \right], & r > a \end{cases} \tag{46}$$

where $\beta = \pi/2\alpha$.

12.5. Transforms Involving Polar Coordinates[1]

In problems involving polar coordinates, one is confronted with transforms of the type $F(p)\sin(p\theta)$ and $F(p)\cos(p\theta)$, as in (45). Suppose that $F(p)$ is the Mellin transform of a real function $f(r)$; then proceeding formally we have[2]

$$\begin{aligned} \mathscr{M}[f(re^{i\theta}); r \to p] &= \int_0^{\infty} f(re^{i\theta})\, r^{p-1}\, dr \\ &= e^{-ip\theta} \int f(s)\, s^{p-1}\, ds \\ &= e^{-ip\theta} F(p), \end{aligned} \tag{47}$$

provided the s integral is equivalent to a Mellin transform. This leads to the useful formulas

$$\begin{aligned} F(p) \cos (p\theta) &= \mathscr{M}[\mathrm{Re}\, f(re^{i\theta}); r \to p], \\ F(p) \sin (p\theta) &= \mathscr{M}[-\mathrm{Im}\, f(re^{i\theta}); r \to p]. \end{aligned} \tag{48}$$

Sector of Validity: In order to carry out the variable change $s = r \exp(i\theta)$ in (47) we must assume that $f(r)$ is the value of an analytic function $f(z)$ defined in some sector $-\alpha < \arg(z) < \alpha$, with $r = |z|$. Replacing the upper limit for r in (47) by R, we can write

$$\int_0^R f(re^{i\theta})\, r^{p-1}\, dr = e^{-ip\theta}\left\{\int_0^R f(s)\, s^{p-1}\, ds + \int_C f(s)\, s^{p-1}\, ds\right\}, \tag{49}$$

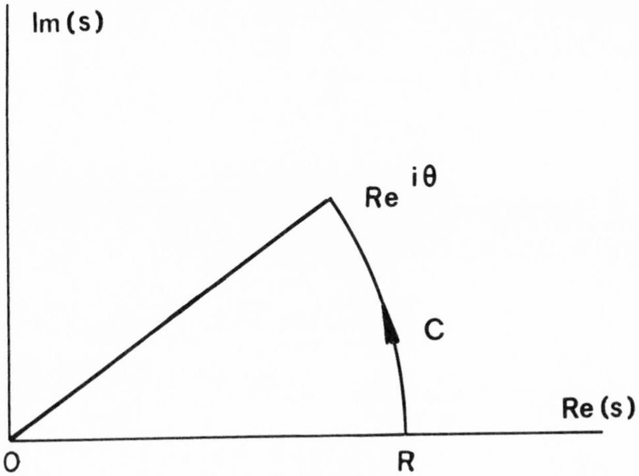

Figure 12.2

(see Figure 2). We need the second integral to become zero as $R \to \infty$. A sufficient condition is

$$z^p\, f(z) \to 0 \quad \text{as} \quad |z| \to \infty, \quad |\arg(z)| < \alpha. \tag{50}$$

We have already assumed that $\mathcal{M}[f(r)]$ exists in some strip $p_1 < \mathrm{Re}(p) < p_2$, thus

$$r^p\, f(r) \to 0 \quad \text{as} \quad r \to \infty, \quad \mathrm{Re}(p) < p_2. \tag{51}$$

The usual situation is that (50) is also valid when $\mathrm{Re}(p) < p_2$ provided that α is suitably chosen; consequently (50) is a restriction on α rather than $f(z)$. For example, if $f(r) = \exp(-r)$, then $f(z) = \exp(-z)$ and $\alpha = \pi/2$.

Applications:

(i) $$\mathcal{M}[\ell n(1+r)] = \frac{\pi}{p \sin p\pi}, \quad -1 < \operatorname{Re}(p) < 0. \tag{52}$$

Hence

$$\mathcal{M}[\tfrac{1}{2}\ \ell n(1+2r\cos\theta + r^2)] = \frac{\pi \cos p\theta}{p \sin p\pi}, \tag{53}$$

and

$$\mathcal{M}[\arctan (\frac{r \sin\theta}{1+r\cos\theta})] = \frac{-\pi \sin p\theta}{p\sin p\pi},$$

$$-\pi/2 < \theta < \pi/2,$$

$$-1 < \operatorname{Re}(p) < 0.$$

(ii) $$\mathcal{M}[\tfrac{\pi}{2} - \arctan r] = \frac{\pi}{2p\cos(p\pi/2)}, \quad 0 < \operatorname{Re}(p) < 1. \tag{54}$$

To check the sector of validity of (50) in this case we write

$$\arctan z = \frac{\pi}{2} + \frac{1}{2i}[\ell n(z-i) - \ell n(z+i)] \tag{55}$$

where the branches of the logarithm are chosen so that $0 < \arctan r < \pi/2$. It follows immediately that $\alpha = \pi/2$; using (55) we find that

$$\operatorname{Re}[\tfrac{\pi}{2} - \arctan z] = \begin{cases} \pi - \arctan\left(\dfrac{2r\cos\theta}{1-r^2}\right), & 0 \le r < 1 \\ \arctan\left(\dfrac{2r\cos\theta}{r^2-1}\right), & r > 1 \end{cases} \tag{56}$$

and this function has as its Mellin transform the function $\cos(p\theta)/p\cos(p\pi/2)$. Equation (46) may be recovered by applying the results of Problems 10 and 11 in succession.

12.6. Hermite Functions

The Mellin transform may sometimes be of use in solving ordinary differential equations with polynomial coefficients, using a technique which was first employed by Barnes in his investigations of the hypergeometric function. We choose here to discuss the Hermite equation

$$H_\nu''(x) - xH_\nu'(x) + \nu H_\nu(x) = 0. \tag{57}$$

On taking the Mellin transform, we get

$$(p-1)(p-2)\ S(p-2) + (\nu+p)\ S(p) = 0, \tag{58}$$

$$S(p) = \mathscr{M}[H_\nu;p].$$

It is apparent that the Mellin transform does not give us $S(p)$ directly, but rather a difference equation for $S(p)$.

Solution of the Difference Equation: We first reduce the difference equation for $S(p)$ to standard form (in which the arguments differ by an integer) by writing $p = -2s$ and $S(p) = T(s)$. Thus we have to solve

$$\frac{T(s+1)}{T(s)} = \frac{(s-\frac{1}{2}\nu)}{2(s+\frac{1}{2})(s+1)}. \tag{59}$$

A particular solution is

$$T(s) = K\ 2^{-s}\ (s-\tfrac{1}{2}\nu-1)!\ (-s-\tfrac{1}{2})!\ (-s-1)!, \tag{60}$$

but it is not unique, since we may multiply $T(s)$ by any function which is a solution of the difference equation

$$\frac{Y(s+1)}{Y(s)} = 1. \tag{61}$$

At this point, we appeal to the fact that $S(p)$ is a Mellin transform, defined only in some strip $\alpha < \mathrm{Re}(p) < \beta$. Therefore, equation (58) is valid only in the overlap of the two strips

$$\begin{aligned} \alpha &< \mathrm{Re}(p) < \beta \\ \alpha &< \mathrm{Re}(p-2) < \beta \end{aligned} \tag{62}$$

and there is no such overlap unless $\beta > \alpha+2$. Thus, $Y(s)$ cannot have poles, since they would give rise to a row of poles in $S(p)$ separated by exactly two units. Also, $Y(s)$ cannot grow faster than $|s|$ as $\mathrm{Im}(s) \to \infty$ in the inversion strip; otherwise the inversion integral would diverge. Therefore, by (61), $Y(s)$ is a bounded entire function, and thus equal to a constant. Hence (60) is the only acceptable solution, and then only if $\mathrm{Re}(\nu) < -2$, so we have

$$H_\nu(x) = \frac{K}{2\pi i}\int_{c-i\infty}^{c+i\infty} (s-\tfrac{1}{2}\nu-1)!\ (-s-\tfrac{1}{2})!(-s-1)!\left[\frac{x^2}{2}\right]^s ds, \tag{63}$$

$$\mathrm{Re}(\nu) < -2.$$

By convention the coefficient of x^ν in $H_\nu(x)$ is unity, and since the pole of $(s-\frac{1}{2}\nu-1)!$ at $s = \frac{1}{2}\nu$ has residue 1, we have

$$K = \frac{2^{\nu/2}}{(-\frac{1}{2}\nu-\frac{1}{2})!\ (-\frac{1}{2}\nu-1)!}\ . \tag{64}$$

<u>Complete Descending Expansion</u>: The poles of $(s-\frac{1}{2}\nu-1)!$ lie at $s = (\frac{1}{2}\nu-r)$, $r = 0,1,\ldots$, with residues $(-1)^r/r!$ Thus

$$\begin{aligned} H_\nu(x) &= \frac{2^{\nu/2}}{(-\frac{1}{2}\nu-\frac{1}{2})!(-\frac{1}{2}\nu-1)!} \\ &\times \sum_{r=0}^{\infty} \frac{(-1)^r}{r!}\ (-\tfrac{1}{2}\nu+r-\tfrac{1}{2})!\ (-\tfrac{1}{2}\nu+r-1)!\ \left[\frac{x^2}{2}\right]^{\frac{1}{2}\nu-r} \\ &= x^\nu - \frac{\nu(\nu-1)}{2}\, x^{\nu-2} + \frac{\nu(\nu-1)(\nu-2)(\nu-3)}{2!\ 2^2}\, x^{\nu-4} + \cdots\ . \end{aligned} \tag{65}$$

The restriction on ν may now be lifted by analytic continuation. If ν is a positive integer or zero, we have a polynomial of degree ν.

Complete Ascending Expansion: Closing the contour to the right leads to the expansion

$$H_\nu(x) = \frac{2^{\nu/2}}{(-\frac{1}{2}\nu-\frac{1}{2})!\,(-\frac{1}{2}\nu-1)!}$$
$$\times \sum_{r=0}^{\infty} \frac{(-1)^r}{r!} \left\{ (r-\frac{1}{2}\nu-\frac{1}{2})!\,(-r+\frac{1}{2})! \left(\frac{x^2}{2}\right)^{r+\frac{1}{2}} \right. \tag{66}$$
$$\left. + \; (r-\frac{1}{2}\nu-1)!\,(-r+\frac{1}{2})! \left(\frac{x^2}{2}\right)^{r} \right\}.$$

If ν is zero or a positive integer, we must first calculate the ratios of the factorials outside and inside the summation before using this formula, which then gives us a polynomial. In other cases, the expansion is an infinite series.

Problems

Prove the following general properties of the Mellin transform.

1. $\mathcal{M}[f(ax);p] = a^{-p}\,F(p)$

2. $\mathcal{M}[f(x^a);p] = a^{-1}\,F(p/a)$

3. $\mathcal{M}[x^{-1}f(x^{-1});p] = F(1-p)$

4. $\mathcal{M}[\ell n\ x\ f(x);p] = \frac{d}{dp}\,F(p)$

5. $\mathcal{M}[(xd/dx)^n f(x);p] = (-1)^n\,p^n F(p)$

6. $\mathcal{M}\left[\int_0^x f(u)du;p\right] = -\frac{1}{p}\,F(p+1)$

7. $\mathcal{M}\left[\int_x^\infty f(u)du;p\right] = \frac{1}{p} F(p+1)$

Verify the following Mellin transforms.

8. $\mathcal{M}[(1+x)^{-a};p] = \frac{(p-1)!(a-p-1)!}{(a-1)!}, \quad 0 < \text{Re}(p) < \text{Re}(a)$

9. $\mathcal{M}[(1+x)^{-1};p] = \frac{\pi}{\sin \pi p}$

10. $\mathcal{M}[(1+x^a)^{-1};p] = \frac{\pi}{a \sin(\pi p/a)}$

11. $\mathcal{M}[e^{i\beta x};p] = \frac{(p-1)!\, e^{i\pi p/2}}{\beta^p}, \quad 0 < \text{Re}(p) < 1$

12. $\mathcal{M}[\cos \beta x;p] = \frac{(p-1)!\, \cos(\pi p/2)}{\beta^p}, \quad 0 < \text{Re}(p) < 1$

13. $\mathcal{M}[\sin \beta x;p] = \frac{(p-1)!\, \sin(\pi p/2)}{\beta^p}, \quad -1 < \text{Re}(p) < 1$

14. $\mathcal{M}[J_\nu(x);p] = \frac{2^{p-1}(\frac{1}{2}\nu+\frac{1}{2}p-1)!}{(\frac{1}{2}\nu-\frac{1}{2}p)!}, \quad -\nu < \text{Re}(p) < \nu+2$

[Hint: use the Poisson integral representation for $J_\nu(x)$.]

15. $\mathcal{M}[e^{-\alpha x^2};p] = \frac{1}{2}\,\alpha^{-p/2}(\frac{1}{2}p-1)!$

16. $\mathcal{M}[(1-x)^{a-1}\, h(1-x);p] = \frac{(p-1)!(a-1)!}{(p+a-1)!}$

17. $\mathcal{M}[e^{-x}\ell n x;p] = (p-1)!\, \psi(p-1)$

18. $\mathcal{M}\left[\frac{1+x\cos\theta}{1+2x\cos\theta+x^2};p\right] = \frac{\pi \cos p\theta}{\sin p\pi}$

19. $\mathcal{M}\left[\frac{x\sin\theta}{1+2x\cos\theta+x^2};p\right] = \frac{\pi \sin p\theta}{\sin p\pi}$

20. $$\mathscr{M}\left[\int_0^1 \frac{u^{2/3}\ (1-u^2)^{1/2}}{(u+x)^{1/3}}\,du;p\right] = \frac{\sqrt{\pi}(p-1)!(-p-2/3)!(\frac{1}{2}p-1/3)!}{4\ (-2/3)!(\frac{1}{2}p+7/6)!}$$

21. $$\mathscr{M}\left[\int_0^\infty u^{2/3}(u+x)^{-1/3}e^{-u^2}du;p\right] = \frac{(p-1)!(-p-2/3)!(\frac{1}{2}p-1/3)!}{2(-2/3)!}$$

22. The complementary error function is

$$\operatorname{erfc}(x) = \frac{2}{\sqrt{\pi}}\int_x^\infty e^{-u^2}\ du;$$

show $\mathscr{M}[\operatorname{erfc}(x);p] = (\frac{1}{2}p-\frac{1}{2})!/p\sqrt{\pi}$.

23. The cosine integral is defined by

$$\operatorname{Ci}(x) = -\int_x^\infty \frac{\cos u}{u}\,du;$$

show $\mathscr{M}[\operatorname{Ci}(x);p] = \dfrac{(p-1)!}{p}\cos(\pi p/2)$.

24. Find the steady state temperature distribution inside a wedge $0 \le r < \infty$, $0 \le \theta \le \alpha$, if the boundary $\theta = 0$ is held at temperature zero, while the other boundary is maintained at

$$u(r,\alpha) = \begin{cases} T_0, & r \le a \\ 0\ , & r > a. \end{cases}$$

25. The boundary $\theta = 0$ of an infinite wedge $0 \le r < \infty$, $0 \le \theta \le \alpha$, is held at zero temperature. Through the other boundary, the concentrated heat flow

$$q(r) = Q\delta(r-a)$$

is maintained. Show that the steady state temperature distribution is

$$u(r,\theta) = \frac{Q}{2\pi\,\kappa\,\cosh[\pi \ln(r/a)/2\alpha]}$$
$$\times \ln\left[\frac{\cosh[\pi \ln(r/a)/2\alpha] + \sin(\pi\theta/2\alpha)}{\cosh[\pi \ln(r/a)/2\alpha] - \sin(\pi\theta/2\alpha)}\right].$$

26. A thin charged wire with charge q per unit length is placed along the line $r = r_0$, $\theta = \theta_0$, inside a wedge shaped region $0 \leq r < \infty$, $0 \leq \theta \leq \alpha$, whose boundaries are held at zero potential. Show that the electrostatic potential may be written as

$$\phi(r,\theta) = \begin{cases} \dfrac{2q}{i}\displaystyle\int_{-i\infty}^{i\infty} \frac{\sin p(\alpha-\theta_0)\,\sin p\theta}{p\,\sin p\alpha}\left(\frac{r_0}{r}\right)^p dp, & \theta \leq \theta_0 \\ \dfrac{2q}{i}\displaystyle\int_{-i\infty}^{i\infty} \frac{\sin p\theta_0\,\sin p(\alpha-\theta)}{p\,\sin p\alpha}\left(\frac{r_0}{r}\right)^p dp, & \theta \geq \theta_0. \end{cases}$$

27. In the preceding problem, show that if $\alpha = 2\pi$, $r_0 = a$, and $\theta_0 = \pi$, then

$$\phi(r,\theta) = q \ln \frac{1+2\sqrt{r/a}\,\sin(\theta/2)+(r/a)}{1-2\sqrt{r/a}\,\sin(\theta/2)+(r/a)}.$$

Calculate the charge density induced on the boundary $\theta = 0$, $r \geq 0$.

28. If in Problem 24 the boundaries are at $\theta = \pm\alpha$, with the boundary conditions

$$u(r,-\alpha) = f(r),$$
$$u(r,\alpha) = g(r),$$

then show that

$$u(r,\theta) = \frac{\beta r^\beta}{\pi}\cos(\beta\theta)\left\{\int_0^\infty \frac{u^{\beta-1} f(u)\,du}{u^{2\beta} - 2r^\beta u^\beta \sin\beta\theta + r^{2\beta}}\right.$$
$$\left. + \int_0^\infty \frac{u^{\beta-1} g(u)\,du}{u^{2\beta}+2r^\beta u^\beta \sin\beta\theta + r^{2\beta}}\right\},$$

where $\beta = \pi/2\alpha$.

29. Solve the Laguerre equation

$$xy''(x) + (\xi+1-x)\, y'(x) + \eta y(x) = 0$$

by the Mellin transform. In particular, derive the Laguerre polynomials corresponding to the choices

$$\eta = m-\ell-1,$$
$$\xi = \ell(\ell+1),$$

where ℓ and m are non-negative integers.

30. Show that if the integral transforms

$$F(x) = \int_0^\infty k(xt)\, f(t)\, dt$$
$$f(t) = \int_0^\infty \ell(tx)\, F(x)\, dx$$

are reciprocal to each other, then

$$L(p)\, K(1-p) = 1.$$

31. By considering the Mellin transform of $J_\nu(kx)$ verify the Hankel transform pair (15.1,2).

Footnotes

1. See W. J. Harrington, SIAM Review (1967), 9, 542.

2. We write $\mathscr{M}[f(r,\theta);\ r \to p]$ to indicate that r is the variable being integrated out to give a function of p.

§13. MELLIN TRANSFORMS IN SUMMATION

13.1. Mellin Summation Formula[1]

Suppose we wish to evaluate the sum

$$S = \sum_{n=1}^{\infty} f(n). \tag{1}$$

If the function $f(n)$, regarded as a function of a continuous variable, has the Mellin transform $F(p)$, then we may write

$$f(n) = \frac{1}{2\pi i} \int_{c-i\infty}^{c+i\infty} F(p)\, n^{-p}\, dp, \tag{2}$$

and consequently

$$\begin{aligned} S &= \frac{1}{2\pi i} \int_{c-i\infty}^{c+i\infty} F(p) \sum_{n=1}^{\infty} n^{-p}\, dp \\ &= \frac{1}{2\pi i} \int_{c-i\infty}^{c+i\infty} F(p)\, \zeta(p)\, dp, \end{aligned} \tag{3}$$

where $\zeta(p)$ is the Riemann zeta function, whose more important properties are discussed in Appendix B.

An Example: We consider the sum

$$S = \sum_{n=1}^{\infty} \frac{\cos \beta n}{n^2}, \quad 0 \leq \beta < 2\pi. \tag{4}$$

From Problem 12.12, we have

$$\int_0^{\infty} \cos(\beta n)\, n^{p-1}\, dn = \frac{(p-1)!}{\beta^p} \cos(\pi p/2), \qquad 0 < \mathrm{Re}(p) < 1, \tag{5}$$

and thus the Mellin transform of $\cos(\beta n)/n^2$ is

$$F(p) = -\frac{(p-3)!}{\beta^{p-2}} \cos(\pi p/2). \tag{6}$$

The sum now becomes

$$S = -\frac{1}{2\pi i}\int_{c-i\infty}^{c+i\infty}\frac{(p-3)!}{\beta^{p-2}}\cos(\pi p/2)\,\zeta(p)\,dp,$$

$$2 < \mathrm{Re}(p) < 3, \tag{7}$$

where the interchange of order of integration and summation is made possible because the sum converges uniformly in p on the inversion contour.

Using the Riemann relation (B11) the integral may be cast in the more convenient form

$$S = -\frac{\beta^2}{4\pi i}\int_{c-i\infty}^{c+i\infty}\left(\frac{2\pi}{\beta}\right)^p\frac{\zeta(1-p)}{(p-1)(p-2)}\,dp. \tag{8}$$

We have three simple poles at $p = 2, 1,$ and 0, from which we obtain the simple result

$$\begin{aligned} S &= -\frac{\beta^2}{2}\left[\frac{(2\pi)^2}{\beta^2}\zeta(-1) - \frac{2\pi}{\beta}\zeta(0) - \frac{1}{2}\right] \\ &= \frac{\pi^2}{6} - \frac{\pi\beta}{2} + \frac{\beta^2}{4}. \end{aligned} \tag{9}$$

<u>A Further Example</u>: We consider the finite series

$$S(x) = \sum_{m=1}^{M}\frac{(1-xm^{2/3})^{1/2}}{m^{2/3}}, \tag{10}$$

$$M = \text{largest integer} \le x^{-3/2}.$$

For small x the series is slowly convergent; we will obtain an asymptotic formula which is rapidly convergent. First we write

$$S(x^{2/3}) = x^{2/3}\sum_{m=1}^{\infty} f(xm), \tag{11}$$

where

$$f(t) = \begin{cases} (1-t^{2/3})^{1/2}/t^{2/3}, & t \le 1 \\ 0, & t > 1, \end{cases} \tag{12}$$

and, using the relation

$$\mathscr{M}[f(t^{\alpha});\ t \to p] = \frac{1}{\alpha} F(\frac{p}{\alpha} - 1) \tag{13}$$

we obtain

$$F(p) = \frac{3}{2}\,\frac{[(3p/2)-2]!\,(1/2)!}{[(3p/2)-(1/2)]!} \tag{14}$$

and

$$S(x^{2/3}) = \frac{1}{2\pi i}\int_{c-i\infty}^{c+i\infty} \frac{3}{2}\,\frac{[(3p/2)-2]!\,(1/2)!}{[(3p/2)-(1/2)]!}\,x^{-p-2/3}\zeta(p), \qquad c > 2/3. \tag{15}$$

The factor $[(3p/2)-2]!$ has poles at $p = 2/3, 0, -2/3, \ldots, -(n-1)2/3$, with residues $(2/3)(-1)^n/n!$ and $\zeta(p)$ has a pole at $p = 1$ with residue 1. Hence

$$S(x) \simeq \frac{3\pi}{4\sqrt{x}} + \zeta(2/3) + \frac{1}{4}x - \frac{1}{8}\zeta(-2/3)x^2 + \ldots, \tag{16}$$

where the remainder term, which is a difficult integral, eventually increases with increasing n. Nevertheless, (16) is a useful asymptotic series for small x.

13.2. A Problem of Ramanujan[2]

Consider the function

$$f(x) = \sum_{n=-\infty}^{\infty} \left[e^{-xe^n} - e^{-e^n}\right]. \tag{17}$$

This infinite sum converges for all positive x, but it cannot be summed by the Mellin summation formula. It obviously satisfies the functional equation

$$f(ex) = f(e) + f(x) \tag{18}$$

which is the same functional equation satisfied by the function $\ell n\ x$. Is our function then identically equal to $\ell n\ x$ (up to a constant factor)? We can show that $f(x) \neq \ell n\ x$ as follows.

The relation

$$\int_0^\infty e^{-y}\, y^{p-1}\, dy = (p-1)!, \quad \mathrm{Re}(p) > 0 \tag{19}$$

allows us to write, by the Mellin inversion formula,

$$e^{-y} = \frac{1}{2\pi i}\int_{c-i\infty}^{c+i\infty} (p-1)!\; y^{-p}\, dp, \quad c > 0. \tag{20}$$

Hence, substituting this integral representation into (17), we can write

$$\begin{aligned} f(x) &= \sum_{n=-\infty}^{\infty} \frac{1}{2\pi i}\int_{c-i\infty}^{c+i\infty} (p-1)!\; e^{-np}\,(x^{-p}-1)\,dp \\ &= \sum_{n=0}^{\infty} \frac{1}{2\pi i}\int_{c-i\infty}^{c+i\infty} (p-1)!\; e^{-np}(x^{-p}-1)\,dp \\ &\quad + \sum_{m=1}^{\infty} \frac{1}{2\pi i}\int_{c-i\infty}^{c+i\infty} (p-1)!\; e^{mp}(x^{-p}-1)\,dp \\ &\equiv S_1 + S_2. \end{aligned} \tag{21}$$

The first sum S_1 converges uniformly with respect to p, so that we can change the order of integration and summation and carry out the sum to write

$$S_1 = \frac{1}{2\pi i}\int_{c-i\infty}^{c+i\infty} (p-1)!\; \frac{(x^{-p}-1)}{1-e^{-p}}\, dp. \tag{22}$$

This interchange is not permissible in the second sum S_2, but we can overcome the problem by observing that the integrand has no pole at $p = 0$, the pole due to $(p-1)!$ being cancelled by the zero of $(x^{-p}-1)$. Thus we can translate the contour to the left to write

$$S_2 = \sum_{m=1}^{\infty} \frac{1}{2\pi i}\int_{c'-i\infty}^{c'+i\infty} (p-1)!\; e^{mp}(x^{-p}-1)\,dp, \quad -1 < c' < 0. \tag{23}$$

On the new contour our sum converges, so bringing the summation inside we have

$$S_2 = -\frac{1}{2\pi i}\int_{c'-i\infty}^{c'+i\infty} (p-1)!\ \frac{(x^{-p}-1)}{(1-e^{-p})}\, dp \tag{24}$$

and

$$f(x) = S_1 + S_2$$

$$= \frac{1}{2\pi i}\int_C (p-1)!\ \frac{(x^{-p}-1)}{(1-e^{-p})}\, dp \tag{25}$$

where the closed contour of integration is indicated in Figure 1. The poles of the integrand are a single pole at

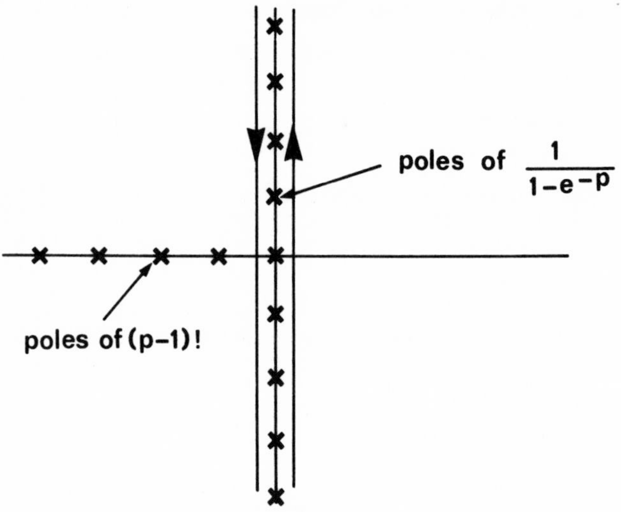

Figure 13.1

$p = 0$, and single poles due to zeros of the function $1-e^{-p}$ at $p = \pm 2\pi in$ along the imaginary axis. The residue of the integrand at $p = 0$ is

$$\lim_{p\to 0} \frac{(x^{-p}-1)}{1-e^{-p}} = -\ell n\ x \tag{26}$$

Addition of the residues at the poles gives us

$$f(x) = -\ell n\ x + \sum_{n=-\infty}^{\infty} (2\pi in-1)!\ (x^{-2\pi in}-1) \tag{27}$$

Our function $f(x)$ then does not coincide with the function $-\ell n\ x$, except at isolated points given by

$$(x^{-2\pi i n}-1) = 0, \quad n = 1,2,3,\ldots . \tag{28}$$

The function in fact wobbles about the function $-\ell n\ x$. A similar problem led Ramanujan to his fallacious proof of the Prime Number Theorem--he forgot the wobbles!

13.3. Asymptotic Behavior of Power Series[3]

A problem which sometimes occurs in that of finding the asymptotic (large z) behavior of a function defined as a power series in z. We have already seen that an integral representation may be a useful starting point for asymptotic analysis; the Mellin transform is the most direct method for a power series because it involves ascending and descending expansions simultaneously.

Consider the integral representation

$$f(z) = \frac{1}{2\pi i}\int_{c-i\infty}^{c+i\infty} F(p)\, z^{-p}\, dp, \quad c > 0. \tag{29}$$

This will equal the power series $\Sigma\, a_n z^n$ provided that the contour may be closed to the left, and that $F(p)$ has simple poles at $0,-1,-2,\ldots$, with residues a_n at $p = -n$. A possible choice of $F(p)$ which satisfies this latter condition is

$$F(p) = \pi(-1)^p\, a_{-p} \operatorname{cosec}(\pi p) \tag{30}$$

where $0 < c < 1$, and we have assumed that the coefficients are expressed by a suitable formula.

Stirling's Series: It is shown in many places[4] that for $|\alpha| < 1$ we can write

$$\ell n(\alpha!) = -\gamma\alpha + \sum_{n=2}^{\infty} (-1)^n \alpha^n \zeta(n)/n. \qquad (31)$$

Using (30), we have

$$\ell n(\alpha!) = -\gamma\alpha + \frac{1}{2\pi i}\int_{c-i\infty}^{c+i\infty} \frac{\pi\alpha^s \zeta(s)}{s \sin(\pi s)} ds, \quad 1 < c < 2. \qquad (32)$$

By the principle of analytic continuation we can replace the restriction $|\alpha| < 1$ by $|\arg(\alpha)| < \pi$; subsequently moving the contour to the left yields the asymptotic expansion

$$\ell n(\alpha!) \sim (\alpha+\tfrac{1}{2})\ell n\, \alpha - \alpha + \frac{1}{2}\ell n(2\pi) + \sum_{n=1}^{\infty} \frac{(-1)^n \zeta(-n)}{n\, \alpha^n}, \qquad (33)$$

which is known as Stirling's series.

Incomplete Factorial Function: As a second example, consider the functions

$$(\alpha,z)! = \int_0^z t^\alpha e^{-t} dt, \quad \text{Re}(\alpha) > -1$$

$$= \sum_{n=0}^{\infty} \frac{(-1)^n z^{n+\alpha+1}}{n!(\alpha+1+n)} \qquad (34)$$

and

$$[\alpha,z]! = \int_z^\infty t^\alpha e^{-t} dt$$

$$= \alpha! - (\alpha,z)! \qquad (35)$$

For $\text{Re}(z) \to \infty$, $(\alpha,z)!$ obviously behaves as $\exp(-z)$, so it is expedient to deal also with the function $f(z) = \exp(z)(\alpha,z)!$ which is defined by the power series

$$f(z) = z^{\alpha+1} \sum_{k=0}^{\infty} \frac{z^k}{k!} \sum_{n=0}^{\infty} \frac{(-z)^n}{n!(\alpha+1+n)}$$

$$= z^{\alpha+1} \sum_{m=0}^{\infty} z^m \left[\sum_{\ell=0}^{m} \frac{(-1)^\ell}{(m-\ell)!\ell!(\alpha+1+\ell)}\right] \qquad (36)$$

$$= z^{\alpha+1} \sum_{m=0}^{\infty} \frac{\alpha!\, z^m}{(\alpha+1+m)!}$$

Using (29) and (30) we tentatively write

$$(\alpha,z)! = \frac{\alpha!\ z^{\alpha+1}\ e^{-z}}{2\pi i}\int_{c-i\infty}^{c+i\infty}\frac{(p-1)!(-p)!}{(\alpha+1-p)!}(-z)^{-p}dp, \tag{37}$$

$$0 < c < 1.$$

The question of the convergence of (37) can be settled by replacing p by $c+i\xi$ and using the asymptotic forms for large ξ for the factorial functions; this gives for the absolute value of the integrand the behavior $\exp[-(\pi\xi/2) + \xi \arg(-z)]$ and thus (37) will converge in the sector

$$|\arg(-z)| < \pi/2. \tag{38}$$

To obtain an integral representation in the sector $|\arg(z)| < \pi/2$, we must work with the power series for $(\alpha,z)!$ directly, to yield

$$(\alpha,z)! = \frac{z^{\alpha+1}}{2\pi i}\int_{c-i\infty}^{c+i\infty}\frac{(p-1)!}{\alpha+1-p}\ z^{-p}\ dp, \tag{39}$$

$$|\arg(z)| < \pi/2.$$

Moving the path of integration in (37) and (39) to the right, we obtain the asymptotic information

$$(\alpha,z)! = \alpha! + O(z^{\alpha}\ e^{-z}), \quad |\arg(z)| < \pi/2,$$

$$= -\alpha!\ z^{\alpha+1}\ e^{-z}\sum_{k=0}^{\infty}\frac{1}{(\alpha-1)!\ z^{k+1}}, \tag{40}$$

$$|\arg(-z)| < \pi/2.$$

A complete descending expansion of $[\alpha,z]!$ may be obtained by a direct application of the Mellin transform; this leads to the complete expansion

$$(\alpha,z)! = \alpha! - \frac{z^{\alpha}\, e^{-z}}{(-\alpha-1)!} \sum_{k=0}^{\infty} \frac{(k-\alpha-1)!}{(-z)^k}, \qquad |\arg (z)| < \pi/2. \tag{41}$$

Further examples of these techniques may be found in the problems.

Problems

1. Evaluate $\sum_{n=1}^{\infty} \frac{\sin \beta n}{n}$.

2. Show that

$$\sum_{n=1}^{\infty} (-1)^{n+1} n^{-s} = (1-2^{1-s})\zeta(s),$$

$$\sum_{n=0}^{\infty} (2n+1)^{-s} = (1-2^{-s})\zeta(s).$$

3. Evalute $\sum_{n=0}^{\infty} \frac{J_1[(2n+1)y]}{2n+1}$.

4. Derive, by the method of Mellin transforms, the Jacobi theta function transformation

$$\sum_{n=1}^{\infty} e^{-\alpha n^2} = -\frac{1}{2} + \frac{1}{2}(\pi/\alpha)^{1/2} + \sum_{m=1}^{\infty} e^{-\pi^2 m^2/\alpha}.$$

5. Show that[5]

$$\sum_{n=1}^{\infty} \ln(1-e^{-nx}) = -\frac{\pi^2}{6x} - \frac{1}{2}\ln(x/2\pi) + \frac{x}{24} + \sum_{n=1}^{\infty} \ln(1-4\pi^2 n/x).$$

6. Show that[5]

$$\sum_{n=1}^{\infty} \ell n[1-e^{-(2n-1)x}] = -\frac{\pi^2}{12x} + \frac{1}{2}\ell n\ 2 - \frac{x}{24} - \sum_{n=1}^{\infty} \ell n[1-e^{-(2n-1)2\pi^2/x}].$$

7. By writing

$$(\alpha+\ell)^{-1} = \int_0^1 u^{\alpha+\ell-1}\,du,$$

show that

$$\sum_{\ell=0}^{m} \frac{(-1)^\ell}{(m-\ell)!\,\ell!\,(\alpha+\ell)} = \frac{(\alpha-1)!}{(m+\alpha)!}\ .$$

8. Let

$$f(x) = \sum_{n=0}^{\infty} \frac{(-x)^n}{n!\,(2n+1)}$$

and

$$g(x) = e^x f(x).$$

Show that

$$g(x) = \frac{\sqrt{\pi}}{2} \sum_{n=0}^{\infty} \frac{x^n}{(n+\frac{1}{2})!}\ .$$

Hence deduce the integral representations

$$f(x) = \begin{cases} \dfrac{1}{2\pi i}\displaystyle\int_{\gamma-i\infty}^{\gamma+i\infty} \frac{(p-1)!}{1-2p} x^{-p}\,dp, & |\arg(x)| < \pi/2 \\ \dfrac{1}{2\pi i}\displaystyle\int_{\gamma-i\infty}^{\gamma+i\infty} \frac{(p-1)!\,(-p)!}{(-p+\frac{1}{2})!}(-x)^{-p}\,dp, & |\arg(-x)| < \pi/2 \end{cases}$$

and the asymptotic series

$$f(x) = \begin{cases} \dfrac{\sqrt{\pi}}{2}\, x^{-1/2} + O(x^{-1}e^{-x}), & |\arg(x)| < \pi/2 \\ \dfrac{e^{-x}}{2\sqrt{\pi}} \displaystyle\sum_{n=0}^{\infty} \frac{(n-1/2)!}{(-x)^{n+1}} + O(x^{-1/2}), & |\arg(-x)| < \pi/2. \end{cases}$$

Footnotes

1. This section is based on G. G. MacFarlane, Phil. Mag. (vii) (1949), 40, 188. MacFarlane considers the more general problem of evaluating sums of the form

$$\sum_{n=0}^{\infty} f[(n+a)^{\nu}].$$

2. This treatment is due to B. W. Ninham.

3. DINGLE (1973), Ch. 2.

4. For example, OLVER (1974), p. 64.

5. This is a transformation in the theory of elliptic modular functions.

§14. INTEGRALS INVOLVING A PARAMETER

14.1. Preliminary Example

Consider the integral[1]

$$B_s(x) = \frac{1}{s!}\int_0^\infty \frac{u^s\,du}{e^{u+x}-1}, \tag{1}$$

and suppose that we require an expansion for small values of the parameter x. When $x = 0$, the integral is simply a zeta function. If we attempt to find an expansion for small x by expanding the integrand in powers of x directly, the expansion will ultimately break down. To see this explicitly, suppose for simplicity that $0 < s < 1$. We have then

$$\begin{aligned} B_s(x) &= \frac{1}{s!}\left\{\int_0^\infty \frac{u^s du}{e^u-1} + \int_0^\infty u^s du\left[\frac{1}{e^{u+x}-1} - \frac{1}{e^u-1}\right]\right\} \\ &= \zeta(s+1) - x\int_0^\infty \frac{u^s du}{(e^u-1)^2} + O(x^2). \end{aligned} \tag{2}$$

The second term in this expansion diverges, as do higher terms, which indicates that our function does not have a power series expansion in x. Therefore, as in the previous section, we consider the Mellin transform of $B_s(x)$

$$S(p) = \frac{1}{s!}\int_0^\infty u^s du \int_0^\infty \frac{x^{p-1}dx}{e^{x+u}-1}. \tag{3}$$

The second integral is

$$\begin{aligned} \int_0^\infty \frac{x^{p-1}dx}{e^{x+u}-1} &= \int_0^\infty e^{-u-x}\, x^{p-1}\,(1 - e^{-u-x})^{-1} dx \\ &= e^{u}\int_0^\infty e^{-x}\, x^{p-1} dx + e^{-2u}\int_0^\infty e^{-2x} x^{p-1} dx \\ &\quad + e^{-3u}\int_0^\infty e^{-3x}\, x^{p-1} dx + \dots \\ &= (p-1)!\left[e^{-u} + \frac{e^{-2u}}{2^p} + \frac{e^{-3u}}{3^p} + \dots\right], \quad \mathrm{Re}(p) > 0. \end{aligned} \tag{4}$$

Inserting this result into (3) we have

$$S(p) = \frac{(p-1)!}{s!}\int_0^\infty u^s du \left[e^{-u} + \frac{e^{-2u}}{2^p} + \frac{e^{-3u}}{3^p} + \cdots\right]$$

$$= (p-1)! \left[1 + \frac{1}{2^{p+s+1}} + \frac{1}{3^{p+s+1}} + \cdots\right] \tag{5}$$

$$= (p-1)!\ \zeta(p+s+1), \quad \mathrm{Re}(p+s) > 0.$$

Hence by the Mellin inversion formula we have

$$B_s(x) = \frac{1}{2\pi i}\int_{c-i\infty}^{c+i\infty} (p-1)!\ \zeta(p+s+1)\ x^{-p}\ dp. \tag{6}$$

The ascending expansion for $B_s(x)$ can now be found by translating the contour to the left. The poles are those of $(p-1)!$, and of the zeta function at $p = -s$ with residue unity. If s is not an integer we have

$$B_s(x) = \sum_{r=0}^{\infty} \frac{(-1)^r}{r!}\ \zeta(s-r+1)\ x^r - \frac{\pi}{s!\ \sin\pi s}\ x^s. \tag{7}$$

14.2. Another Example[2]

Consider the function $g(\gamma)$ defined by

$$g(\gamma) = 2\pi^{1/2}\ \gamma^{3/2} \int_0^\infty \frac{e^{-\gamma k^2}}{e^{\pi/k}-1}\ k\ dk. \tag{8}$$

Its Mellin transform is simply

$$G(p) = 2\pi^{1/2} \int_0^\infty \gamma^{p-1}\ d\gamma\ \gamma^{3/2} \int_0^\infty \frac{e^{-\gamma k^2}}{e^{\pi/k}-1}\ k\ dk$$

$$= 2\pi^{1/2} \int_0^\infty \frac{(p+\frac{1}{2})!\ k^{-2p-2}\ dk}{e^{\pi/k}-1} \tag{9}$$

$$= 2\pi^{-2p-\frac{1}{2}}\ (p+\tfrac{1}{2})!\ (2p)!\zeta(2p+1).$$

On using the inversion integral and closing the contour to the left, we obtain the convergent expansion

$$2g(\gamma) = 1 - (\pi\gamma)^{1/2} + \sum_{n=0}^{\infty} \frac{(-1)^n}{n!} (\tfrac{1}{2}n)!\, \zeta(n+2)\gamma^{\frac{1}{2}n+1}, \qquad (10)$$

which may be used to compute the value of g up to $\gamma = 10$ quite easily. In the right-hand half-plane $G(p)$ has no poles, so the Mellin transform does not yield a descending expansion in this case.

14.3. Ascending Expansions for Fourier Integrals[3]

Consider the integral

$$f(x) = \int_0^{\infty} A(k)e^{ikx}\, dk \qquad (11)$$

which is a typical Fourier integral representation for the function $f(x)$. To obtain an expansion for small x, we might try to replace the exponential term by its Taylor series and integrate term by term, which gives

$$f(x) = \sum_{n=0}^{\infty} \frac{(ix)^n}{n!} \int_0^{\infty} A(k)\, k^n\, dk. \qquad (12)$$

This will be a useful asymptotic series if at least the first few terms converge, but in most cases it is not particularly valuable.

Suppose now that $A(k)$ has the asymptotic form

$$A(k) \sim k^{-\nu} \sum_{\ell=0}^{\infty} a_\ell k^{-\ell}, \qquad 0 < \nu < 1. \qquad (13)$$

The analysis of the cases $\nu = 0$ and $\nu = 1$ requires some modifications to the following argument, and is left to the problems. In the present case, we note that if $A(k)$ satisfies some rather general conditions, then $f(x) \to 0$ as $x \to \infty$ (see Section 2.1). Furthermore, it is easily shown that the asymptotic form (13) implies that $f(x) = O(x^{\nu-1})$ as $x \to 0$. For $\nu < \mathrm{Re}(p) < 1$, we consider the Mellin transform of $f(x)$, viz.

$$F(p) = \int_0^\infty x^{p-1}\,dx \int_0^\infty A(k)\,e^{ikx}\,dk$$
$$= (p-1)!\,e^{i\pi p/2} \int_0^\infty A(k)\,k^{-p}\,dk, \qquad (14)$$
$$1 - \nu < \mathrm{Re}(p) < 1.$$

In order to obtain an ascending expansion for $f(x)$, we need to know the analytic structure of $F(p)$ for $\mathrm{Re}(p) < 1-\nu$. To this end, we define the functions

$$R_n(k) = A(k) - k^{-\nu} \sum_{\ell=0}^{n} a_\ell k^{-\ell}. \qquad (15)$$

Now for $\mathrm{Re}(p) > 1-\nu$ we can write

$$\int_0^\infty A(k)k^{-p}\,dk$$
$$= \int_0^1 A(k)k^{-p}\,dk + \int_1^\infty R_n(k)k^{-p}\,dk + \sum_{\ell=0}^{n} a_\ell \int_1^\infty k^{-p-\nu-\ell}dk$$
$$= \Omega_n(p) + \sum_{\ell=0}^{n} \frac{a_\ell}{p+\nu+\ell-1}, \qquad (16)$$

where $\Omega_n(p)$ is the sum of the first two integrals, which is an analytic function for $-\nu-n < \mathrm{Re}(p) < 1$. Thus (16) is an analytic continuation into this larger region. Moreover we note that if we restrict p to the strip $-\nu-n < \mathrm{Re}(p) < 1-\nu-n$, then

$$\int_0^1 A(k)k^{-p}\,dp = \int_0^1 R_n(k)k^{-p}\,dp - \sum_{\ell=0}^{n} \frac{a_\ell}{p+\nu+\ell-1}, \qquad (17)$$

so that one analytic continuation of $F(p)$ into this strip is obtained by replacing $A(k)$ by $R_n(k)$ in (14).

Returning to $F(p)$, we see that there are poles at $p = -n$, $n = 0,1,2,\ldots$, due to the factor $(p-1)!$. The residues in the Mellin inversion are

$$\frac{(ix)^n}{n!} \int_0^\infty R_n(k) k^n \, dk. \tag{18}$$

There are also poles from the integral (16) at $p = 1-\nu-n$, $n = 0,1,2,\ldots$, with corresponding residues

$$(-\nu-n)! \, e^{i\pi(1-\nu-n)/2} \, a_n \, x^{\nu+n-1}$$
$$= \frac{\pi}{\sin(\pi\nu)} \, x^{\nu-1} \, e^{i\pi(1-\nu)/2} \, \frac{a_n (ix)^n}{(\nu+n-1)!}. \tag{19}$$

Consequently, moving the contour of the inversion integral to the left gives the asymptotic series

$$\int_0^\infty A(k) \, e^{ikx} \, dk$$
$$\sim \sum_{m=0}^\infty \frac{(ix)^m}{m!} \int_0^\infty R_m(k) \, k^m \, dk \tag{20}$$
$$+ \frac{\pi}{\sin \pi\nu} \, x^{\nu-1} \, e^{i\pi(1-\nu)/2} \sum_{m=0}^\infty \frac{a_m (ix)^m}{(m+\nu-1)!}$$

It is instructive to compare this with (12). In particular, if the first N coefficients a_n are zero, then (20) and (12) coincide to the first N terms.

14.4. Multidimensional Integrals[4]

Many research problems degenerate at some point or another to the evaluation of rather complicated multidimensional integrals containing a parameter. Normally one is interested in only the leading terms in an asymptotic expansion of such an integral; we give below some typical examples.

Example 1. Partition Function of an Electron Gas: An important correction to the equation of state of a classical electron gas is given by the integral

$$f(\lambda) = \int_0^\infty \left\{e^{-\beta q(x)} - 1 + \beta q(x) - \tfrac{1}{2}|\beta q(x)|^2\right\}x^2\,dx, \tag{21}$$

where

$$q(x) = \varepsilon^2 \frac{e^{-Kx}}{x}. \tag{22}$$

We require an asymptotic expansion for small values of the parameter $\lambda = K\beta\varepsilon^2$. Consider the integral representation of the exponential function (12.8); we can translate the contour there to the left by writing

$$e^{-y} - 1 + y - \frac{y^2}{2} = \frac{1}{2\pi i}\int_{c-i\infty}^{c+i\infty} (p-1)!\; y^{-p}dp, \quad -3 < c < -2. \tag{23}$$

Now we replace y by $\beta q(x) = \beta\varepsilon^2 e^{-Kx}/x$ and integrate, giving

$$f(\lambda) = \int_0^\infty x^2\,dx\,\frac{1}{2\pi i}\int_{c-i\infty}^{c+i\infty} (p-1)!\left[\beta\varepsilon^2\,\frac{e^{-Kx}}{x}\right]^{-p} dp. \tag{24}$$

A change of variable to $t = x/\beta\varepsilon^2$ and the substitution $K\beta\varepsilon^2 = \lambda$ give

$$f(\lambda) = (\beta\varepsilon^2)^3 \int_0^\infty t^2 dt\,\frac{1}{2\pi i}\int_{c-i\infty}^{c+i\infty} (p-1)!\; t^p\, e^{\lambda pt}\,dp, \tag{25}$$

$$-3 < c < -2.$$

For the contour above the orders of integration may be interchanged, and the t integration performed at once. Thus

$$\int_0^\infty t^{p+2}\, e^{\lambda pt}dt = (p+2)!\;(-\lambda p)^{-3-p} \tag{26}$$

so that

$$\frac{f(\lambda)}{(\beta\varepsilon^2)^3} = \frac{1}{2\pi i}\int_{c-i\infty}^{c+i\infty} (p-1)!\;(p+2)!\;\lambda^{-3-p}(-p)^{-3-p}\,dp. \tag{27}$$

The integrand has double poles at $p = -3,-4,-5,\ldots$; evaluating the residues at these poles, we obtain the expansion

$$\frac{f(\lambda)}{(\beta\varepsilon^2)^3} = \sum_{n=3}^{\infty} \frac{(\lambda n)^{n-3}}{n!(n-3)!} \left[\ell n\ \lambda + \ell n\ n - \frac{(n-3)}{n} - \psi(n) - \psi(n-3) \right]. \tag{28}$$

The first term of this expansion

$$\frac{1}{6}\left[\ell n\ 3 + 2\gamma - (11/6) + \ell n\ \lambda \right], \tag{29}$$

where γ is Euler's constant, was first obtained by Abel by a very much more cumbersome calculation. Note that the Mellin transform method yields the complete expansion of the integral $f(\lambda)$ almost trivially.

Example 2. Correlation Energy of a Degenerate Electron Gas: Gellmann and Brueckner first gave the leading corrections to the correlation energy of a degenerate electron gas in the form

$$J = \int_0^{\infty} du \int_0^1 \frac{dq}{q} \sum_{n=2}^{\infty} \frac{(-1)^n}{n} [R(u)]^n \left(\frac{x}{q^2}\right)^{n-2}, \tag{30}$$

where

$$R(u) = 1 - u \arctan (1/u) \tag{31}$$

and x is a parameter. We require the leading terms for both large and small x. In the original paper the integral J was evaluated numerically. Writing $y = (x/q^2)R(u)$, we see that the sum over n is just the formal expansion of

$$-\left(\frac{x}{q^2}\right)^{-2} [\ell n\ (1+y) - y]. \tag{32}$$

We can give this function an integral representation by observing that

$$\int_0^\infty \ell n(1+y)\; y^{p-1} dy = \frac{y^p}{p} \ell n(1+y) \Big|_0^\infty - \frac{1}{p}\int_0^\infty \frac{y^p}{(1+y)}\, dy$$

$$= -\frac{1}{p}\int_1^\infty (z-1)^p dz = -\frac{1}{p}\int_0^1 (1-\eta)^p \eta^{-p-1} d\eta$$

$$= \frac{\pi}{p \sin \pi p}, \quad -1 < \mathrm{Re}(p) < 0, \tag{33}$$

so we have

$$\ell n(1+y) = \frac{1}{2\pi i}\int_{c-i\infty}^{c+i\infty} \frac{\pi}{p \sin \pi p}\, y^p\, dp, \quad 0 < \mathrm{Re}(p) < 1. \tag{34}$$

Hence, translating the contour to the right, the sum which occurs in (30) may be written as

$$\sum_{n=2}^\infty \frac{(-1)^n}{n} [R(u)]^n \left(\frac{x}{q^2}\right)^{n-2}$$

$$= -\frac{1}{2\pi i}\int_{c-i\infty}^{c+i\infty} \frac{\pi}{p \sin \pi p} [R(u)]^p \left(\frac{x}{q^2}\right)^{p-2} dp, \tag{35}$$

$$1 < c < 2.$$

After substitution of this expression into (30), the q integration can be carried out immediately. The q integral is

$$\int_0^1 \frac{dq}{q}\left(\frac{1}{q^2}\right)^{p-2} = \int_0^1 q^{3-2p}\, dq = \frac{1}{4-2p}, \tag{36}$$

and our original integral J reduces to

$$J = \frac{1}{2\pi i}\int_{c-i\infty}^{c+i\infty} \frac{\pi}{2p \sin \pi p} \frac{\Phi(p)}{(p-2)} x^{p-2}\, dp, \qquad 1 < c < 2, \tag{37}$$

where

$$\Phi(p) = \int_0^\infty [R(u)]^p\, du. \tag{38}$$

A new feature here is the appearance in the integrand of a function $\Phi(p)$ which cannot be evaluated explicitly in terms of elementary functions. However, in order to find an expansion for J it is sufficient to know only the poles and residues of $\Phi(p)$. We can find these as follows.

Consider the integral (38). As $u \to 0$, $R(u) \sim 1 - (u\pi/2) \to 1$, so the integrand is well behaved at the lower limit for all values of p. On the other hand, for large values of u, we have

$$\arctan(1/u) = (1/u) - (1/3u^3) + \ldots, \tag{39}$$

hence

$$R(u) = (1/3u^2) + O(u^{-4}). \tag{40}$$

Now choose L sufficiently large so that (40) is a good approximation to $R(u)$ for $u > L$, and split the integral (38) as follows:

$$\Phi(p) = \int_0^L [R(u)]^p\,du + \int_L^\infty [R(u)]^p\,du. \tag{41}$$

The first integral is well-behaved (has no poles) as a function of p for all p. The second has a pole, since when we use (40) we have

$$\int_L^\infty [R(u)]^p\,du = \int_L^\infty \frac{u^{-2p}}{3^p}\,du + (\text{an integral convergent at } p = \tfrac{1}{2})$$

$$= \frac{L^{1-2p}}{2(p-\frac{1}{2})3^p} + (\text{remainder}).$$

Thus as $p \to \frac{1}{2}$ from above, $\Phi(p)$ has a single pole at $p = \frac{1}{2}$ with residue $1/2\sqrt{3}$. The leading terms in the descending expansion are thus

$$J = -\frac{2\pi}{3\sqrt{3}} + \frac{\pi}{4x} + O(x^{-2}). \tag{42}$$

The ascending expansion is obtained from (37) by translating the contour to the right. The result is

$$J = -\frac{1}{2}\,[\Phi(2)\,\ell n x - \tfrac{1}{2}\Phi(2) + \Phi'(2)] + O(x\,\ell n x). \tag{43}$$

Problems

1. Obtain expansions corresponding to (7) for s not equal to an integer.

Obtain ascending and descending expansions for the following integrals.

2. $\operatorname{erfc}(x)$ [See Problem (12.22).]

3. $\operatorname{Ci}(x)$ [See Problem (12.23).]

4. $\int_0^1 \frac{u^{2/3}(1-u^2)^{1/2}}{(u+x)^{1/3}}\,du$ [See Problem (12.20).]

5. $\int_0^\infty \frac{u^{2/3}\,e^{-u^2}}{(u+x)^{1/3}}\,du$ [See Problem (12.21).]

6. $\int_0^\infty u^2\,\ell n(1-e^{-\sqrt{u^4+\alpha}})\,du$

7. If $u(y) = \int_0^1 t^3(1-t^4)^{1/2}\,e^{-ty}\sin(ty)\,dt$,

   show that for large y

$$u(y) = 2^{-3/4}\sqrt{\pi}\,e^{-y}\left\{y^{-3/2}(y+3\pi/8) + O(y^{-5/2})\right\}.$$

8. By taking the Mellin transform with respect to x in the Poisson integral representation of $J_\nu(x)$, deduce the asymptotic expansion

$$J_\nu(x) \sim \left(\frac{2}{\pi x}\right)^{1/2}\Bigg\{\cos[x-(\nu\pi/2)-(\pi/4)]$$

$$\times\left[1+\sum_{r=1}^{\infty}(-1)^r\frac{(4\nu^2-1)(4\nu^2-3^2)\cdots(4\nu^2-(4r-1)^2)}{(2r)!\;2^{6r}\;x^{2r}}\right]$$

$$+\sin[x-(\nu\pi/2)-(\pi/4)]\left[\sum_{r=1}^{\infty}(-1)^r\frac{(4\nu^2-1)(4\nu^2-3^2)\cdots(4\nu^2-(4r-3)^2)}{(2r-1)!\;2^{6r-3}\;x^{2r-1}}\right]\Bigg\}.$$

9. For integer s, show that

$$B_s(x) = \sum_{\substack{r\neq s\\ r=0}}^{n} x^r\,\frac{\zeta(s-r+1)}{r!} - \frac{x^s}{s!}\{\ell n|x|-\psi(s)+\psi(0)\}$$

$$+ O(x^{n+1}).$$

10. If

$$f(\alpha) = \int_0^1 dx \int_0^\infty dy\,\frac{1-\exp[-\alpha y^2x(1-x)]}{1+y^2},$$

show directly that for small α

$$f'(\alpha) \simeq \frac{\pi^{3/2}}{16\alpha^{1/2}} - \frac{\pi}{12},$$

$$f(\alpha) \simeq \frac{\pi^{3/2}\alpha^{1/2}}{8} - \frac{\pi\alpha}{12}.$$

Use the Mellin transform to find complete ascending and descending expansions.

11. If (13) is replaced by

$$A(k) = \sum_{\ell=1}^{n} a_\ell k^{-\ell} + O(k^{-n-1}),$$

then show that the ascending expansion for $f(x)$ is

$$f(x) = \sum_n \frac{(ix)^n}{n!}\Bigg\{\int_0^\infty R_n(k)k^n\,dk$$

$$+ a_{n+1}\left[-\ell n|x|+\ell n\gamma+(i\pi\,\mathrm{sgn}(x)/2)+\sum_{r=1}^{n}\frac{1}{r}\right]\Bigg\},$$

where

$$R_n(k) = A(k) - \sum_{r=1}^{n} a_r k^{-r} - a_{n+1} k^{-n-1} h(k-1),$$

and h is the Heaviside step function.

Footnotes

1. A large number of useful expansions for the related Fermi-Dirac integral

$$F_s(x) = \frac{1}{s!} \int_0^\infty \frac{u^s \, du}{e^{x+u}+1}$$

have been obtained by this technique by A. Wasserman, T. J. Buckholtz, and H. E. DeWitt, J. Math. Phys. (1970), 11, 477.

2. B. Davies and R. G. Storer, Phys. Rev. (1968), 171, 150.

3. These results were obtained by H. C. Levey and J. J. Mahoney, Q. Appl. Math. (1967), 26, 101, by a direct analysis. It is interesting to compare the two methods of derivation.

4. Based on material written by B. W. Ninham.

§15. HANKEL TRANSFORMS

15.1. The Hankel Transform Pair

Bessel functions have frequently occurred in our investigations of the Laplace and Fourier transforms; indeed, we could rewrite most of the formulas we have derived in terms of Bessel functions of order $\pm\frac{1}{2}$, since $(2x/\pi)^{1/2}K_{1/2}(x) = \exp(-x)$, with similar relations for $\sin(x)$ and $\cos(x)$. Furthermore, we noted in Problem 12.31 that the integral transform

$$F_\nu(k) = \int_0^\infty f(x)\, J_\nu(kx)x\, dx \tag{1}$$

has for its inverse the reciprocal formula

$$f(x) = \int_0^\infty F_\nu(k)J_\nu(kx)\, k\, dk. \tag{2}$$

These formulas constitue the Hankel transform pair. Proof of the validity of these results for various classes of functions $f(x)$ (such as functions satisfying Dirichlet conditions) may be found in various places;[1] we will be content here to reproduce a rather elegant treatment due to MacRobert which is sufficient to cover many situations occuring in practice.

MacRobert's Proof: We consider the integral

$$\int_0^\infty J_\nu(kt)\, k\, dk \int_a^b J_\nu(kx)\, f(x)\, x\, dx, \quad 0 < a < b, \tag{3}$$

where we assume that $f(x)$ is analytic in some region of the complex plane containing the line $a \le x \le b$. Now we split up one of the Bessel functions into Hankel functions, and deform the x contour onto the contours C_1 and C_2

shown in Figure 1. Thus (3) becomes

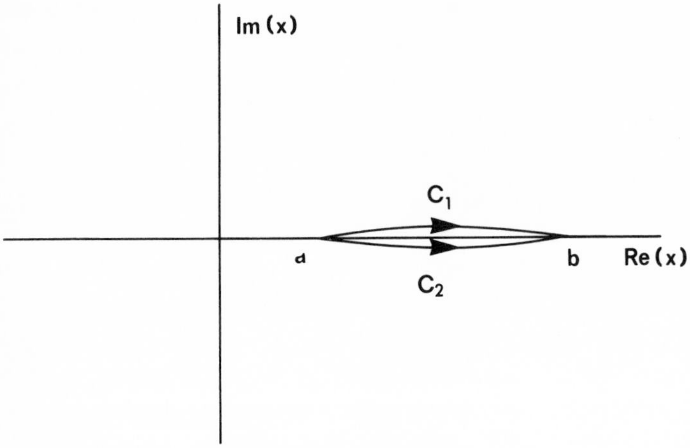

Figure 15.1

$$\frac{1}{2}\int_0^\infty J_\nu(kt)\ k\ dk \int_{C_1} H_\nu^{(1)}(kx)\ f(x)\ x\ dx$$
$$+\ \frac{1}{2}\int_0^\infty J_\nu(kt)\ k\ dk \int_{C_2} H_\nu^{(2)}(kx)\ f(x)\ x\ dx. \tag{4}$$

We may reverse the order of integration because the Hankel functions fall off exponentially on the contours C_1 and C_2 as $k \to \infty$; using Lommel's integral[2] this yields

$$\frac{i}{\pi}\int_{C_1} \frac{(t/x)^\nu}{x^2-t^2}\ f(x)\ x\ dx - \frac{i}{\pi}\int_{C_2} \frac{(t/x)^\nu}{x^2-t^2}\ f(x)\ x\ dx$$
$$= \frac{1}{2\pi i}\int_C (t/x)^{\nu-1}\left\{\frac{1}{x-t} - \frac{1}{x+t}\right\} f(x)\ dx \tag{5}$$
$$= \begin{cases} f(t), & a < t < b \\ 0\ , & x < a \quad \text{or} \quad x > b \end{cases}$$

where C is the loop C_2-C_1. From this the transform pair (1) and (2) follow, provided that either $f(x)$ or $F_\nu(k)$ has the necessary analytic properties.[3] This is usually the case for practical applications, although (1) and (2) are valid for a wider class of functions.[4]

Connection with the Fourier Transform: Consider the two-variable Fourier transform pair[5]

$$\begin{aligned} F(\underset{\sim}{k}) &= \frac{1}{2\pi} \int f(\underset{\sim}{r})\, e^{i\underset{\sim}{k}\cdot\underset{\sim}{r}}\, d^2\underset{\sim}{r}, \\ f(\underset{\sim}{r}) &= \frac{1}{2\pi} \int F(\underset{\sim}{k})\, e^{-i\underset{\sim}{k}\cdot\underset{\sim}{r}}\, d^2\underset{\sim}{k}. \end{aligned} \tag{6}$$

Suppose that we introduce polar coordinates

$$\begin{aligned} \underset{\sim}{r} &\to (r,\theta), \\ \underset{\sim}{k} &\to (k,\phi), \end{aligned} \tag{7}$$

and expand both $f(r,\theta)$ and $F(k,\phi)$ in Fourier series:

$$\begin{aligned} f(r,\theta) &= \sum_{n=-\infty}^{\infty} f_n(r)\, e^{in\theta}, \\ F(k,\phi) &= \sum_{n=-\infty}^{\infty} F_n(k)\, e^{in\phi}, \end{aligned} \tag{8}$$

where

$$\begin{aligned} f_n(r) &= \frac{1}{2\pi} \int_0^{2\pi} f(r,\theta)\, e^{-in\theta}\, d\theta, \\ F_n(k) &= \frac{1}{2\pi} \int_0^{2\pi} F(k,\phi)\, e^{-in\phi}\, d\phi. \end{aligned} \tag{9}$$

On substituting (6a) into (9b) and using (8a) to represent $f(r,\theta)$, we obtain

$$\begin{aligned} F_n(k) &= \frac{1}{(2\pi)^2} \int_0^{2\pi} e^{-in\phi}\, d\phi \int_0^{\infty} r\, dr \int_0^{2\pi} d\theta\, e^{ikr\cos(\theta-\phi)} \\ &\quad \times \sum_{m=-\infty}^{\infty} f_m(r)\, e^{im\theta} \\ &= \frac{1}{2\pi} \int_0^{\infty} r\, dr \int_0^{2\pi} d\alpha\, e^{in\alpha}\, e^{ikr\cos\alpha}\, f_n(r) \\ &= \int_0^{\infty} r\, dr\, J_n(kr)\, f_n(r). \end{aligned} \tag{10}$$

Similarly we may derive the relation

$$f_n(r) = \int_0^\infty k\ dk\ J_n(kr)\ F_n(kr). \tag{11}$$

Equations (10) and (11) are the Hankel transform pair with $\nu = n$.

15.2. Elementary Properties

Because of their increased generality over the Laplace and Fourier transforms, Hankel transforms do not have as many elementary properties as do the former. We will recount here those elementary properties which correspond to Sections 1.2, 1.3, and 7.2.

Derivatives: Suppose that $F_\nu(k)$ is the Hankel transform of order ν of the function $f(x)$; then the Hankel transform of the function $g(x) = f'(x)$ is

$$\begin{aligned} G_\nu(k) &= \int_0^\infty f'(x)\ J_\nu(kx)\ x\ dx \\ &= [xf(x)\ J_\nu(kx)]_0^\infty \\ &\qquad - \int_0^\infty f(x)\ \frac{d}{dx}\ [x\ J_\nu(kx)]\ dx \end{aligned} \tag{12}$$

We assume that the behavior of $f(x)$ at 0 and ∞ makes the bracket zero, and use (20.46) and (20.47) to write

$$\frac{d}{dx}\left[x\ J_\nu(kx)\right] = \frac{kx}{2\nu}\left[(\nu+1)J_{\nu-1}(kx) - (\nu-1)J_{\nu+1}(kx)\right]. \tag{13}$$

Hence we have

$$G_\nu(k) = -k\left[\frac{\nu+1}{2\nu}\ F_{\nu-1}(k) - \frac{\nu-1}{2\nu}\ F_{\nu+1}(k)\right]. \tag{14}$$

Formulas for transforms of higher derivatives may be obtained by repeated application of this result.

Bessel's Equation: Let $f(x)$ be an arbitrary function, and consider the transform of the combination

$$g(x) = \frac{d^2}{dx^2} f(x) + \frac{1}{x}\frac{d}{dx} f(x) - \frac{\nu^2}{x^2} f(x). \tag{15}$$

Integrating by parts, assuming at each stage that the contributions from $x = 0$ and $x = \infty$ are zero, we have

$$\begin{aligned}
G_\nu(k) &= \int_0^\infty \left[\frac{d}{dx} x \frac{df}{dx} - \frac{\nu^2}{x} f(x)\right] J_\nu(kx)\, dx \\
&= -\int_0^\infty \left[x \frac{df}{dx} kJ'_\nu(kx) + \frac{\nu^2}{x} f(x)\, J_\nu(kx)\right] dx \\
&= \int_0^\infty \left[k^2 J''_\nu(kx) + \frac{k}{x} J'_\nu(kx) - \frac{\nu^2}{x^2} J_\nu(kx)\right] f(x)\, x\, dx \qquad (16) \\
&= -k^2 \int_0^\infty f(x)\, J_\nu(kx)\, x\, dx \\
&= -k^2 F_\nu(k).
\end{aligned}$$

Thus Hankel transforms may lead to significant simplification in problems involving Bessel's equation.

Parseval's Theorem: There is no simple addition formula for Bessel functions such as exist for the exponential and trigonometric functions, thus the Hankel transform does not satisfy any simple convolution relation. However, a simple relation of Parseval type can be derived as follows. Let $F_\nu(k)$ and $G_\nu(k)$ be Hankel transforms of order ν; then

$$\begin{aligned}
&\int_0^\infty F_\nu(k)\, G_\nu(k)\, k\, dk \\
&\quad = \int_0^\infty F_\nu(k)\, k\, dk \int_0^\infty g(x)\, J_\nu(kx)\, x\, dx \\
&\quad = \int_0^\infty g(x)\, x\, dx \int_0^\infty F_\nu(k)\, J_\nu(kx)\, k\, dk \qquad (17) \\
&\quad = \int_0^\infty f(x)\, g(x)\, x\, dx.
\end{aligned}$$

The similarity with (7.32) and (7.34) is obvious.

15.3. Some Examples

Let

$$f(x) = x^{\nu}(a^2-x^2)^{\mu}\ h(a-x), \quad \mu > -1; \tag{18}$$

then, on expanding $J_\nu(kx)$ by (20.45), we have

$$\begin{aligned} F_\nu(k) &= \int_0^a x^{\nu+1}\ (a^2-x^2)^{\mu}\ J_\nu(kx)\ dx \\ &= \sum_{m=0}^{\infty} \frac{(-1)^m (k/2)^{\nu+2m}}{(\nu+m)!m!} \int_0^a x^{2\nu+2m+1}(a^2-x^2)^{\mu}\ dx. \end{aligned} \tag{19}$$

The latter integral is a Beta function (Appendix A); expressing it in terms of factorials we get

$$\begin{aligned} F_\nu(k) &= \sum_{m=0}^{\infty} \frac{(-1)^m \mu!\,(k/2)^{\nu+2m} a^{2\mu+2\nu+2m+2}}{2(\mu+\nu+m+1)!m!} \\ &= 2^{\mu} a^{\mu+\nu+1}\ k^{-\mu-1}\ \mu!\ J_{\nu+\mu+1}(ak). \end{aligned} \tag{20}$$

Using the reciprocal Hankel transform, and replacing $\mu+\nu+1$ by μ, we obtain another useful integral, namely

$$\int_0^\infty x^{1-\mu+\nu}\ J_\mu(ax)\ J_\nu(bx)\ dx = \frac{b^{\nu}(a^2-b^2)^{\mu-\nu-1}}{2^{\mu-\nu-1}\ a^{\mu}(\mu-\nu-1)!}\ h(a-b), \tag{21}$$

where the restriction $\mu > \nu$ is needed to make the integral converge. A further result is obtained by setting $\mu = 0$ in (20) and using the Parseval relation (17), so that

$$\begin{aligned} &\int_0^\infty \frac{J_{\nu+1}(ax)\ J_{\nu+1}(bx)}{x}\ dx \\ &= (ab)^{-\nu-1} \int_0^\infty k^{2\nu+1}\ h(k-a)\ h(k-b)\ dk \\ &= \begin{cases} \dfrac{1}{2(\nu+1)} \left(\dfrac{a}{b}\right)^{\nu+1}, & a < b \\[2ex] \dfrac{1}{2(\nu+1)} \left(\dfrac{b}{a}\right)^{\nu+1}, & a > b. \end{cases} \end{aligned} \tag{22}$$

Many other similar and related results may be obtained; some of them are stated in the problems.

15.4. Boundary-value Problems

The Hankel transform can be used to solve numerous boundary-value problems in a relatively straightforward way, using various properties of Bessel functions. We solve two illustrative problems here; others are found in the problem section.

Heat Conduction: Suppose that heat enters a semi-infinite body of thermal conductivity κ through a disc of radius a, at a constant rate Q. The remainder of the surface at $z = 0$ is insulated. We will find the steady-state temperature distribution of the body, u, which satisfies Laplace's equation with appropriate boundary conditions. Using cylindrical polar coordinates, we can write for $u(r,z)$ the equations

$$u_{rr} + \frac{1}{r} u_r + u_{zz} = 0,$$
$$-\kappa u_z = \begin{cases} Q/a^2, & r < a \\ 0, & r > a. \end{cases} \tag{23}$$

Taking the Hankel transform of order 0 with respect to r, the equations become

$$U_{zz}(k,z) - k^2 U(k,z) = 0,$$
$$-\kappa U_z(k,0) = QJ_1(ka)/ka. \tag{24}$$

The solution of these equations which remains finite as $z \to \infty$ is

$$U(k,z) = \frac{Q}{\kappa a} \frac{J_1(ka)}{k} e^{-kz}, \tag{25}$$

leading to a temperature distribution given by the integral representation

$$u(r,z) = \frac{Q}{\kappa a} \int_0^\infty \frac{J_0(kr)J_1(ka)}{k} e^{-kz} dk. \tag{26}$$

An Electrostatic Problem: We will find the electrostatic potential generated in the space between two grounded plates at $z = \pm a$ by a point charge q at $r = 0$, $z = 0$. The potential ϕ satisfies Laplace's equation except at the origin where it has the singular behavior $\phi(r,z) \simeq q/\sqrt{r^2+z^2}$. Writing $\phi(r,z) = q/\sqrt{r^2+z^2} + \psi(r,z)$, we are faced with the equations

$$\begin{aligned} &\psi_{rr}(r,z) + \frac{1}{r}\psi_r(r,z) + \psi_{zz}(r,z) = 0, \\ &\psi(r,\pm a) + \frac{q}{\sqrt{r^2+a^2}} = 0. \end{aligned} \tag{27}$$

The Hankel transform of order zero turns these into the simpler equations

$$\begin{aligned} &\Psi_{zz}(k,z) - k^2\Psi(k,z) = 0, \\ &\Psi(k,\pm a) = -qe^{-ka}/k, \end{aligned} \tag{28}$$

and the solutions follow immediately, viz.

$$\Psi(k,z) = -q\frac{\cosh(kz)}{\cosh(ka)} \frac{e^{-ka}}{k}, \tag{29}$$

$$\phi(r,z) = \frac{q}{\sqrt{r^2+z^2}} - q\int_0^\infty \frac{\cosh(kz)}{\cosh(ka)} e^{-ka} J_0(kr)\, dk. \tag{30}$$

15.5. Weber's Integral

For some applications, a generalization of the Hankel transform using Weber's integral may be useful. We sketch a few salient points here, relegating most of the details to the problems. We commence by considering the cylinder functions

$$Z_\nu(kr) = J_\nu(kr)\, Y_\nu(ka) - Y_\nu(kr)\, J_\nu(ka), \tag{31}$$

chosen[6] so that $Z_\nu(ka) = 0$. It can be shown that if $Z_\nu(kr)$ rather than $J_\nu(kr)$ is used in (1), we obtain the transform pair

$$F_\nu(k) = \int_a^\infty f(x)\, Z_\nu(kx)\, x\, dx, \tag{32}$$

$$f(x) = \int_0^\infty F_\nu(k)\, \frac{Z_\nu(kx)}{J_\nu^2(ka) + Y_\nu^2(ka)}\, k\, dk. \tag{33}$$

A Simple Application: We consider an infinite slab of uniform solid material of thickness 2ℓ, through which there is a circular hole of radius a. If the plane faces are held at temperature zero, while the circular surface is heated to the temperature T_0, then the steady-state temperature $u(r,z)$ will satisfy the equations

$$\begin{aligned} &u_{rr}(r,z) + \frac{1}{r}\, u_r(r,z) + u_{zz}(r,z) = 0, \\ &u(r,\pm\ell) = 0, \\ &u(a,z) = T_0. \end{aligned} \tag{34}$$

Taking the transform (32) of (34) with $\nu = 0$, and using the result of Problem 18, we obtain the ordinary differential equation

$$U_{zz}(k,z) - k^2U(k,z) = \frac{2}{\pi} T_0. \tag{35}$$

The solution, chosen to satisfy the boundary conditions $U(k,\pm\ell) = 0$, is

$$U(k,z) = -\frac{2T_0}{\pi k^2}\left[1 - \frac{\cosh kz}{\cosh k\ell}\right]. \tag{36}$$

Use of the inversion integral (33) now yields an integral representation of the solution.

Connection with Fourier Series: This problem may also be solved by expanding in a Fourier cosine series in the z variable, and we are led to seek the connection between the two solutions. Expressing the Bessel functions in (33) as Hankel functions, we can write $u(r,z)$ as

$$\begin{aligned} u(r,z) = -\frac{T_0}{\pi i}\int_0^\infty &\left[\frac{H_0^{(1)}(kr)}{H_0^{(1)}(ka)} - \frac{H_0^{(2)}(kr)}{H_0^{(2)}(ka)}\right] \\ &\times\left[1 - \frac{\cosh(kz)}{\cosh(k\ell)}\right]\frac{dk}{k} \end{aligned} \tag{37}$$

The functions $H_0^{(1,2)}(z)$ have no zeros for $\mathrm{Re}(z) > 0$, hence we may deform the contour of integration as follows:

(i) The term multiplied by $H_0^{(1)}(kr)/H_0^{(1)}(ka)$ is integrated along the contour C_1 of Figure 2, which is chosen to coincide with the imaginary axis except for indentations around the poles of $U(k,z)$ at $\ell k = i\pi(n+\frac{1}{2})$, $n = 0,1,2,\ldots$. After this change of contour, we introduce the new variable $\xi = -ik$, so that the contour in ξ is C_2.

(ii) The term multiplied by $H_0^{(2)}(kr)/H_0^{(2)}(ka)$ is integrated along C_3, and subsequently we write $\xi = ik$ to bring the ξ contour to C_4.

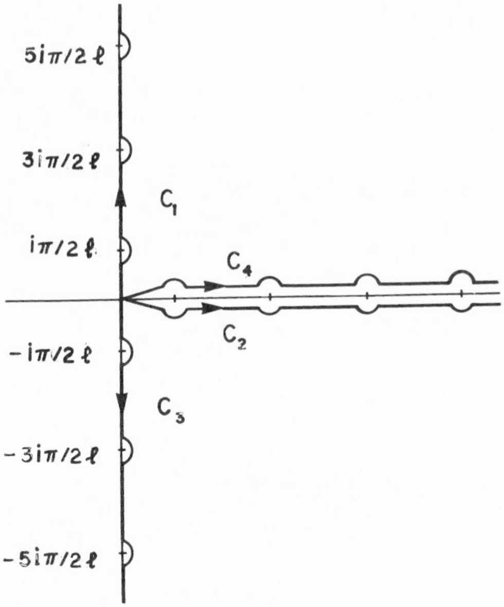

Figure 15.2

It is evident that the contours C_2 and C_4 coincide, except for the indentations at $\xi_n = \pi(n+\frac{1}{2})/\ell$, where they pass on either side. Hence (37) becomes

$$\begin{aligned} u(r,z) &= -\frac{T_0}{\pi i}\int_{C_2} \frac{K_0(\xi r)}{K_0(\xi a)}\left|1 - \frac{\cos(\xi z)}{\cos(\xi \ell)}\right| \frac{d\xi}{\xi} \\ &+ \frac{T_0}{\pi i}\int_{C_4} \frac{K_0(\xi r)}{K_0(\xi a)}\left|1 - \frac{\cos(\xi z)}{\cos(\xi \ell)}\right| \frac{d\xi}{\xi} \\ &= -2T_0 \sum_{n=0}^{\infty} \frac{(-1)^n}{\xi_n}\frac{K_0(\xi_n r)}{K_0(\xi_n a)}\cos(\xi_n z), \end{aligned} \tag{38}$$

which is the Fourier series solution.

Problems

Prove the following general properties of the Hankel transform of order ν. (Here we denote the integral (1) by $H_\nu[f(x);k]$.)

1. $H_\nu[f(ax);k] = a^{-2}F_\nu(k/a)$

2. $H_\nu[x^{-1}f(x);k] = \frac{k}{2\nu}[F_{\nu-1}(k) + F_{\nu+1}(k)]$

3. $H_\nu[x^{\nu-1}\frac{d}{dx}\{x^{1-\nu}f(x)\};k] = -kF_{\nu-1}(k)$

4. $H_\nu[x^{-\nu-1}\frac{d}{dx}\{x^{\nu+1}f(x)\};k] = kF_{\nu+1}(k)$

5. $H_\nu[e^{-px}f(x);k] = \mathscr{L}[xf(x)J_\nu(kx);p]$

Verify the following Hankel transforms.

6. $H_\nu[x^{\nu-1}e^{-px};k] = \dfrac{2^\nu k^\nu(\nu-\frac{1}{2})!}{\sqrt{\pi}(p^2+k^2)^{\nu+1/2}}$

7. $H_\nu[x^{-1}e^{-px};k] = (p^2+k^2)^{-1/2}\left[\dfrac{k}{p+\sqrt{(p^2+k^2)}}\right]^\nu$

8. $H_\nu[x^{p-1};k] = \dfrac{2^p(\frac{1}{2}\nu+\frac{1}{2}p-\frac{1}{2})!}{k^{p+1}(\frac{1}{2}\nu-\frac{1}{2}p-\frac{1}{2})!}$

9. Sonine's First Integral: Show that

$$J_{\mu+\nu+1}(x) = \frac{x^{\nu+1}}{2^\nu\nu!}\int_0^{\pi/2} J_\mu(x\sin\theta)\sin^{\mu+1}\theta\cos^{2\nu+1}\theta\, d\theta.$$

[Use (21).]

10. Show that

$$H_0[e^{-ax^2}J_0(bx);k] = \frac{1}{2}a\exp\left(-\frac{-k^2+b^2}{4a}\right)I_0\left(\frac{bk}{2a}\right).$$

11. <u>Sonine's Second Integral:</u> Use the result of Problem 20.24 to show that

$$\int_0^\infty t^{\mu+1} (t^2+a^2)^{-\nu/2} J_\nu(b\sqrt{t^2+a^2}) J_\mu(xt)\, dt$$

$$= b^{-\nu} x^\mu a^{-\nu+\mu+1} (b^2-x^2)^{(\nu-\mu+1)/2} J_{\nu-\mu-1}(a\sqrt{b^2-x^2})\, h(b-x).$$

12. Show that

$$H_\nu[x^{-\nu}(x^2-a^2)^{\nu-n-1}\, h(x-a);\, k]$$

$$= 2^{\nu-n-1} (\nu-n-1)!\, k^{n-\nu} a^{-n} J_n(ka).$$

(Set $u^2 = t^2 + a^2$ in Sonine's second integral and let $x \to 0$.)

13. Let ϕ be a solution of Laplace's equation in the half-space $z \geq 0$. The boundary $z = 0$ is held at the potential $\phi = f(r)$. Show that the potential elsewhere is given by the expression

$$\phi(r,z) = \int_0^\infty J_0(kr)e^{-kz}k\, dk \int_0^\infty J_0(ks)f(s)\, s\, ds.$$

Examine the special case $f(r) = V\, h(a-r)$ and show that

$$\phi(r,z) = Va \int_0^\infty e^{-kz} J_1(ka)J_0(kr)\, dk.$$

14. The initial temperature distribution of an infinite uniform region is $u(\underset{\sim}{r},0) = f(|\underset{\sim}{r}|)$. Show that

$$u(r,t) = \frac{\kappa}{2t} \int_0^\infty e^{-\kappa(r^2+s^2)/4t} I_0(\kappa rs/2t)\, f(s)\, s\, ds.$$

(Use Problem 10.)

15. The vibration of a thin elastic plate is described by the equation

$$c^2\nabla^4 w + \frac{\partial^2 w}{\partial t^2} = 0,$$

where c is the ratio of the rigidity of the plate (against bending) and its mass per unit area. Show that the motion of an infinite plate, starting from the axially symmetric initial conditions

$$w(r,0) = f(r),$$
$$w_t(r,0) = 0,$$

subsequently is described by the expression

$$w(r,t) = \int_0^\infty F_0(k)\ J_0(kr)\ \cos\ (ctk^2)\ kdk.$$

Derive the alternative formula

$$w(r,t) = \frac{1}{2ct}\int_0^\infty J_0(rs/2ct)\sin[(r^2+s^2)/4ct]\ f(s)\ s\ ds.$$

[Extend Problem 10 to verify the relation

$$\int_0^\infty J_0(kr)\ J_0(ks)\ \cos\ (ctk^2)\ kdk$$
$$= \frac{1}{2ct}\ J_0(rs/2ct)\ \sin\{(r^2+s^2)/4ct\}.]$$

16. Two point charges $+q$ and $-q$ are placed in vacuum on either side of a slab of material of dielectric constant ε. The geometry is shown in Figure 3. Find an expression for the electrostatic potential in each of the three regions.

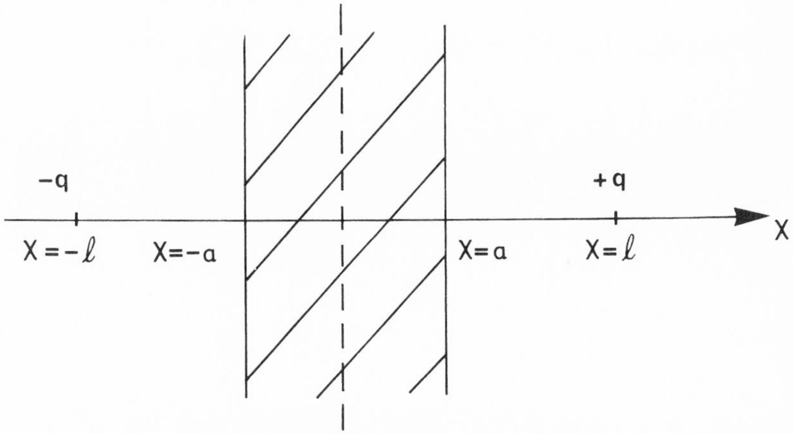

Figure 15.3

(Hint: in each vacuum region, separate off the singular part of the solution.)

17. A direct current J enters a semi-infinite region $z \geq 0$ of conductivity σ through an electrode of radius a making contact at the surface $z = 0$. Show that the current density $\underset{\sim}{j}$ is given by

$$\underset{\sim}{j} = -\underset{\sim}{\nabla} u$$

where

$$u(r,z) = \frac{J}{\pi a \sigma} \int_0^\infty e^{-kz} J_1(ka)\, J_0(kr) \frac{dk}{k}.$$

Show that as $a \to 0$

$$u(r,z) \simeq \frac{J}{2\pi\sigma\sqrt{r^2+z^2}}.$$

18. If

$$f(x) = g''(x) + \frac{1}{x} g'(x) - \frac{\nu^2}{x^2} g(x),$$

show that the Weber transforms (32) of f and g are related by

$$F_\nu(k) = -k^2 G_\nu(k) - \frac{2}{\pi} g(a).$$

19. By considering the integral

$$\int_0^\infty Z_\nu(kx)\ x\ dx \int_b^c \frac{Z_\nu(px)}{J_\nu^2(pa) + Y_\nu^2(pa)} F_\nu(p)\ pdp$$

and using Lommel's integral,[2] construct a proof of the inversion theorem for Weber's transform.

20. Construct an integral transform using the functions

$$Z_\nu(kr) = J_\nu(kr)\ Y_\nu'(ka) - Y_\nu(kr)\ J_\nu'(ka)$$

and show how the transforms of functions $f(x)$ and $g(x)$, where

$$f(x) = g''(x) + \frac{1}{x} g'(x) - \frac{\nu^2}{x^2} g(x)$$

and $g(x)$ is arbitrary, are related.

21. A cylindrical hole of radius a is drilled in an infinite body, and the walls of the hole are maintained at temperature T_1 from time $t = 0$. If the body is initially at temperature T_0, show that the temperature distribution is given by

$$u(r,t) = T_0 + \frac{2(T_1 - T_0)}{\pi} \int_0^\infty \frac{Z_0(kr)(1-e^{-\kappa k^2/t})}{J_0^2(ka) + Y_0^2(ka)} \frac{dk}{k},$$

where κ is the thermal conductivity and Z_ν is defined in (31).

Footnotes

1. For example, SNEDDON (1972).

2. Lommel's integral is

$$(\lambda^2-\mu^2) \int_a^b U_\nu(\lambda x) \; V_\nu(\mu x) x \; dx$$

$$= [U_\nu(\lambda x)\mu x V_\nu'(\mu x) - U_\nu'(\lambda x)\lambda x \; V_\nu(\mu x)]_a^b,$$

for any pair of cylinder functions U_ν and V_ν [WATSON (1958), p. 134]. It may be used to obtain results such as

$$(\lambda^2-\mu^2) \int_0^\infty J_\nu(\lambda x) H_\nu^{(1)}(\lambda x) x \; dx = -\frac{2i}{\pi} (\lambda/\mu)^\nu, \quad \mathrm{Im}(\mu) > 0.$$

3. If $F_\nu(k)$ is analytic in a region of the complex plane containing $a \leq k \leq b$, then we replace (3) by

$$\int_0^\infty J_\nu(kx) \; x \; dx \int_a^b J_\nu(px) \; F_\nu(p) \; p \; dp.$$

4. In particular the case $b \to \infty$ is easy to handle. Also, if the interval $0 \leq x < \infty$ can be split up into a finite number of subintervals in each of which the condition of MacRobert's proof applies, then the proof is easily generalized. This covers most functions which arise in applications.

5. We have chosen the constants 2π in a more symmetrical way than in (11.1) and (11.2).

6. Another transform is obtained from the choice $Z_\nu'(ka) = 0$.

§16. DUAL INTEGRAL EQUATIONS[1]

16.1. The Electrified Disc

To motivate this section, we first solve a classical problem of electrostatics. We wish to find the electrostatic potential ϕ created by an isolated thin conducting disc of radius a, whose potential is V. Noting the symmetry of the problem about the axis of the disc and introducing cylindrical polar coordinates r, θ, and z, we reduce the problem to that of satisfying the equations

$$\phi_{rr} + \frac{1}{r}\phi_r + \phi_{zz} = 0 \tag{1}$$

and

$$\begin{aligned} \phi(r,0) &= V, & r < a, \\ \phi_z(r,0+) &= \phi_z(r,0-), & r > a. \end{aligned} \tag{2}$$

Applying the Hankel transform of order zero, we easily find from (1) that

$$\Phi(k,z) = A(k)e^{-k|z|}, \tag{3}$$

and that the boundary conditions (2) reduce to the "dual integral equations"

$$\begin{aligned} &\int_0^\infty A(k)\, J_0(kr)kdk = V, \quad r < a, \\ &\int_0^\infty kA(k)\, J_0(kr)\, kdk = 0, \quad r > a. \end{aligned} \tag{4}$$

If we differentiate (4a) with respect to r, we obtain an alternative pair of equations, namely

$$\begin{aligned} \phi_r(r,0) &= -\int_0^\infty A(k)\, J_1(kr)\, k^2 dk \\ &= 0, \quad r < a, \end{aligned} \tag{5}$$

$$\phi_z(r,0+) = \int_0^\infty A(k)\, J_0(kr)\, k^2 dk$$
$$= 0, \quad r > a. \tag{6}$$

From (15.21) we see that the function

$$A(k) = C(ka)^{-3/2}\, J_{1/2}(ka) \tag{7}$$

satisfies both of these equations; furthermore, with this form for $A(k)$, (15.21) gives

$$\phi_r(r,0) = -\frac{C}{a}\sqrt{\frac{2}{\pi}}\; \frac{h(r-a)}{r\sqrt{r^2-a^2}} \tag{8}$$

and thus

$$\phi(r,0) = -\int_0^\infty \phi_t(t,0)dt$$
$$= \begin{cases} \dfrac{C}{a^2}\sqrt{\dfrac{2}{\pi}}\; \sin^{-1}(a/r), & r > a \\ \dfrac{C}{a^2}\sqrt{\dfrac{\pi}{2}}, & r < a. \end{cases} \tag{9}$$

Finally, this implies $C = Va^2\sqrt{2/\pi}$, so the solution is

$$\phi(r,z) = \frac{2V}{\pi}\int_0^\infty \frac{\sin(ka)e^{-k|z|}}{k}\, J_0(kr)dk\,. \tag{10}$$

16.2. Dual Integral Equations of Titchmarsh Type

Equations of the type

$$\int_0^\infty k^{-2\alpha} A(k)\, J_\mu(kx)\, kdk = f(x), \quad x < a,$$
$$\int_0^\infty k^{-2\beta} A(k)\, J_\nu(kx)\, kdk = g(x), \quad x > a, \tag{11}$$

where $f(x)$ and $g(x)$ are only known over part of the range $0 < x < \infty$ and $A(k)$ is sought, occur in certain mixed boundary value problems of which the electrified disc is a simple example. A convenient formalism for the solution of these equations can be developed using a modified Hankel transform defined by

$$S_{\nu,\alpha} f = (2/k)^{\alpha} \int_0^{\infty} x^{-\alpha} f(x) J_{2\nu+\alpha}(kx) x \, dx. \tag{12}$$

It is readily verified that the inversion formula for this transform is given by

$$S_{\nu,\alpha}^{-1} = S_{\nu+\alpha,-\alpha}. \tag{13}$$

The dual integral equations (11) can be written

$$\begin{aligned} S_{\frac{1}{2}\mu-\alpha,2\alpha} A(k) &= (2/x)^{2\alpha} f(x), \quad x < a, \\ S_{\frac{1}{2}\nu-\beta,2\beta} A(k) &= (2/x)^{2\beta} g(x), \quad x > a. \end{aligned} \tag{14}$$

Suppose now that we can find two operators L_1 and L_2 with the following properties:

(i)
$$\begin{aligned} L_1 S_{\frac{1}{2}\mu-\alpha,2\alpha} &= S_{\gamma,\delta}, \\ L_2 S_{\frac{1}{2}\nu-\beta,2\beta} &= S_{\gamma,\delta}; \end{aligned} \tag{15}$$

(ii) $L_1 f(x)$ only involves values of $f(x)$ for $x \leq a$;

(iii) $L_2 g(x)$ only involves values of $g(x)$ for $x \geq a$.

Then (14) will become

$$S_{\gamma,\delta} A(k) = \begin{cases} L_1 (2/x)^{2\alpha} f(x), & x < a \\ L_2 (2/x)^{2\beta} g(x), & x > a \end{cases} \tag{16}$$

and $A(k)$ can be found by applying the inverse operator $S_{\gamma+\delta,-\delta}$ to the right-hand side, which is a known function.

Choice of Operators: Using the inversion operators on (14), we find that L_1 and L_2 must satisfy

$$\begin{aligned} L_1 &= S_{\gamma,\delta}\, S_{\frac{1}{2}\mu+\alpha,-2\alpha}, \\ L_2 &= S_{\gamma,\delta}\, S_{\frac{1}{2}\nu+\beta,-2\beta}. \end{aligned} \tag{17}$$

We will deal here with L_1, leaving the corresponding calculations for L_2 to the reader. Written as a double integral, $L_1 f$ is

$$\begin{aligned} L_1 f = (2/x)^{\delta} \int_0^{\infty} k^{-\delta} J_{2\gamma+\delta}(kx)\,k\,dk \\ \times \left\{ (k/2)^{2\alpha} \int_0^{\infty} u^{2\alpha} J_{\mu}(ku)\ f(u)u\,du \right\}, \end{aligned} \tag{18}$$

and, if we interchange the order of integration, this appears as

$$L_1 f = \int_0^{\infty} \omega(x,u)\ f(u)\ du, \tag{19}$$

$$\omega(x,u) = 2^{\delta-2\alpha} x^{-\delta} u^{1+2\alpha} \times \int_0^{\infty} k^{2\alpha-\delta} J_{2\gamma+\delta}(kx) J_{\mu}(ku)\,k\,dk. \tag{20}$$

This equation represents the first of three conditions which we want the operators L_1 and L_2 to satisfy. The second condition requires that $\omega(x,u) = 0$ when $u > x$; reference to (15.21) shows that this is easily satisfied by choosing $\gamma = \frac{1}{2}\mu-\alpha$, for which

$$L_1 f = \frac{2x^{2\alpha-2\delta-\mu}}{(\delta-2\alpha-1)!} \int_0^{x} u^{1+2\alpha+\mu} (x^2-u^2)^{\delta-2\alpha-1}\ f(u)\ du. \tag{21}$$

In a similar way, the choice $\delta = \frac{1}{2}(\nu-\mu) + \alpha-\beta$ gives

$$L_2 g = \frac{2x^{\mu-2\alpha}}{(\frac{1}{2}\nu-\frac{1}{2}\mu+\alpha-\beta-1)!} \int_x^{\infty} u^{1+2\beta-\nu} (u^2-x^2)^{\alpha-\beta+\frac{1}{2}\nu-\frac{1}{2}\mu-1}\, g(u)\,du. \tag{22}$$

With these expressions in (16), the problem is formally solved.

Restrictions on Parameters: If $f(u)$ and $g(u)$ tend to finite (non-zero) limits as $u \to 0$ and $u \to \infty$ respectively, then we need

$$-\mu-2 < 2\alpha < \mu \tag{23}$$

for both integrals (21) and (22) to converge at these limits. This causes no difficulty, since we can choose a new α by redefining $A(k)$ by

$$k^{-2\alpha}A(k) = k^{-2\alpha'} A'(k). \tag{24}$$

In order for the integrals to converge at $u = x$, we also need the restrictions

$$\frac{1}{2}(\nu-\mu) \pm \alpha-\beta > 0. \tag{25}$$

However, by extending the definitions of the operators suitably, we can lift this restriction. We turn to this task in the next section.

16.3. Erdelyi-Kober Operators

The operators L_1 and L_2 are usually known as the Erdelyi-Kober operators of fractional integration, and in the conventional notation are defined as

$$I_{\eta,\alpha}f = \frac{2x^{-2\alpha-2\eta}}{(\alpha-1)!}\int_0^x u^{2\eta+1} (x^2-u^2)^{\alpha-1} f(u)\, du, \tag{26}$$

$$K_{\eta,\alpha}f = \frac{2x^{2\eta}}{(\alpha-1)!}\int_x^\infty u^{-2\alpha-2\eta+1} (u^2-x^2)^{\alpha-1} f(u)\, du. \tag{27}$$

These definitions are restricted by $\mathrm{Re}(\eta) > -\frac{1}{2}$, $\mathrm{Re}(\alpha) > 0$.

Properties: We will investigate here the operators $I_{\eta,\alpha}$ only, relegating the derivations of the corresponding properties of $K_{\eta,\alpha}$ to the problems. First we note that

$$I_{\eta,\alpha} x^{2\beta} f(x) = x^{2\beta} I_{\eta+\beta,\alpha} f(x). \tag{28}$$

Secondly, consider

$$I_{\eta,\alpha} I_{\eta+\alpha,\beta} f(x) = \frac{2x^{-2\eta-2\alpha}}{(\alpha-1)!} \int_0^x u^{2\eta+1} (x^2-u^2)^{\alpha-1} du \\ \times \frac{2u^{-2\eta-2\alpha-2\beta}}{(\beta-1)!} \int_0^u t^{2\eta+2\alpha+1} (u^2-t^2)^{\beta-1} f(t)\, dt. \tag{29}$$

Interchanging the order of integration and evaluating the inner (u) integral by the variable change $s = [1-(t^2/u^2)]/[1-(t^2/x^2)]$, which transforms it to a Beta function, we find that

$$I_{\eta,\alpha} I_{\eta+\alpha,\beta} f(x) = I_{\eta,\alpha+\beta} f(x). \tag{30}$$

A similar treatment shows that

$$I_{\eta+\alpha,\beta} I_{\eta,\alpha} f(x) = I_{\eta,\alpha+\beta} f(x). \tag{31}$$

Connection with Differentiation: We introduce the differential operator

$$D_x = \frac{1}{2x} \frac{d}{dx} . \tag{32}$$

Then, using integration by parts, we see

$$I_{\eta,\alpha} x^{-2\eta} D_x x^{2\eta} f(x)$$

$$= \frac{2x^{-2\alpha-2\eta}}{(\alpha-1)!} \int_0^x (x^2-u^2)^{\alpha-1} \frac{1}{2} \frac{d}{du} [u^{2\eta} f(u)]\, du \tag{33}$$

$$= x^{-2} I_{\eta,\alpha-1} f(x),$$

and recursive use of this formula yields

$$x^{2m} I_{\eta,\alpha} x^{-2\eta} D_x^m x^{2\eta} = I_{\eta,\alpha-m}, \qquad (34)$$

$$\mathrm{Re}(\alpha-m) > 0.$$

Similarly

$$x^{-2\eta-2\alpha} D_x x^{2\eta+2\alpha} I_{\eta,\alpha} f(x)$$

$$= \frac{2x^{-2\eta-2\alpha-1}}{(\alpha-1)!} \frac{d}{dx} \int_0^x u^{2\eta+1}(x^2-u^2)^{\alpha-1} f(u)\, du \qquad (35)$$

$$= x^{-2} I_{\eta,\alpha-1} f(x),$$

and in this case, repeated application yields

$$x^{2m-2\eta-2\alpha} D_x^m x^{2\eta+2\alpha} I_{\eta,\alpha} = I_{\eta,\alpha-m}, \qquad (36)$$

$$\mathrm{Re}(\alpha-m) > 0.$$

Analytic Continuation: We will now lift the restriction $\mathrm{Re}(\alpha) > 0$ from the definition of $I_{\eta,\alpha}$ and formulas involving $I_{\eta,\alpha}$. First we use (34) to define $I_{\eta,\alpha}$ when $\mathrm{Re}(\alpha) < 0$ by choosing an integer m such that $\mathrm{Re}(\alpha+m) > 0$ and writing

$$I_{\eta,\alpha} f(x) = x^{-2\eta-2\alpha} D_x^m x^{2\eta+2\alpha+2m} I_{\eta,\alpha+m} f(x). \qquad (37)$$

It is trivial to show that with this definition equations (28), (34) and (36) hold without the restriction on $\mathrm{Re}(\alpha)$. Moreover,

$$I_{\eta,0} f(x) = f(x). \qquad (38)$$

Now on setting $\beta = -\alpha$ in (30), we find that if $\mathrm{Re}(\alpha) < 0$ then

$$
\begin{aligned}
I_{\eta,\alpha}\, I_{\eta+\alpha,-\alpha}
&= x^{-2\eta-2\alpha} D_x^m\, x^{2\eta+2\alpha+2m}\, I_{\eta,\alpha+m}\, I_{\eta+\alpha,-\alpha} \\
&= x^{-2\eta-2\alpha} D_x^m\, x^{2\eta+2\alpha+2m}\, I_{\eta+\alpha,m} \qquad (39)\\
&= I_{\eta+\alpha,0}.
\end{aligned}
$$

Thus, from (38), another possible definition of $I_{\eta,\alpha}$ for $\mathrm{Re}(\alpha) < 0$ is

$$I_{\eta,\alpha} = I^{-1}_{\eta+\alpha,-\alpha}. \qquad (40)$$

Now let $\mathrm{Re}(\alpha) > 0$ and $\mathrm{Re}(\beta) > 0$; then on taking the inverse of (30) we see

$$I^{-1}_{\eta+\alpha,\beta}\, I^{-1}_{\eta,\alpha} = I^{-1}_{\eta,\alpha+\beta}, \qquad (41)$$

which, using (40) and making the substitutions $\eta+\alpha+\beta \to \eta$, $-\beta \to \alpha$, and $-\alpha \to \beta$, becomes (30) again, except that now $\mathrm{Re}(\alpha) < 0$ and $\mathrm{Re}(\beta) < 0$. By similar arguments, (30) and (31) may be extended to all values of α and β.

Connection with Modified Hankel Transform: Consider

$$
\begin{aligned}
I_{\eta+\alpha,\beta}\, S_{\eta,\alpha}\, f(x)
&= \frac{2x^{-2\eta-2\alpha-2\beta}}{(\beta-1)!} \int_0^x du\, u^{2\eta+2\alpha+1}\, (x^2-u^2)^{\beta-1} \\
&\quad \times 2^{\alpha} u^{-\alpha} \int_0^\infty t^{1-\alpha} J_{2\eta+\alpha}(ut)\, f(t)\, dt \qquad (42)\\
&= \frac{2^{1+\alpha} x^{-2\eta-2\alpha-2\beta}}{(\beta-1)!} \int_0^\infty t^{1-\alpha} f(t)\, dt \\
&\quad \times \int_0^x du\, u^{2\eta+\alpha+1} (x^2-u^2)^{\beta-1}\, J_{2\eta+\alpha}(ut).
\end{aligned}
$$

The inner (u) integral can be evaluated by using (15.19) and (15.20), to give

$$I_{\eta+\alpha,\beta}S_{\eta,\alpha} = S_{\eta,\alpha+\beta}. \tag{43}$$

By similar methods, we can also derive the identities

$$S_{\eta+\alpha,\beta}\, I_{\eta,\alpha} = S_{\eta,\alpha+\beta}, \tag{44}$$

$$K_{\eta,\alpha}\, S_{\eta+\alpha,\beta} = S_{\eta,\alpha+\beta}, \tag{45}$$

$$S_{\eta,\alpha}\, K_{\eta+\alpha,\beta} = S_{\eta,\alpha+\beta}, \tag{46}$$

$$S_{\eta+\alpha,\beta}\, S_{\eta,\alpha} = I_{\eta,\alpha+\beta}, \tag{47}$$

$$S_{\eta,\alpha}\, S_{\eta+\alpha,\beta} = K_{\eta,\alpha+\beta}. \tag{48}$$

Finally, we note that in the notation of the Erdelyi-Kober operators, the solution of the dual integral equations (11) is given by

$$S_{\frac{1}{2}\mu-\alpha,\lambda-\mu+2\alpha}\, A(k) = h(x), \tag{49}$$

where

$$\lambda = \tfrac{1}{2}\mu+\tfrac{1}{2}\nu-\alpha+\beta \tag{50}$$

and

$$h(x) = \begin{cases} 2^{2\alpha}\, x^{-2\alpha}\, I_{\frac{1}{2}\mu,\lambda-\mu}\, f, & x < a \\ 2^{2\beta}\, x^{-2\beta}\, K_{\frac{1}{2}\mu-\alpha+\beta,\nu-\lambda}\, g, & x > a. \end{cases} \tag{51}$$

Problems

1. Show that the solution of the dual integral equations

$$\int_0^\infty A(k)\, J_1(kx)\, dk = -1, \quad x < 1$$

$$\int_0^\infty kA(k)\, J_1(kx)\, dk = 0, \quad x > 1$$

is

$$A(k) = \frac{\cos(k) - 1}{k} .$$

2. Show that the solution of the dual integral equations

$$\int_0^\infty A(k) \sin(kx)\, dk = f(x), \quad x < 1$$

$$\int_0^\infty kA(k) \sin(kx)\, dk = 0 \quad , \quad x > 1$$

is given by

$$A(k) = \frac{2}{\pi} \int_0^1 J_0(kx) \left\{ \frac{d}{dx} \int_0^x \frac{y\, f(y)\, dy}{\sqrt{x^2 - y^2}} \right\} dx.$$

3. Show that the solution of the dual integral equations

$$\int_0^\infty k\, A(k)\, J_0(kx)\, dk = f(x), \quad x < 1$$

$$\int_0^\infty A(k)\, J_0(kx)\, dk = 0 \quad , \quad x > 1$$

is

$$A(k) = \frac{2}{\pi} \int_0^1 \sin\,(kx)\, dx \int_0^x \frac{y\, f(y)\, dy}{\sqrt{x^2 - y^2}} .$$

4. Consider the problem of finding a function $u(x,y)$ which is harmonic in the half-plane $y \geq 0$, and satisfies the mixed boundary conditions

$$u_y(x,0) = v(x), \quad |x| < 1,$$
$$u(x,0) = 0 \quad , \quad |x| > 1.$$

Obtain a pair of dual integral equations by writing

$$v(x) = v^+(x) + v^-(x),$$

$$v^+(x) = \frac{1}{2}[v(x) + v(-x)],$$

$$v^-(x) = \frac{1}{2}[v(x) - v(-x)],$$

with similar definitions of u^+ and u^-, and using the Fourier cosine transform on u^+ and Fourier sine transform on u^-. Show that, if $v^- = 0$, then

$$u(x,y) = \frac{2}{\pi}\int_0^\infty A(k) \cos(kx) e^{-ky}\, dk,$$

where

$$A(k) = -\int_0^1 t\, J_0(kt)\, dt \int_0^t \frac{v(s)\, ds}{\sqrt{t^2-s^2}} .$$

Find the remainder of the solution when $v^- \neq 0$.

5. Show that the solution of Problem 4 with $v = -1$ is

$$u(x,y) = \int_0^\infty J_1(k) \cos(kx)\, e^{-ky} \frac{dk}{k} .$$

6. Verify the relations

$$K_{\eta,\alpha}\, x^{2\beta} f(x) = x^{2\beta}\, K_{\eta-\beta,\alpha} f(x),$$

$$K_{\eta,\alpha} K_{\eta+\alpha,\beta} = K_{\eta,\alpha+\beta},$$

$$x^{2\eta+2n} D_x^n\, x^{-2\eta} K_{\eta,\alpha+n}\, f(x) = K_{\eta+n,\alpha} f(x).$$

7. Consider the dual integral equations

$$\int_0^\infty G(k)\, A(k)\, J_\nu(kx)\, dk = f(x), \quad x < 1,$$

$$\int_0^\infty k\, A(k)\, J_\nu(kx)\, dk = 0 \quad , \quad x > 1,$$

where $G(k)$ is a given function with the asymptotic form

$$G(k) \sim k^{1-2\alpha}, \quad k \to \infty.$$

By defining the functions

$$g(x) = \int_0^\infty k\, A(k)\, J_\nu(kx)\, dk, \quad x > 0$$

and

$$H(x) = \frac{x^{\nu-\beta+1}}{2^{\beta-1}(\beta-1)!} \int_x^1 y^{1-\nu}(y^2-x^2)^{\beta-1}\, g(y)\, dy,$$

show that $H(x)$ is determined by the Fredholm integral equation

$$H(x) + x\int_0^1 K(x,y)\, H(y)\, dy$$

$$= \frac{2^\beta(\beta-1)!\,\sin(1-\beta)\pi}{\pi} x^{\beta-\nu} \frac{d}{dx} \int_0^x \frac{y^{\nu+1}\, f(y)}{(x^2-y^2)^\beta}\, dy,$$

where

$$K(x,y) = \int_0^1 [k^{2\beta}G(k) - k]\, J_{\nu-\beta}(kx)\, J_{\nu-\beta}(ky)\, dk.$$

8. Consider a condenser made of two equal, coaxial, parallel, circular metal discs of unit radius and separation ℓ, which are charged to potentials V_0 and $-V_0$. Show that the potential $\phi(r,z)$ can be represented as (see Figure 1)

$$\phi(r,z) = \begin{cases} V_0\displaystyle\int_0^\infty \{e^{-kz} - e^{-k(z-\ell)}\}\, A(k)\, J_0(kr)\, dk, & z > \ell \\ V_0\displaystyle\int_0^\infty \{e^{-kz} - e^{k(z-\ell)}\}\, A(k)\, J_0(kr)\, dk, & 0 < z < \ell \\ V_0\displaystyle\int_0^\infty \{e^{kz} - e^{k(z-\ell)}\}\, A(k)\, J_0(kr)\, dk, & z < 0 \end{cases}$$

provided that the function $A(k)$ satisfies the dual integral equations

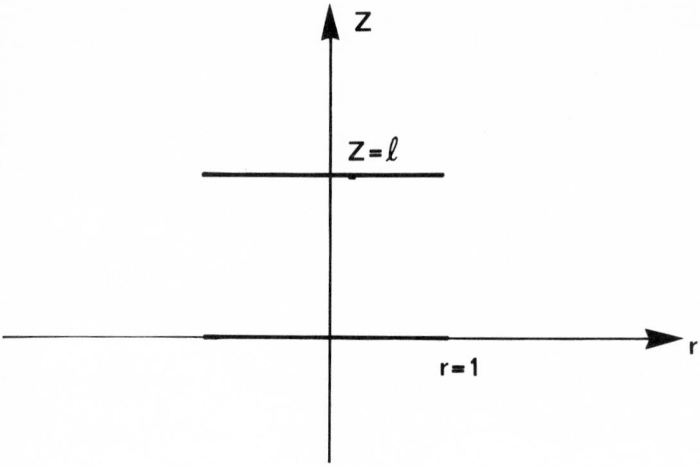

Figure 16.1

$$\int_0^\infty (1 - e^{-k\ell})\, A(k)\, J_0(kr)\, dk = 1, \qquad r < 1,$$

$$\int_0^\infty k\, A(k)\, J_0(kr)\, dk = 0, \qquad r > 1.$$

Show that an application of Problem 7 leads to

$$A(k) = \frac{2k^2}{\pi} \int_0^1 g(t) \cos(kt)\, dt$$

where $g(r)$ satisfies the Fredholm equation

$$g(r) - \frac{\ell}{\pi} \int_{-1}^1 \frac{g(t)dt}{\ell^2 + (r-t)^2} = 1.$$

Show also that the capacity of the condenser is

$$C = \frac{1}{\pi} \int_0^1 g(r)\, dr.$$

Footnotes

1. The most comprehensive reference on mixed boundary-value problems and dual integral equations is SNEDDON (1966).

§17. INTEGRAL TRANSFORMS GENERATED BY GREEN'S FUNCTIONS

17.1. The Basic Formula

In this section we will investigate (in a purely formal manner) some properties of the self-adjoint differential operator [see (10.15)]

$$L[u] = [p(x)\, u'(x)]' + q(x)\, u(x), \tag{1}$$

where $p(x)$ and $q(x)$ are given functions on the interval $a \leq x \leq b$, and the functions $u(x)$ under consideration all satisfy homogeneous boundary conditions of the type [see (10.2)]

$$\begin{aligned} a_1 u(a) + a_2 u'(a) &= 0, \\ b_1 u(b) + b_2 u'(b) &= 0. \end{aligned} \tag{2}$$

We shall not take special note of the cases where a and/or b are infinite, although this is crucial to a rigorous analysis.

We first recall the following results from Section 10.1. Let ϕ_λ and ψ_λ be solutions of the equation

$$L[u] = \lambda r(x) u(x), \tag{3}$$

where λ is a constant and $r(x)$ a given function such that ϕ_λ and ψ_λ satisfy the boundary conditions

$$\begin{aligned} a_1 \phi_\lambda(a) + a_2 \phi_\lambda'(a) &= 0, \\ b_1 \psi_\lambda(b) + b_2 \psi_\lambda'(b) &= 0. \end{aligned} \tag{4}$$

Then the Green's function for the operator $L-\lambda r$ which satisfies the complete boundary conditions (2) is

$$g(x,x';\lambda) = \frac{\phi_\lambda(x^<)\psi_\lambda(x^>)}{\Delta(\lambda)}, \tag{5}$$

$$\Delta(\lambda) = p(x)\, W[\phi_\lambda,\psi_\lambda],$$

where $x^<$ is the smaller of x and x', and $x^>$ the larger. As we showed in Section 10, $\Delta(\lambda)$ is independent of x, although in the present problem it is a function of λ. The Green's function is undefined when $\Delta(\lambda) = 0$, that is, when the functions ϕ_λ and ψ_λ are linearly dependent, making each one a solution of the eigenvalue problem given by (2) and (3) together. Thus there is a close connection between Green's functions and eigenfunctions; we refer the reader to one of the many excellent texts for relevant details.[1]

Consider the partial differential equation

$$ir(x)\,\frac{\partial\phi(x,t)}{\partial t} = L[\phi(x,t)] \tag{6}$$

together with the initial conditions

$$\phi(x,0) = f(x) \tag{7}$$

and the boundary conditions (2). Taking the Laplace transform with respect to t, we obtain

$$[L - isr(x)]\,\Phi(x,s) = -ir(x)f(x) \tag{8}$$

where s is the transform variable. In terms of the Green's function (5), the Laplace transform $\Phi(x,s)$ is

$$\Phi(x,s) = -i\int_a^b g(x,x';-is)\, f(x')r(x')dx'. \tag{9}$$

If we apply the inverse transform to $\Phi(x,s)$, we recover a function which is zero for $t < 0$ and equal to $\phi(x,t)$ for $t > 0$; hence on setting $t = 0$ in the inversion integral we recover the average. Explicitly,

$$f(x) = -\frac{1}{\pi}\int_{c-i\infty}^{c+i\infty} ds \int_a^b dx'\ r(x')g(x,x';-is)f(x'). \quad (10)$$

The contour in s must pass to the right of all the singularities of $g(x,x';-is)$ in the s-plane; these are at the points $s = i\lambda$ corresponding to eigenvalues λ of the operator $L-\lambda r$. Using the standard result[2] that the eigenvalues of a self-adjoint operator of the present type are real numbers, we find that c may be any positive number. In our subsequent use of (10), we shall replace s by $i\lambda$ and write

$$f(x) = \frac{1}{\pi i}\int_{ic-\infty}^{ic+\infty} d\lambda \int_a^b dx'\ r(x')g(x,x';\lambda)f(x'),\ c > 0. \quad (11)$$

Similarly, by considering the equation $ir\phi_t = -L[\phi]$, we obtain

$$f(x) = -\frac{1}{\pi i}\int_{ic-\infty}^{ic+\infty} \int_a^b dx' r(x')g(x,x';\lambda)f(x'),\ c < 0. \quad (12)$$

We have derived these formulas without regard to a rigorous justification of the steps involved. Such a justification can be provided[3] when appropriate conditions are applied to the functions $p(x)$, $q(x)$, and $r(x)$, although we will not attempt this here. Alternatively, the formulas may be used to generate useful particular results whose validity must be checked by some other method.

17.2. Finite Intervals

If a and b are finite numbers, and $p(x)$, $p'(x)$, $q(x)$, and $r(x)$ are all continuous on the closed interval

$a \le x \le b$, and if in addition $p(x) > 0$ and $r(x) > 0$ for $a \le x \le b$, then the eigenvalue problem defined by (2) and (3) is a regular Sturm-Liouville problem.[4] It is shown in many texts on mathematical physics that the eigenvalues λ_n and eigenfunctions $\phi_n(x)$ for such a problem have the following properties:

(i) The eigenvalues are real and denumerable.

(ii) There is no point of accumulation of the eigenvalues; i.e., there are only a finite number of eigenvalues in any finite interval.

(iii) There is only one eigenfunction (to within an arbitrary multiplicative constant) for each eigenvalue.

(iv) Different eigenfunctions are orthogonal in the sense that

$$\int_a^b \phi_n(x)\phi_m(x)r(x)dx = 0, \quad n \neq m. \tag{13}$$

<u>Fourier Sine Series</u>: The simplest regular problem corresponds to $p(x) = -1$, $q(x) = 0$, $r(x) = 1$, $a = 0$, and $b = \ell$. It is trivial to show that the eigenvalues and eigenfunctions are

$$\begin{aligned} \lambda_n &= (n\pi/\ell)^2, \\ \phi_{\lambda_n} &= \sin(n\pi x/\ell), \qquad n = 1,2,3,\ldots . \end{aligned} \tag{14}$$

The Green's function (5) is also easy to construct; it is

$$g(x,x';\lambda) = -\frac{\sin(kx^<)\sin[k(\ell-x^>)]}{k \sin k\ell}, \quad k^2 = \lambda. \tag{15}$$

Despite the appearance of $\sqrt{\lambda}$ in these formulas, the Green's function does not have a branch point at $\lambda = 0$, but merely a series of simple poles at $\lambda = (n\pi/\ell)^2$. Adding (11) and (12), we find that

$$f(x) = \frac{1}{2\pi i}\int_C d\lambda \int dx' \frac{\sin(kx^<)\sin[k(\ell-x^>)]}{k\sin k\ell} f(x'), \tag{16}$$

where the contour is shown in Figure 1.

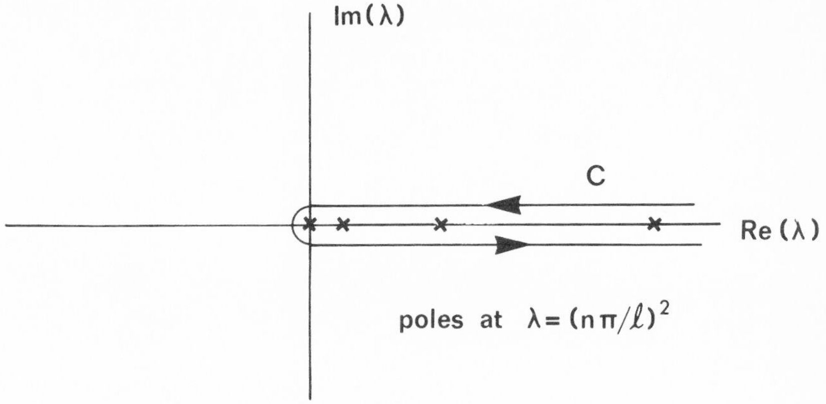

Figure 17.1

Interchanging the orders of integration, we can readily evaluate the λ integral as a sum of residues. This gives the familiar formula

$$f(x) = \frac{2}{\ell}\sum_{n=1}^{\infty} \sin(n\pi x/\ell)\int_0^{\ell} \sin(n\pi x'/\ell) f(x')dx'. \tag{17}$$

Eigenfunction Expansion: Now consider the general case. The zeros of $\Delta(\lambda)$, which determine the poles of $g(x,x';\lambda)$, are all discrete and simple; hence we can proceed by adding (11) and (12) and evaluating residues as for the Fourier series. To find the residues, we need the value of $d\Delta(\lambda)/d\lambda$ at $\lambda = \lambda_n$, i.e., we need to calculate

$$p(x)\frac{d}{d\lambda}[\phi_\lambda(x)\psi'_\lambda(x) - \phi'_\lambda(x)\psi_\lambda(x)]. \tag{18}$$

To get $d\phi_\lambda/d\lambda$ and $d\psi_\lambda/d\lambda$, we differentiate (3) with respect to λ:

$$[L - \lambda r(x)]\frac{\partial\phi_\lambda}{\partial\lambda} = r(x)\phi_\lambda(x), \tag{19}$$

and solve using the Green's function. After some straightforward algebra, we get

$$\frac{d\Delta(\lambda)}{d\lambda} = -\int_a^b r(x)\phi_\lambda(x)\psi_\lambda(x)\,dx. \tag{20}$$

When we put $\lambda = \lambda_n$, ϕ_λ and ψ_λ become the same function, apart from a normalizing factor, and in evaluating the residue this factor cancels, so that the residues are

$$-\frac{\phi_n(x)\phi_n(x')}{\int_a^b r(x)\phi_n^2(x)dx}, \tag{21}$$

where we have written $\phi_n(x)$ for $\phi_{\lambda_n}(x)$. The complete expansion of a function $f(x)$ is

$$f(x) = \sum_n \frac{\phi_n(x)\int_a^b \phi_n(x')f(x')r(x')dx'}{\int_a^b \phi_n^2(x')r(x')dx'} . \tag{22}$$

Note that $f(x)$ satisfies the boundary conditions (2), as is evident from its definition, and so there is no conflict caused by the fact that every term in (22) also satisfies these boundary conditions.

17.3. Some Singular Problems

It is evident from the foregoing that we must consider singular problems if we are to obtain integral transforms. Before generating any new transforms, we will show how the Fourier, Mellin, and Hankel transforms are related to Green's functions.

Fourier Transform: Let $p(x) = -1$, $r(x) = 1$, $q(x) = 0$, $a = -\infty$, and $b = +\infty$. We require as our boundary conditions that $\phi_\lambda(x)$ and $\psi_\lambda(x)$ remain finite as $x \to \mp\infty$, respectively, remembering that $\text{Im}(\lambda) > 0$ in (11). The functions we need are

$$\begin{aligned} \phi_\lambda(x) &= e^{-ikx}, \\ \psi_\lambda(x) &= e^{ikx}, \\ \Delta(\lambda) &= 2ik, \\ k^2 &= \lambda, \qquad 0 < \arg(k) < \pi/2. \end{aligned} \tag{23}$$

and so (11) reads

$$f(x) = \frac{1}{2\pi}\int_{ic-\infty}^{ic+\infty} d\lambda \int_{-\infty}^{\infty} \frac{e^{-ik(x^<-x^>)}}{k} f(x')\, dx', \quad c > 0. \tag{24}$$

Similarly, we may evaluate the functions relevant to (12) to get

$$f(x) = \frac{1}{2\pi}\int_{ic-\infty}^{ic+\infty} d\lambda \int_{-\infty}^{\infty} \frac{e^{ik(x^<-x^>)}}{k} f(x')dx', \quad c > 0, \tag{25}$$

$$-\pi/2 < \arg(k) < 0.$$

When λ is on the negative real axis, these two expressions differ only in sign. Adding them and letting $c \to 0$, we get

$$f(x) = \frac{1}{2\pi}\int_0^\infty \frac{d\lambda}{\sqrt{\lambda}} \int_{-\infty}^{\infty} \cos[k(x-x')]\, f(x')\, dx'. \tag{26}$$

If we replace $\sqrt{\lambda}$ by k and break up the cosine into complex exponentials we obtain the exponential Fourier transform.

<u>An Alternative Formula</u>: In each of our examples above, the eigenvalues of the operator have had a lower bound and the Green's function $g(x,x';\lambda)$ has been analytic across the real axis to the left of the lowest eigenvalue. Whenever this is the case, we can add (11) and (12), to obtain the formula

$$f(x) = \frac{1}{2\pi i}\int_C d\lambda \int_a^b g(x,x';\lambda)f(x')r(x')dx', \tag{27}$$

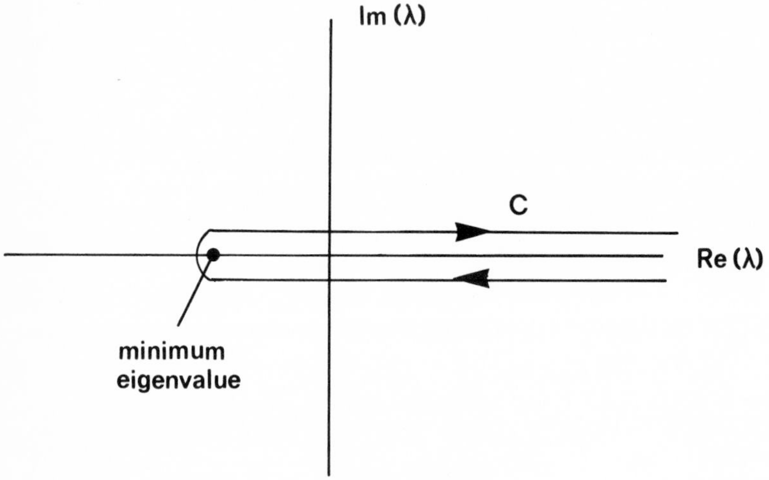

Figure 17.2

where the contour C is shown in Figure 2. In the regular case, the eigenvalues are discrete, and the value of the integral is given by a sum of residues. With singular problems, however, the Green's function generally has a branch cut along the real axis, and other appropriate methods of evaluating the integral must be found.

Mellin Transform: If we set $p(x) = x$, $q(x) = 0$ and $r(x) = x^{-1}$ in the interval $0 \le x < \infty$, then we again have a singular problem. The equation for ϕ_λ and ψ_λ is

$$(xu')' = \lambda u/x \tag{28}$$

with the solutions $x^{\pm i\sqrt{\lambda}}$. On the contour C we choose $\sqrt{\lambda}$ by $0 < \arg\sqrt{\lambda} < \pi$; then $\mathrm{Re}(i\sqrt{\lambda}) < 0$. Thus

$$g(x,x';\lambda) = \frac{-i}{2\sqrt{\lambda}} (x^<)^{-i\sqrt{\lambda}} (x^>)^{i\sqrt{\lambda}}, \tag{29}$$

and the integral reduces to

$$\begin{aligned}&\frac{1}{2\pi i}\int_0^\infty d\lambda \int_0^\infty [g(x,\xi;\lambda+ic) - g(x,\xi;\lambda-ic)]f(\xi)\frac{d\xi}{\xi}\\ &= \frac{1}{4\pi}\int_0^\infty \frac{d\lambda}{\sqrt{\lambda}}\int_0^\infty \left[x^{-i\sqrt{\lambda}}\xi^{i\sqrt{\lambda}} + x^{i\sqrt{\lambda}}\xi^{-i\sqrt{\lambda}}\right] f(\xi)\frac{d\xi}{\xi}.\end{aligned} \tag{30}$$

Now write $q^2 = \lambda$ in the first term and $(-q)^2 = \lambda$ in the second; this gives

$$f(x) = \frac{1}{2\pi}\int_{-\infty}^\infty x^{-iq}\, dq \int_0^\infty \xi^{iq} f(\xi)\frac{d\xi}{\xi}, \tag{31}$$

which is the Mellin transform formula with $p = iq$.

Hankel Transform: Proceeding as above, we consider the interval $0 \le x < \infty$ with $p(x) = -x$, $q(x) = \nu^2/x$ and $r(x) = x$, where $\text{Re}(\nu) > -\frac{1}{2}$. The Green's function, finite as $x \to 0$ or ∞, is easily constructed as

$$\begin{aligned}&g(x,x';\lambda) = \frac{\pi i}{2} J_\nu(kx^<)H_\nu^{(1)}(kx^>),\\ &k^2 = \lambda, \quad 0 < \arg(k) < \pi/2.\end{aligned} \tag{32}$$

Using the relations

$$\begin{aligned}&J_\nu(ze^{i\pi}) = e^{i\pi\nu} J_\nu(z),\\ &H_\nu^{(1)}(ze^{i\pi}) = -e^{-i\pi\nu} H_\nu^{(2)}(z),\\ &-\pi < \arg(z) < 0,\end{aligned} \tag{33}$$

we readily reduce the integral around the branch cut $0 \le \lambda < \infty$ to the Hankel transform formula

$$f(x) = \int_0^\infty k\,dk \int_0^\infty J_\nu(kx)J_\nu(k\xi)f(\xi)\ \xi d\xi. \tag{34}$$

17.4. Kontorovich - Lebedev Transform

The Hankel transform is the first example we considered for which $q(x) \neq 0$. In fact, the Hankel transform, with ν replaced by $\nu + \frac{1}{2}$, may be obtained from the choice $p(x) = 1$, $q(x) = \nu^2/x^2$, $r(x) = 1$ on the interval $0 \leq x < \infty$. Setting $\nu = 0$ gives the Fourier sine transform as a special case, thus we can regard the Hankel transform as a generalization of the Fourier sine transform to nonzero $q(x)$. In a similar way, we can generalize the Mellin transform by the choices

$$\begin{aligned} p(x) &= -x, \\ q(x) &= k^2 x, \\ r(x) &= 1/x, \\ 0 &\leq x < \infty. \end{aligned} \tag{35}$$

The functions $\phi_\lambda(x)$ and $\psi_\lambda(x)$ must satisfy the differential equation

$$\phi''_\lambda + \frac{1}{x}\phi'_\lambda - \left(k^2 - \frac{\lambda}{x^2}\right)\phi_\lambda = 0, \tag{36}$$

whose solutions are modified Bessel functions of order $\sqrt{-\lambda}$. If we choose $\alpha = \sqrt{-\lambda}$ by $\mathrm{Re}(\alpha) > 0$ for λ on the contour C, then the Green's function which goes to zero as $x \to 0$ or $x \to \infty$ is

$$g(x,x';\lambda) = I_\alpha(kx^<)K_\alpha(kx^>). \tag{37}$$

To evaluate the integral (27) we need the value of $g(x,x';\lambda+i\varepsilon) - g(x,x';\lambda-i\varepsilon)$ for real positive λ and $\varepsilon \to 0$. Using the results of Problem 20.17 we can write (with $\nu = \sqrt{\lambda}$)

$$\begin{aligned} I_{-i\nu}(kx) - I_{i\nu}(kx) &= \frac{2i}{\pi} \sinh(\pi\nu)\, K_{i\nu}(kx), \\ K_{-i\nu}(kx) - K_{i\nu}(kx) &= 0, \end{aligned} \tag{38}$$

and hence

$$\begin{aligned} &g(x,x';\lambda+i0) - g(x,x';\lambda-i0) \\ &\quad = \frac{2i}{\pi} \sinh(\pi\nu)\, K_{i\nu}(kx)\, K_{i\nu}(kx'). \end{aligned} \tag{39}$$

Using this result in (27), and subsequently changing the integration variable to ν, we obtain a form of the Kontorovich-Lebedev transform, viz.

$$f(x) = \frac{2}{\pi^2} \int_0^\infty \nu\sinh(\pi\nu)\, d\nu \int_0^\infty K_{i\nu}(kx) K_{i\nu}(kx') f(x') \frac{dx'}{x'}. \tag{40}$$

<u>An Alternative Formula</u>: Equation (40) is the original transform given by Kontorovich and Lebedev; however, there is an alternative formula which demonstrates the close connection with the Mellin transform. To obtain it, we note that (38) demonstrates that the I_α functions may be written in terms of K_α, so that the distinction between $x^<$ and $x^>$ could be dropped, provided we maintain convergence of the integrals. We therefore set $x^< = x$ and $x^> = x'$ in (37), and substitute into (27). Using α as a new variable, this gives

$$f(x) = \frac{1}{\pi i} \int_{-i\infty}^{i\infty} \alpha\, d\alpha\ \ I_\alpha(kx) \int_0^\infty K_\alpha(kx') f(x') \frac{dx'}{x'}. \tag{41}$$

<u>Regions of Convergence</u>: Consider the Kontorovich-Lebedev transform of a function $f(x)$, defined as

$$F(k,\nu) = \int_0^\infty K_\nu(kx) f(x)\, \frac{dx}{x}. \tag{42}$$

If we assume that f(x) has the asymptotic form

$$f(x) \sim \begin{cases} x^{\beta}, & x \to 0 \\ x^{\alpha}, & x \to \infty \end{cases} \tag{43}$$

then, using the asymptotic forms of the modified Bessel functions

$$\begin{aligned} I_{\nu}(x) &\sim \frac{x^{\nu}}{2^{\nu}\nu!}, && x \to 0, \\ K_{\nu}(x) &\sim \frac{2^{\nu-1}(\nu-1)!}{x^{\nu}}, && x \to 0, \\ K_{\nu}(x) &\sim \left(\frac{\pi}{2x}\right)^{1/2} e^{-x}, && x \to \infty, \end{aligned} \tag{44}$$

we see immediately that if f(x) is a "reasonable" function (for example, if it satisfies Dirichlet's conditions) then the integral (42) converges in the region $\mathrm{Re}(\nu) < \beta$ for all $k > 0$. Thus the inversion integral is defined whenever $\beta > 0$; it can be shown that the transform pair may be extended to functions for which $\beta \leq 0$ by moving the inversion contour.

Relation to Mellin Transform: If $\alpha < \beta$, then in the strip $\alpha < \mathrm{Re}(\nu) < \beta$ we may let $k \to 0$ in (42) to obtain

$$F(0,\nu) = \frac{1}{2^{\nu}\nu!} \mathscr{M}[f(x);\nu]. \tag{45}$$

Furthermore, in this limit the inversion integral (41) is the Mellin inversion.

17.5. Boundary-value Problems in a Wedge

To illustrate a simple use of the Kontorovich-Lebedev transform, we consider the problem of determining a function $u(r,\theta,z)$ which is harmonic in the wedge

$$\begin{aligned} 0 &\le r < \infty, \\ 0 &\le \theta \le \alpha, \\ 0 &\le z \le \ell, \end{aligned} \tag{46}$$

and which satisfies the boundary condition

$$u(r,\alpha,z) = f(r,z) \tag{47}$$

on one boundary and is zero on all the other boundaries. First, we introduce a Fourier series in z, using the functions $\sin(n\pi z/\ell)$, which satisfy the required boundary conditions at $z = 0$ and $z = \ell$. Thus we write

$$\begin{aligned} u(r,\theta,z) &= \sum_{n=1}^{\infty} u_n(r,\theta) \sin(n\pi z/\ell), \\ f(r,z) &= \sum_{n=1}^{\infty} f_n(r) \sin(n\pi z/\ell), \end{aligned} \tag{48}$$

and the coefficients $u_n(r,\theta)$ are determined by

$$\begin{aligned} \left\{ \frac{\partial^2}{\partial r^2} + \frac{1}{r}\frac{\partial}{\partial r} + \frac{1}{r^2}\frac{\partial^2}{\partial \theta^2} - \left(\frac{n\pi}{\ell}\right)^2 \right\} u_n(r,\theta) &= 0, \\ u_n(r,0) &= 0, \\ u_n(r,\alpha) &= f_n(r). \end{aligned} \tag{49}$$

We denote the Kontorovich-Lebedev transforms of $u_n(r,\theta)$ and $f_n(r)$ with respect to r by $U_n(k,\nu,\theta)$ and $F_n(k,\nu)$ respectively; on multiplying the differential equation for u_n by $rK_\nu(n\pi r/\ell)$ and using integration by parts twice[5] we reduce (49) to

$$\begin{aligned} \left\{ \frac{d^2}{d\theta^2} + \nu^2 \right\} U_n(n\pi/\ell,\nu,\theta) &= 0, \\ U_n(n\pi/\ell,\nu,0) &= 0, \\ U_n(n\pi/\ell,\nu,\alpha) &= F_n(n\pi/\ell,\nu). \end{aligned} \tag{50}$$

These equations are readily solved to yield the expression

$$U_n(n\pi/\ell,\nu,\theta) = \frac{\sinh(\nu\theta)}{\sinh(\nu\alpha)} F_n(n\pi/\ell,\nu), \tag{51}$$

from which an explicit integral representation of the solution can be constructed.

17.6. Diffraction of a Pulse by a Two-Dimensional Half-Plane[6]

As a more difficult example of the use of the Kontorovich-Lebedev transform, we will construct an explicit representation of the Green's function for the two-dimensional scalar wave equation,

$$[\nabla^2 - \frac{1}{c^2}\frac{\partial^2}{\partial t^2}]\ g(\underset{\sim}{r},\underset{\sim}{r}_0,t) = \delta(\underset{\sim}{r}-\underset{\sim}{r}_0)\ \delta(t) \tag{52}$$

subject to the boundary condition $g = 0$ on the positive x-axis and the initial conditions $g = \partial g/\partial t = 0$ at $t = 0$. This Green's function represents the wave pattern generated by a pulse at $\underset{\sim}{r} = \underset{\sim}{r}_0$, $t = 0$, including the effects of diffraction by a semi-infinite barrier. We have solved the free-space problem in Section 10; we must now find the effect of the barrier on this solution. Introduce polar coordinates, and also the variables

$$\begin{aligned} R &= [r^2 + r_0^2 - 2rr_0\cos(\theta-\theta_0)]^{1/2}, \\ R_0 &= [x^2 + r_0^2 - 2\ x\ r_0\ \cos\ \theta_0]^{1/2}. \end{aligned} \tag{53}$$

All of these quantities are depicted in Figure 3.

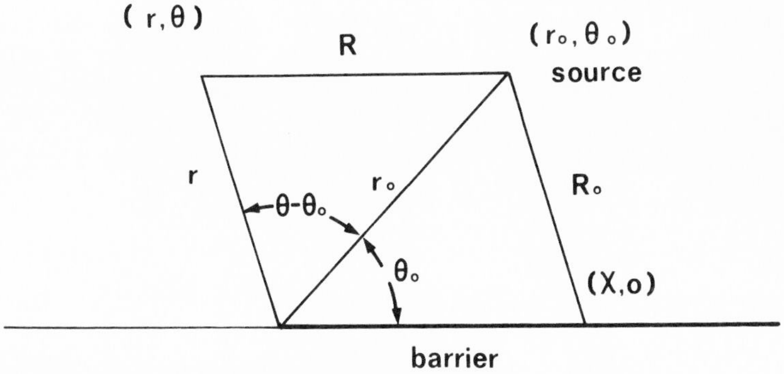

Figure 17.3

Laplace Transform: We introduce a function ϕ by $g = g_0+\phi$, where g_0 is the free-space solution.[7] Then the equations which determine ϕ are

$$[\nabla^2 - \frac{1}{c^2}\frac{\partial^2}{\partial t^2}]\ \phi(r,\theta,r_0,\theta_0,t) = 0,$$

$$\phi(r,0,r_0,\theta_0,t) = \begin{cases} 0, & t < R_0/c \\ -1/2\pi\sqrt{t^2-R_0^2/c^2}, & t > R_0/c, \end{cases} \tag{54}$$

$$\phi(r,\theta,r_0,\theta_0,0) = 0,$$

$$\phi_t(r,\theta,r_0,\theta_0,0) = 0.$$

We denote the Laplace transform of ϕ with respect to t by Φ; then the Laplace transform of these equations is

$$\nabla^2\Phi - (p^2/c^2)\Phi = 0,$$
$$\Phi = -\frac{1}{2\pi}K_0(pR_0/c), \quad \theta = 0. \tag{55}$$

Kontorovich-Lebedev Transform: It is convenient to introduce another new function

$$\psi = \Phi + \frac{1}{2\pi} K_0(pr_0/c) \tag{56}$$

which goes to zero as $r \to 0$. After multiplying by r^2, the equation for ψ is

$$\left\{ r^2 \frac{\partial^2}{\partial r^2} + r \frac{\partial}{\partial r} + \frac{\partial^2}{\partial \theta^2} - \frac{p^2 r^2}{c^2} \right\} \psi(r,\theta,r_0,\theta_0,p) = - \frac{p^2 r^2}{c^2} \frac{K_0(pr_0/c)}{2\pi} \tag{57}$$

and the boundary conditions are[8]

$$\psi(r,\theta,r_0,\theta_0,p) = \frac{1}{2\pi} [K_0(pr_0/c) - K_0(pR/c)] \quad \text{at} \quad \theta = 0 \quad \text{and} \quad \theta = 2\pi. \tag{58}$$

These equations are now transformed by the Kontorovich-Lebedev transform with respect to r. The transform of ψ is

$$\Psi(\nu,\theta,r_0,\theta_0,p) = \int_0^\infty \psi(r,\theta,r_0,\theta_0,p) K_\nu(r) \frac{dr}{r}, \tag{59}$$

and (57) and (58) transform to

$$\left\{ \frac{d^2}{d\theta^2} + \nu^2 \right\} \Psi = - \frac{\nu K_0(pr_0/c)}{4 \sin(\pi\nu/2)}, \tag{60}$$

$$\Psi = \frac{K_\nu(pr_0/c) \cos \nu(\pi - \theta_0)}{2\nu \sin(\pi\nu)} - \frac{K_0(pr_0/c)}{4\nu \sin(\pi\nu/2)} \quad \text{at} \quad \theta = 0 \quad \text{and} \quad \theta = 2\pi. \tag{61}$$

The solution to these last two equations is

$$\Psi(\nu,\theta,r_0,\theta_0,p) = K_\nu(pr_0/c) \frac{\cos(\pi - \theta_0)\nu \cos(\pi - \theta)\nu}{\nu \sin 2\pi\nu} - \frac{K_0(pr_0/c)}{4\nu \sin(\pi\nu/2)}. \tag{62}$$

<u>The Solution</u>: The function ψ is obtained by evaluating the inversion integral

$$\psi(r,\theta,r_0,\theta_0,p) = \frac{1}{\pi i}\int_{-i\infty}^{i\infty} \Psi(\nu,\theta,r_0,\theta_0,p)\, I_\nu(pr/c)\, \nu\, d\nu, \tag{63}$$

and subsequently we must invert the Laplace transform to get ϕ and g. The technical details of this inversion, which are given in Turner's paper, are quite complicated; we merely quote the result here, which is

$$g = u_1 + u_2 + u_3, \tag{64}$$

$$u_1 = \begin{cases} 0, & t < R/c \\ \dfrac{1}{4\pi\sqrt{t^2-R^2/c^2}}, & t > R/c \end{cases} \tag{65}$$

$$u_2 = \begin{cases} 0, & t < R_1/c \\ \dfrac{-1}{4\pi\sqrt{t^2-R_1^2/c^2}}, & t > R_1/c \end{cases} \tag{66}$$

$$u_3 = \begin{cases} \dfrac{c}{2\pi\sqrt{rr_0}} \displaystyle\sum_{n=0}^{\infty} P_n\left[\frac{r^2+r_0^2-c^2t^2}{2rr_0}\right] \sin(n+\tfrac{1}{2})\theta \sin(n+\tfrac{1}{2})\theta_0, \\ \qquad\qquad |r-r_0| < ct < r+r_0 \\ 0, \quad \text{otherwise} \end{cases} \tag{67}$$

where $R_1 = [r^2+r_0^2-2rr_0\cos(\theta+\theta_0)]^{1/2}$ is the distance to the image of r_0,θ_0 in the plane $y = 0$. For a further discussion of this solution, the reader should consult Turner's original paper.

Problems

1. By setting $p(x) = -1$, $q(x) = 0$, and $r(x) = 1$ over the interval $a \leq x < \infty$, with the boundary conditions $g = 0$ at $x = a$, g finite as $x \to \infty$, derive the transform pair

$$F(k) = \int_a^\infty \{\sin(kx)\cos(ka) - \cos(kx)\sin(ka)\} f(x)\,dx,$$

$$f(x) = \int_0^\infty \{\sin(kx)\cos(ka) - \cos(kx)\sin(ka)\} F(k)\,dk.$$

Consider the limit $a \to 0$.

2. Repeat Problem 1 with $g = 0$ replaced by $dg/dx = 0$ at $x = a$.

3. Consider the Green's function obtained over the interval $0 \leq x < \infty$ by setting $p(x) = -1$, $g(x) = 0$, $r(x) = 1$, and

$$\frac{dg}{dx} + h\,g = 0, \qquad x = a.$$

Show that if $h < 0$ the resulting integral transform is

$$F(k) = \int_0^\infty \phi(k,x)\, f(x)\,dx,$$

$$f(x) = \frac{2h^2}{\pi}\int_0^\infty \frac{\phi(k,x)\,F(k)}{h^2 + k^2}\,dk,$$

where

$$\phi(k,x) = \sin(kx) - (k/h)\cos(kx),$$

and that if $h > 0$, there is an extra contribution from a pole at $\lambda = -h^2$, giving

$$f(x) = \frac{2h^2}{\pi}\int_0^\infty \frac{\phi(k,x)\,F(k)}{h^2 + k^2}\,dk + 2Ahe^{-hx},$$

$$A = \int_0^\infty e^{-hx} f(x)\,dx,$$

with the other quantities defined as before.

4. Recover the Weber transform (Section 15.5) by using Green's functions.

5. By considering the Hermite equation (see Section 20), recover the eigenfunction expansion

$$f(x) = \sum_{n=0}^{\infty} \frac{H_n(x)}{2^n\, n!\sqrt{\pi}} \int_{-\infty}^{\infty} e^{-x} H_n(x)\, f(x)\, dx.$$

6. A quadrant-shaped slab $0 \le x < \infty$, $0 \le y < \infty$, $0 \le z \le \ell$, has the face $x = 0$ held at temperature T_0, while the other faces are held at temperature zero. Show that the temperature distribution is

$$u(r,\theta,z) = \frac{8T_0}{\pi^2} \sum_{n=0}^{\infty} \frac{\sin[(2n+1)\pi z/\ell]}{2n+1}$$

$$\times \int_0^{\infty} \frac{\cosh(\pi\nu/2)\, \sinh(\theta\nu)}{\sinh\,(\pi\nu/2)} K_{i\nu}[(2n+1)\pi r/\ell]\, d\nu.$$

Using the integral representation

$$K_{i\nu}(x) = \frac{1}{\cosh(\pi\nu/2)} \int_0^{\infty} \cos(x \sinh t)\, \cos(\pi\nu)\, dt,$$

reduce the result to the simpler form

$$u = \frac{8T_0}{\pi^2} \sin 2\theta \sum_{n=0}^{\infty} \frac{\sinh[(2n+1)\pi z/\ell]}{2n+1}$$

$$\times \int_0^{\infty} \frac{\cos[(2n+1)\pi r.\, \sinh(t/\ell)]}{\cosh 2t + \cos 2\theta}\, dt.$$

7. A point charge q is placed near the edge of a conductor of rectangular shape held at zero potential (see Figure 4). Find expressions for the potential and the density of charge induced on the boundary.

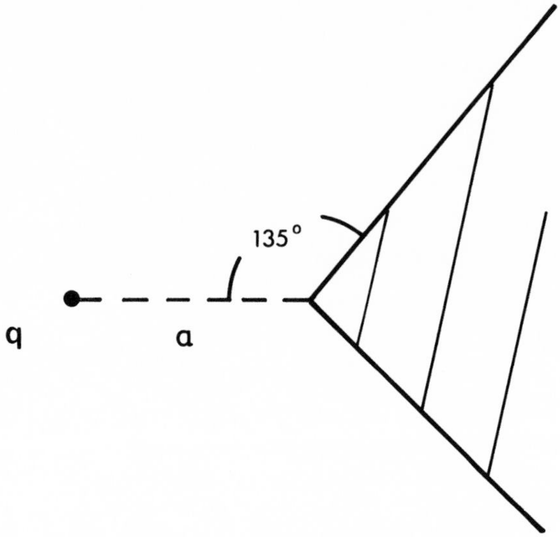

Figure 17.4

8. A line source of current, $J = J_0 \exp(i\omega t)$, is placed parallel to the edge of a thin conducting sheet $0 \le x < \infty$, $-\infty < y < \infty$, in the plane of the sheet at a distance a from the edge. Show that the density of current induced in the sheet is

$$j = \frac{J_0}{2\pi}\left(\frac{a}{x}\right)^{1/2} \frac{e^{-ik(x+a)}}{x+a},$$

$k = \omega/c.$

9. Plane waves whose propagation is governed by the Helmholtz equation are incident on a screen in the form of a half-plane $r \ge 0$, $\theta = \alpha$, on which the boundary condition is $\partial\phi/\partial n = 0$. The incident wave is

$$\phi_{inc} = e^{i(\omega t - kx)}.$$

Show that the total field ϕ is given by

$$\phi = \frac{1}{2} e^{-ikr\cos\theta}[1 + e^{i\pi/4} \operatorname{erf}\{\sqrt{2kr}\sin(\theta/2)\}]$$

$$+ \frac{1}{2} e^{-ikr\cos(\theta-2\alpha)}[1 - e^{i\pi/4} \operatorname{erf}\{\sqrt{2kr}\sin(\theta/2-\alpha)\}].$$

Footnotes

1. For example, STAKGOLD (1968).

2. STAKGOLD (1968), Ch. 4.

3. TITCHMARSH (1953), Ch. 6.

4. If any one of these conditions is not satisfied, we have a singular problem.

5. These manipulations involve assumptions about the solution which can only be verified a posteriori. Alternatively, we could work with a suitable set of generalized functions from the outset.

6. R. D. Turner, Q. Appl. Math. (1956), 14, 63.

7. MORSE & FESHBACH (1953), p. 842.

8. The evaluation of the right-hand side (58) is discussed at length in Turner's paper.

Part IV: Special Techniques

§18. THE WIENER-HOPF TECHNIQUE[1]

The solution of boundary value problems using integral transforms is comparatively easy for certain simple regions. There are many important problems, however, where the boundary data is of such a form that although an integral transform may be sensibly taken, it does not lead directly to an explicit solution. A typical problem involves a semi-infinite boundary, and may arise in such fields as electromagnetic theory, hydrodynamics, elasticity, and others. The Wiener-Hopf technique, which gives the solution to many problems of this kind, was first developed systematically by Wiener and Hopf in 1931, although the germ of the idea is contained in earlier work by Carleman. While it is most often used in conjunction with the Fourier transform, it is a significant and natural tool for use with the Laplace and Mellin transforms also. As usual, we develop the method in relation to some illustrative problems.

18.1. The Sommerfeld Diffraction Problem

In this section we will study a problem involving the reflection and diffraction of waves in two space dimensions, commonly known as the Sommerfeld diffraction problem.[2] We commence with the wave equation

$$\frac{\partial^2\phi}{\partial t^2} = c^2\nabla^2\phi \tag{1}$$

in the unbounded region $-\infty < x < \infty$, $-\infty < y < \infty$. We will not investigate the initial value problem, but rather look for particular steady-state solutions with the time dependence $\exp(-i\Omega t)$. Then the wave equation becomes the Helmholtz equation in two dimensions, namely

$$(\nabla^2+k^2)\ \phi(x,y) = 0, \qquad k = \Omega/c. \tag{2}$$

We impose three conditions on (2) to complete the specification of the problem:

(i) We suppose that the motion is caused by a steady incident plane wave

$$\phi_{inc} = e^{-ik(x\cos\theta + y\sin\theta)}, \tag{3}$$

which represents plane waves proceeding in a direction making angle θ with the positive x-axis.

(ii) We assume that the positive x-axis is a barrier to the waves. Specifically, we impose the boundary condition $\partial\phi/\partial y = 0$ for $y = 0$, $x \geq 0$. If we introduce as the new unknown function $\psi = \phi - \phi_{inc}$, this amounts to the boundary condition

$$\psi_y(x,0) = i\,k\sin\theta\ e^{-ik(x\cos\theta + y\sin\theta)}, \quad x \geq 0. \tag{4}$$

Because of this ψ may be discontinuous across the positive x-axis. However, we must have continuity for negative x, giving the further boundary condition

$$\psi(x,0+) - \psi(x,0-) = 0, \quad x < 0. \tag{5}$$

(iii) In choosing the inversion contour, we must ensure that the resulting solution is the steady-state component of the (more complicated) initial-value problem which we ought to have solved. We saw in Section 8.4 that one way to do this is to replace Ω by $\Omega + i\delta$, where $\delta > 0$ if $\Omega > 0$; in the present case this amounts to replacing k by $k + i\varepsilon$. After the problem is solved we allow ε to become zero.

Preliminary Considerations: The Wiener-Hopf technique relies on the use of Liouville's theorem,[3] and hence on having some information about the analytic properties of the Fourier transforms involved. It is obvious that in using an integral transform to solve any problem we are making some assumptions about the unknown function. In the present case, we will need information about the analytic properties of the transform of ψ, and this comes from physical considerations. Referring to Figure 1, there are three regions in which we expect ψ to behave quite differently, which we have labeled I, II, and III. In region I, ψ should consist of the reflection of the incident plane wave plus an outgoing diffracted wave coming from the edge of the barrier. In region II, we expect ψ to be only a diffracted wave. Region III is in the 'shadow' of the barrier, and here the complete solution ϕ must be only a diffracted wave. Hence $\psi = \phi - \phi_{inc}$ consists of a diffracted component and the

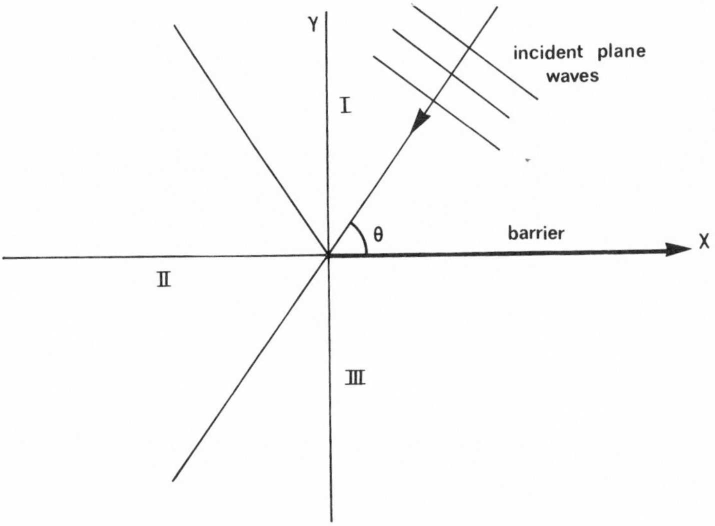

Figure 18.1

negative of ϕ_{inc}.

We are particularly concerned with the behavior of these functions on the x-axis, since this is where the boundary conditions are applied. The incident waves have amplitude $\exp(\varepsilon x \cos\theta)$, and we can estimate the amplitude of the diffracted wave by the following argument: the diffraction is caused by the edge of the barrier, and the strength of this term at a distance r from the origin must be proportional to the strength of the incident wave at the origin at time r/c previously. However, ϕ_{inc} is increasing in time as $\exp(\delta t)$, where $\delta = \varepsilon c$, so the diffracted wave must decrease as $\exp(-\varepsilon r)$ for large r. This gives us the estimates

$$\psi(x,0) \sim \begin{cases} e^{\varepsilon x \cos\theta}, & x > 0 \\ e^{\varepsilon x}, & x < 0 \end{cases}. \tag{6}$$

From (6) we expect that the Fourier transform will converge

in the strip $\varepsilon \cos \theta < \operatorname{Im}(\omega) < \varepsilon$, so we confine the inversion contour to this region.

Basic Procedure: We take the Fourier transform with respect to x of the equation for ψ, obtaining

$$\left[\frac{d^2}{dy^2} + k^2 - \omega^2\right] \Psi(\omega,y) = 0, \tag{7}$$

with the independent solutions

$$\Psi(\omega,y) = \exp(\pm\, y \sqrt{\omega^2 - k^2})\, f(\omega). \tag{8}$$

The execution of the method requires that we consider only transforms which are analytic in a strip containing the inversion contour. From physical considerations, moreover, we must choose from the solutions (8) a function which is bounded as $|y| \to \infty$, and this requires that we have knowledge of the sign of $\operatorname{Re}(\sqrt{\omega^2 - k^2})$ on the contour. It is easy to show that if we choose the branch of (8) so that $\sqrt{\omega^2 - k^2} = ik$ for $\omega = 0$, and cut the ω-plane as indicated in Figure 2, then

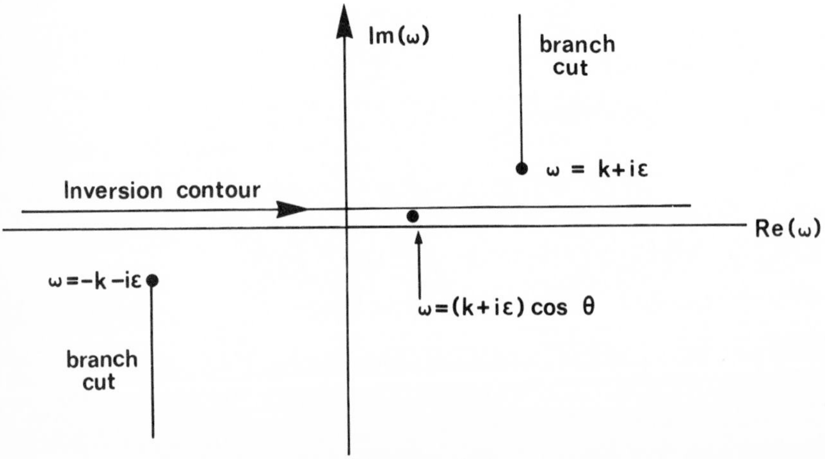

Figure 18.2

$\mathrm{Re}\,(\sqrt{\omega^2 - k^2}) \geq 0$ in the strip $-\varepsilon < \mathrm{Im}(\omega) < \varepsilon$, and suitable solutions of (7) are given by

$$\Psi(\omega,y) = \Psi(\omega,0\pm)\,\exp\,(-|y|\,\sqrt{\omega^2-k^2}), \tag{9}$$

where $\Psi(\omega,0\pm)$ are still to be determined.

We must now incorporate the boundary conditions at $y = 0$, recognizing the fact that they are different for positive and negative x, which is the cause of the complication. We therefore split the Fourier transforms of the boundary values into two parts, defining the four functions

$$\begin{aligned} A_-(\omega) &= \tfrac{1}{2}\int_{-\infty}^{0} \{\psi(x,0+) - \psi(x,0-)\}\, e^{i\omega x}\, dx \\ &= 0, \\ A_+(\omega) &= \tfrac{1}{2}\int_{0}^{\infty} \{\psi(x,0+) - \psi(x,0-)\}\, e^{i\omega x}\, dx, \\ B_-(\omega) &= \int_{-\infty}^{0} \psi_y(x,0)\, e^{i\omega x}\, dx, \\ B_+(\omega) &= \int_{0}^{\infty} \psi_y(x,0)\, e^{i\omega x}\, dx \\ &= \frac{-k\sin\theta}{\omega - k\cos\theta}, \quad \mathrm{Im}(\omega) > \varepsilon\cos\theta. \end{aligned} \tag{10}$$

The whole difficulty with the problem is that while we know the functions $A_-(\omega)$ and $B_+(\omega)$, we do not know $A_+(\omega)$ and $B_-(\omega)$, so that there is not enough explicit information to write down $\Psi(\omega,0\pm)$ immediately from the boundary conditions. We do know that ψ_y and hence Ψ_y is continuous at $y = 0$, and on differentiating (9) and setting $y = 0-$ and $y = 0+$, this gives the relation

$$\begin{aligned} 2\,\Psi_y(\omega,0) &= \Psi_y(\omega,0+) + \Psi_y(\omega,0-) \\ &= -\sqrt{\omega^2-k^2}\,[\,\Psi(\omega,0+) - \Psi(\omega,0-)], \end{aligned} \tag{11}$$

or in terms of the functions defined above

$$[\, B_+(\omega) + B_-(\omega) \,] = -\sqrt{\omega^2 - k^2}\, A_+(\omega). \tag{12}$$

Further progress can only be made by appealing to the analytic properties of the functions $A_+(\omega)$ and $B_-(\omega)$. It follows from (6) that $A_+(\omega)$ is analytic in the region $\mathrm{Im}(\omega) > \varepsilon \cos\theta$, and $B_-(\omega)$ is analytic in the region $\mathrm{Im}(\omega) < \varepsilon$. The factor $\sqrt{\omega^2 - k^2}$ has branch cuts in both of these regions, so we write it as the product $\sqrt{\omega - k}\,\sqrt{\omega + k}$, which separates the two branch points. Using this factorization, we can rearrange (12) as

$$\frac{1}{\sqrt{\omega - k}}\,[B_+(\omega) + B_-(\omega)] = -\sqrt{\omega + k}\, A_+(\omega). \tag{13}$$

There are three combinations here. $B_-(\omega)/\sqrt{\omega - k}$, which is a new unknown function, is analytic in the region $\mathrm{Im}(\omega) < \varepsilon$, and $-\sqrt{\omega + k}\, A_+(\omega)$, which is also unknown, is analytic in the region $\mathrm{Im}(\omega) > \varepsilon \cos\theta$. We therefore examine the third function, $B_+(\omega)/\sqrt{\omega - k}$, for which we have an explicit formula, and write it as the sum of two functions, each analytic in one or other of the two regions mentioned. By trivial algebraic manipulations, we can write

$$\begin{aligned} \frac{B_+(\omega)}{\sqrt{\omega - k}} &= \frac{B_+(\omega)}{\sqrt{k(\cos\theta - 1)}} + B_+(\omega)\left\{ \frac{1}{\sqrt{\omega - k}} - \frac{1}{\sqrt{k(\cos\theta - 1)}} \right\} \\ &= F_+(\omega) + F_-(\omega). \end{aligned} \tag{14}$$

We have here denoted the first term, $B_+(\omega)/\sqrt{k(\cos\theta - 1)}$, by $F_+(\omega)$; it is obviously analytic in the region $\mathrm{Im}(\omega) > \varepsilon \cos\theta$ since the denominator is independent of ω. For the second term, we have removed the only singularity in $B_+(\omega)$, a pole at $\omega = k \cos\theta$, by arranging for the terms in braces to have a simple zero there. Consequently $F_-(\omega)$ is analytic in the region $\text{Im}(\omega) < \varepsilon$. Using this decomposition we can again rearrange (13) to define a new function $E(\omega)$ by

$$E(\omega) = \frac{B_-(\omega)}{\sqrt{\omega - k}} + F_-(\omega) = -\sqrt{\omega+k}\; A_+(\omega) - F_+(\omega). \tag{15}$$

The point of this is that $E(\omega)$ is an entire function, since it is defined in two overlapping half-planes by functions which are analytic in those half-planes and which coincide in the strip of overlap, $\varepsilon \cos\theta < \text{Im}(\omega) < \varepsilon$. Therefore each function is the analytic continuation of the other and $E(\omega)$ is entire. Under rather weak assumptions (see Problem 1 for further details) we can show that $B_-(\omega)$ and $A_+(\omega)$ tend to zero for large ω in the respective regions $\text{Im}(\omega) < \varepsilon$ and $\text{Im}(\omega) > \varepsilon \cos\theta$, so that the entire function $E(\omega)$ is bounded and tends to zero for large ω. Hence by Liouville's theorem,[3] we conclude that $E(\omega) \equiv 0$. Equation (15) now gives explicit formulas for the unknown functions $A_+(\omega)$ and $B_-(\omega)$, and by working backwards through the definitions we obtain for $\Psi(\omega, y)$ the explicit form

$$\Psi(\omega, y) = \frac{-i\, \text{sgn}(y)\, \sqrt{2k}\, \cos(\theta/2)\, \exp(-|y|\sqrt{\omega^2 - k^2})}{(\omega - k\cos\theta)\sqrt{\omega - k}}. \tag{16}$$

<u>The Solution</u>: We insert (16) into the inverse Fourier transform and allow ε to become zero, which also involves moving the contour off the real axis. The solution can then be written as the integral

$$\psi(x,y) = \frac{\mathrm{sgn}(y)\ \sqrt{k/2}\ \cos(\theta/2)}{\pi i} \int_{C_1} \frac{e^{-i\omega x - |y| \sqrt{\omega^2 - k^2}}}{(\omega - k \cos \theta) \sqrt{\omega - k}}\, d\omega, \tag{17}$$

where k is real and positive and the contour C_1 is shown in Figure 3.

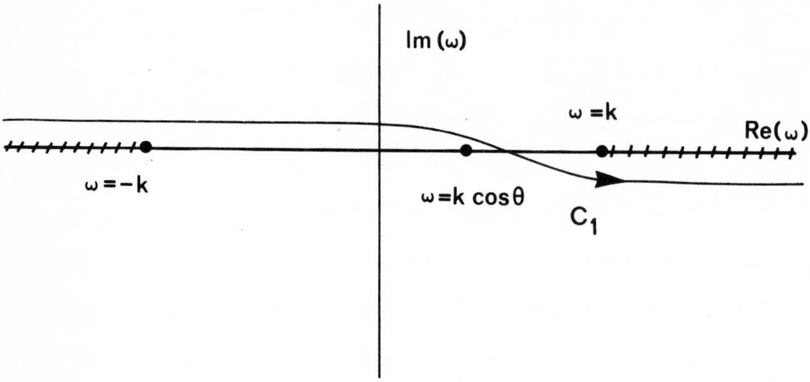

Figure 18.3

We will demonstrate that the integral (17) does indeed describe a solution having the general properties which we discussed in connection with Figure 1. For this purpose, we introduce the variables r and χ by[4]

$$\begin{aligned} x &= r \cos \chi, \\ |y| &= r \sin \chi, \end{aligned} \tag{18}$$

so that the regions of Figure 1 correspond to

$$\begin{aligned} &\text{I:} && 0 < \chi < \pi-\theta, && y > 0, \\ &\text{II:} && \pi-\theta < \chi < \pi, && y \gtrless 0, \\ &\text{III:} && 0 < \chi < \pi-\theta, && y < 0. \end{aligned} \tag{19}$$

Next we introduce the change of variable

$$\omega = -k \cos(\chi + it), \quad -\infty < t < \infty. \tag{20}$$

Elimination of t shows that the contour described by (20) is a hyperbola whose major axis is the real axis in the ω-plane. The vertex is at the point $\omega = -k \cos \chi$, and the asymptotes make an angle $\pi - \chi$ with the real axis (see Figure 4). It can be shown that the integrals along the arcs Γ_1 and Γ_2 tend to zero as their radius $R \to \infty$, hence we can deform the contour C_1 in (17) to this new contour, provided we pick up the residue at the pole at $\omega = k \cos \theta$ if the new contour is on the opposite side of the pole from the original contour. Temporarily denoting this new integral by J, we have the following results for the solution ϕ:

(i) In region I, $k \cos \chi > k \cos(\pi - \theta) = -k \cos \theta$, so that the two contours enclose the pole. Thus

$$\phi = \phi_{inc} + e^{ik(x \cos \theta - y \sin \theta)} + J. \tag{21}$$

Here the second term, which is the residue at the pole, is a reflected plane wave as expected.

(ii) In region II, the contours are on the same side of the pole, and we have

$$\phi = \phi_{inc} + J. \tag{22}$$

(iii) In region III, we again have a contribution from the pole, but because of the different sign of y, it exactly cancels ϕ_{inc}, and we obtain

$$\phi = J. \tag{23}$$

The integral J can be written by straightforward

substitution of (20) into (17) as

$$J = \frac{1}{\pi}\,\mathrm{sgn}(y)\,\sin(\theta/2)\int_{-\infty}^{\infty}\frac{e^{ikr\cosh t}\sin[(\chi+it)/2]}{\cos\theta+\cos(\chi+it)}dt. \quad (24)$$

It is possible to perform further manipulations on this integral which reduce it to the Fresnel integral, but we will not do that here. What we will note is that for large r the major contribution comes from the region $t \simeq 0$, since the

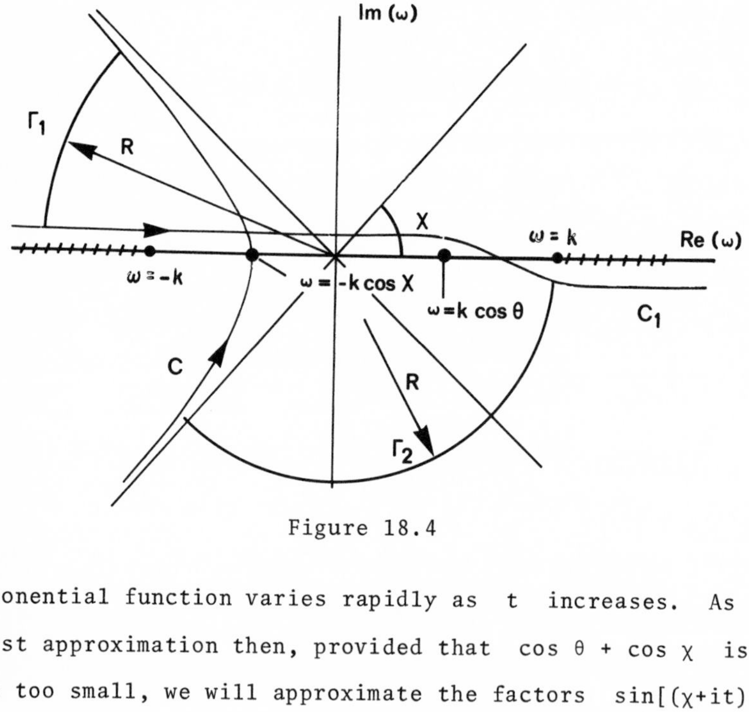

Figure 18.4

exponential function varies rapidly as t increases. As a first approximation then, provided that $\cos\theta + \cos\chi$ is not too small, we will approximate the factors $\sin[(\chi+it)/2]$ and $\cos(\chi+it)$ by $\sin(\chi/2)$ and $\cos\chi$ respectively. Comparison with (20.66) shows that we then have a Hankel function, and since we have already assumed that r is large, we replace this Hankel function by its asymptotic form (Problem 20.11) to obtain the approximation

$$J \simeq e^{i\pi/4}\ \mathrm{sgn}(y) \left\{\frac{\sin(\theta/2)\ \sin(\chi/2)}{\cos\theta + \cos\chi}\right\} \left(\frac{2}{\pi k r}\right)^{1/2} e^{ikr}. \qquad (25)$$

Thus J represents an outgoing diffracted wave whose amplitude is proportional to the factor $\sin(\theta/2)\ \sin(\chi/2)/[\cos\theta + \cos\chi]$. Analysis of J using the method of steepest descents confirms this conclusion, and also shows how it behaves asymptotically when $\cos\theta + \cos\chi \simeq 0$, where it is evident that (25) is invalid. (See also Problem 3.)

18.2. The Wiener-Hopf Procedure: Half-Plane Problems.

The typical problem which may be solved by the Wiener-Hopf technique involves the solution of equations which only give explicit information over a semi-infinite range of a variable. In a mixed boundary-value problem, for instance, we may know the boundary value of one combination of the unknown functions for $x \geq 0$, and of a different combination for $x \leq 0$. After taking the Fourier transform, and finding the general relationship between the partially specified but still unknown functions, we are then faced with the following problem: find unknown functions $\Phi_+(\omega)$ and $\Psi_-(\omega)$ satisfying

$$A(\omega)\ \Phi_+(\omega) + B(\omega)\ \Psi_-(\omega) + C(\omega) = 0, \qquad (26)$$

where this equation holds in a strip $\alpha < \mathrm{Im}(\omega) < \beta$, $\Phi_+(\omega)$ is analytic in the half-plane $\mathrm{Im}(\omega) > \alpha$, and $\Psi_-(\omega)$ is analytic in the half-plane $\mathrm{Im}(\omega) < \beta$. The functions $A(\omega)$, $B(\omega)$ and $C(\omega)$ are analytic in the strip. The fundamental step is to find a 'factorization' of $A(\omega)/B(\omega)$, i.e., to find functions $K_+(\omega)$ analytic in $\mathrm{Im}(\omega) > \alpha$ and $K_-(\omega)$

analytic in $\mathrm{Im}(\omega) < \beta$ such that

$$\frac{A(\omega)}{B(\omega)} = \frac{K_+(\omega)}{K_-(\omega)} . \tag{27}$$

Then we can rewrite (26) as

$$K_+(\omega)\ \Phi_+(\omega) + K_-(\omega)\ \Psi_-(\omega) + K_-(\omega)\ C(\omega)/B(\omega) = 0. \tag{28}$$

For some problems, this decomposition can be found by inspection; a contour integral method which we give in Section 19 may be useful with other problems, and more techniques may be found in the literature.[5] Assuming that the factorization (28) has been performed, we must now effect the further decomposition

$$K_-(\omega)\ C(\omega)/B(\omega) = F_+(\omega) + F_-(\omega), \tag{29}$$

where again $F_+(\omega)$ is analytic for $\mathrm{Im}(\omega) > \alpha$ and $F_-(\omega)$ is analytic for $\mathrm{Im}(\omega) < \beta$. Taking (28) and (29) together, we can now define an entire function $E(\omega)$ by

$$\begin{aligned} E(\omega) &= K_+(\omega)\ \Phi_+(\omega) + F_+(\omega) \\ &= -K_-(\omega)\ \Psi_-(\omega) - F_-(\omega). \end{aligned} \tag{30}$$

At first sight it may seem that we have merely defined a function which is analytic in the strip $\alpha < \mathrm{Im}(\omega) < \beta$, but in fact each of the two ways of defining $E(\omega)$ makes it analytic in a semi-infinite region, and since the two regions overlap we can use the principle of analytic continuation to show that $E(\omega)$ is entire.

Now suppose that we can show that as $|\omega| \to \infty$

$$|K_+(\omega)\ \Phi_+(\omega) + F_+(\omega)| = \mathcal{O}(|\omega|^r), \quad \mathrm{Im}(\omega) > \alpha,$$
$$|K_-(\omega)\ \Psi_-(\omega) + F_-(\omega)| = \mathcal{O}(|\omega|^s), \quad \mathrm{Im}(\omega) < \beta; \qquad (31)$$

then from Liouville's theorem we can conclude that $E(\omega)$ is a polynomial of degree no higher than the largest integer smaller than both r and s. This may reduce the problem to that of determining the coefficients of a polynomial, or it may be that the solution to the problem is not unique, in which case the coefficients play the role of arbitrary constants.

18.3. Integral and Integro-differential Equations.

The original work of Wiener and Hopf was in conjunction with the integral equation

$$\phi(x) = \frac{1}{2}\int_0^\infty E(x-t)\ \phi(t)\ dt,$$
$$E(u) = -E_1(-|u|), \qquad (32)$$

where $E_1(z)$ is the exponential integral. This equation occurs in the study of radiative processes in astrophysics, and is known as Milne's equation. More generally we may consider the problem of solving the equation

$$\lambda\phi(x) + \int_0^\infty k(x-t)\ \phi(t)\ dt = f(x), \quad x > 0. \qquad (33)$$

The first move is to extend the range of the equation to all x, which can be achieved by writing

$$\lambda\phi(x) + \int_0^\infty k(x-t)\ \phi(t)\ dt = f(x) + \psi(x), \qquad (34)$$

where

$$\begin{aligned} \phi(t) &= 0, \quad t < 0, \\ f(x) &= 0, \quad x < 0, \\ \psi(x) &= 0, \quad x > 0. \end{aligned} \tag{35}$$

The Fourier transform of (34) is

$$\lambda\Phi_+(\omega) + K(\omega)\,\Phi_+(\omega) = F_+(\omega) + \Psi_-(\omega), \tag{36}$$

where we have added subscripts to the transforms to indicate the regions of the complex ω-plane in which they should be analytic. This equation is a special case of (26), and should therefore be amenable to the Wiener-Hopf technique. The more general case of an integro-differential equation obtained from (33) by replacing λ by a linear differential operator can be analyzed in a similar manner (see below for an example).

Example 1: To illustrate, we put $k(x) = \exp(-a|x|)$, $a > 0$, and consider the homogeneous problem

$$\int_0^\infty e^{-a|x-y|}\phi(y)\,dy = \phi(x), \quad x > 0. \tag{37}$$

Equation (36) now becomes

$$\left(\frac{2a}{a^2+\omega^2}\right)\Phi_+(\omega) = \Phi_+(\omega) + \Psi_-(\omega), \tag{38}$$

which can be immediately factored as

$$\begin{aligned} \left(\frac{2a - a^2 - \omega^2}{\omega + ia}\right)\Phi_+(\omega) &= (\omega - ia)\,\Psi_-(\omega) \\ &= E(\omega). \end{aligned} \tag{39}$$

If $\phi(x)$ and $\psi(x)$ are bounded as $|x| \to \infty$, then $\Phi_+(\omega)$ and $\Psi_-(\omega)$ are of order $|\omega|^{-1}$ for large $|\omega|$ in the upper and lower half-planes respectively, and (39) defines a bounded entire function $E(\omega)$. By Liouville's theorem, $E(\omega) = A$, where A is an undetermined constant. Thus

$$\Phi_+(\omega) = A\left\{\frac{\omega + ia}{2a - a^2 - \omega^2}\right\}, \qquad \Psi_-(\omega) = \frac{A}{\omega - ia}. \tag{40}$$

Note that the inversion contour must pass above the poles of $\Phi_+(\omega)$ and below the poles of $\Psi_-(\omega)$ so as to satisfy (35). Inversion now yields

$$\phi(x) = A[\cos(bx) + (a/b)\sin(bx)], \qquad b = \sqrt{2a - a^2}. \tag{41}$$

It is instructive to reflect on the fact that there is only one arbitrary constant in this solution, while the integral operator in (37) is the Green's function for the second-order differential operator $(d^2/dx^2 - a^2)$. Acting on (37) with this operator we get the differential equation

$$\phi'' + (2a-a^2)\phi = 0, \tag{42}$$

which has two independent solutions. However, the integral equation also contains the boundary condition

$$\phi'(0) = a\int_0^\infty e^{-ay}\,\phi(y)\,dy = a\,\phi(0), \tag{43}$$

and when this is included with (42) it leads to the solution (41) again.

Example 2: We consider again an equation solved in Section 5.2, namely

$$\lambda \int_0^\infty e^{-a|x-y|}\phi(y)dy = f(x). \tag{44}$$

Here we must factor the equation

$$\left(\frac{2a\lambda}{a^2 + \omega^2}\right)\Phi_+(\omega) = F_+(\omega) + \Psi_-(\omega). \tag{45}$$

Proceeding as for (38) we obtain

$$\Phi_+(\omega) = \left(\frac{\omega^2 + a^2}{2a\lambda}\right)F_+(\omega) + A\left(\frac{\omega + ia}{2a\lambda}\right) \tag{47}$$

and

$$\phi(x) = \frac{a^2 f(x) - f''(x)}{2a\lambda} + B[a\delta(x) + \delta'(x)], \tag{48}$$

where $B = iA/2\lambda$ is an arbitrary constant. This solution involves generalized functions in two ways: explicitly in the combination $a\delta(x) + \delta'(x)$, and implicitly through the appearance of $f''(x)$, the second derivative of a function which may be discontinuous at $x = 0$. As was observed in Section 5.2, (44) implies the boundary condition

$$a\ f(0+) - f'(0+) = 0, \tag{49}$$

so that if we use the notation

$$f''_r(x) = \begin{cases} 0 &, \quad x \leq 0 \\ f''(x), & x > 0 \end{cases} \tag{50}$$

we can write [see (9.24)]

$$f''(x) = f(0)[a\delta(x) + \delta'(x)] + f''_r(x). \tag{51}$$

Hence we can replace $f''(x)$ by $f''_r(x)$ in (48) by adjusting the value of B. In particular, the choice $B = -f(0)$ is the only one for which the solution is an ordinary function rather than a generalized function. We leave it to the reader to show by direct substitution that the constant B

is indeed arbitrary if we allow the solution to be a generalized function.

Example 3: We continue to use the same integral operator to illustrate the variety of phenomena which it may contain, and consider the integro-differential equation

$$\phi''(x) + \frac{a}{2}\int_0^\infty e^{-a|x-y|}\phi(y)\,dy = 0, \quad x \geq 0. \tag{52}$$

Proceeding with the Wiener-Hopf method, we obtain from this

$$-\omega^2\,\Phi_+(\omega) - \phi'(0) + i\omega\phi(0) + \frac{a^2\,\Phi_+(\omega)}{\omega^2 + a^2} = \Psi_-(\omega), \tag{53}$$

and factorization yields

$$\begin{aligned}\frac{(a^2-a^2\omega^2-\omega^4)\Phi_+(\omega) - (\omega^2+a^2)\,\phi'(0) + i\omega\,(\omega^2+a^2)\,\phi(0)}{\omega + ia} \\ = (\omega-ia)\,\Psi_-(\omega) \\ = E(\omega).\end{aligned} \tag{54}$$

We may now examine, in retrospect, the conditions necessary for the validity of our procedure. From the fact that $\phi(x) = 0$ for $x < 0$, we obtain for $\psi(x)$ the simple expression $\psi(x) = (\text{constant})\exp(ax)$, so that its transform converges if $\text{Im}(\omega) < a$. This behavior is reflected in (54), from which we see that $\Psi_-(\omega)$ has a pole at $\omega = ia$. We need an overlapping strip to ensure that $E(\omega)$ is an entire function, so the pole in $\Psi_-(\omega)$ forces us to assume that $\phi(x)$ grows at a rate less than $\exp(ax)$ for large positive x. Applying this restriction we conclude that $E(\omega)$ is an entire function which is of order at most ω^2 in the upper half-plane, and bounded in the lower half-plane. Consequently $E(\omega) = A = \text{constant}$, and we obtain

$$\Phi_+(\omega) = \frac{A(\omega+ia) + (\omega^2+a^2)[\phi'(0) - i\omega\phi(0)]}{a^2 - a^2\omega^2 - \omega^4} . \tag{55}$$

The poles of $\Phi_+(\omega)$ occur at the zeros of the denominator, namely $2\omega^2 = -a^2 \pm \sqrt{4a^2+a^4}$. Three of these lie in the region $\text{Im}(\omega) \leq 0$, and the other in the region $\text{Im}(\omega) > a$. This latter pole clearly violates our original conditions on $\Phi_+(\omega)$. The way out of this difficulty is to choose A so that the numerator of (55) is zero at the awkward point, making $\Phi_+(\omega)$ analytic there. Thus A is not an arbitrary constant, but is determined by our assumption regarding the rate of growth of $\phi(x)$. Inversion of $\Phi_+(\omega)$ gives for $\phi(x)$ a linear combination of three exponential functions depending on two arbitrary constants, namely $\phi(0)$ and $\phi'(0)$.

Boundary Conditions: To investigate the significance of these findings, we use the fact that the integral operator in (52) is a Green's function to convert the problem into a differential equation. Acting on the original equation with the operator $d^2/dx^2 - a^2$ yields the fourth-order equation

$$\phi^{(4)}(x) - a^2\phi''(x) - a^2\phi(x) = 0, \tag{56}$$

whose solution is

$$\phi(x) = \sum_{j=1}^{\infty} c_j\, e^{r_j x}, \tag{57}$$

where the r_j are roots of $r^4 - a^2r^2 - a^2 = 0$. If we impose the condition that $\phi(x)$ grow more slowly than $\exp(ax)$, one of the exponential functions is disallowed, and we recover the solution found above, except that it appears to depend on three arbitrary constants. In fact there is a boundary condition implicit in the original integro-differential equation, namely

$$\phi^{(3)}(0) = a\,\phi''(0), \tag{58}$$

and this reduces the number of independent constants to two.

Problems

1. Show that if the function $\psi(x,0)$ of (6) has the behavior

$$\psi(x,0) \sim x^{\mu}, \qquad x \to 0,$$

where $\mu > -1/2$, then the entire function $E(\omega)$ of (15) is identically zero. Investigate the solution obtained for the Sommerfeld diffraction problem under the weaker assumption that $\mu = -1/2$.

2. By using a suitable free-space Green's function for the Helmholtz equation in a half-plane, show that the solution of the Sommerfeld diffraction problem may be written as

$$\phi = \begin{cases} e^{-ik(x\cos\theta+y\sin\theta)} + e^{-ik(x\cos\theta-y\sin\theta)} \\ \qquad - \frac{1}{2}i \int_{-\infty}^{0} H_0^{(1)}(kR)\, h(\xi)\, d\xi, & y \geq 0 \\ \frac{1}{2}i \int_{-\infty}^{0} H_0^{(1)}(kR)\, h(\xi)\, d\xi, & y \leq 0, \end{cases}$$

$$R^2 = (x-\xi)^2 + y^2,$$

where the unknown function $h(\xi)$ is determined by the integral equation

$$i \int_{-\infty}^{0} H_0^{(1)}(k|x-\xi|)\, h(\xi)\, d\xi = 2\, e^{-ikx\cos\theta}.$$

Solve these equations using the Wiener-Hopf technique.

3. Derive an asymptotic expansion for the function defined by (24) by writing $u = \cosh t$ and deforming the contour so as to employ Watson's lemma for loop integrals.

4. Show that the solution of the mixed boundary-value problem

$$(\nabla^2 + k^2)\,\phi(x,y) = 0, \quad -\infty < x < \infty, \quad y \geq 0,$$

$$\phi(x,0) = 0, \quad x > 0,$$

$$\phi_y(x,0) = g(x), \quad x < 0,$$

is given by

$$\phi(x,y) = \frac{1}{2\pi}\int_C \Phi(\omega,0)\, e^{-i\omega x - |y|\sqrt{\omega^2-k^2}}\, d\omega,$$

where

$$\Phi(\omega,0) = \frac{e^{3i\pi/4}}{\sqrt{\pi(\omega-k)}}\int_{-\infty}^{0} e^{i\omega u}\, du \int_0^\infty \xi^{-1/2}\, e^{ik\xi}\, g(u-\xi)\, d\xi.$$

5. If the boundary conditions in Problem 4 are replaced by

$$\phi(x,0) = f(x), \quad x > 0,$$

$$\phi_y(x,0) = 0\ , \quad x < 0,$$

then show that

$$\Phi(\omega,0) = \frac{e^{3i\pi/4}}{\sqrt{\pi(\omega-k)}}\int_0^\infty e^{i(\omega-k)u} du\, \frac{d}{du}\int_0^\infty \xi^{-1/2} e^{ik(u+\xi)} f(u+\xi)\, d\xi.$$

6. Solve the mixed boundary-value problem

$$(\nabla^2 - k^2)\,\phi(x,y) = 0, \quad -\infty < x < \infty, \quad -\infty < y < \infty;$$

$$\phi_y(x,0) = e^{i\beta x}, \quad x \geq 0,$$

$$\phi(x,y) \to 0, \quad x^2 + y^2 \to \infty.$$

7. Investigate the Sommerfeld diffraction problem when the boundary condition on the screen is replaced by

$$\phi(x,0\pm) = \pm\, i\delta\, \phi_y(x,0\pm), \quad x \geq 0,$$

and show that this leads to the Wiener-Hopf problem

$$\Psi'_-(\omega,0) = -(1 + i\delta\gamma)\,[\Psi'_+(\omega,0+) - \Psi'_+(\omega,0-)]$$

$$+ 2i\delta\gamma k \sin\theta/(\omega - k\sin\theta)$$

where $\psi = \phi - \phi_{inc}$, $\Psi'(\omega,y)$ is the Fourier transform of $\psi_y(x,y)$, and

$$\gamma^2 = \omega^2 - k^2.$$

8. Solve the integro-differential equation

$$\phi''(x) + \alpha^2\phi(x) + \int_0^\infty e^{-|x-y|}\phi(y)\,dy = 0, \quad x \geq 0,$$

subject to

$$\phi(0) = 1,$$

$$\phi(x) \to 0, \quad x \to \infty.$$

9. Find the Green's function for the equations

$$\{\partial^2/\partial x^2 + \alpha^2\}\, G(x,x')$$

$$+ \int_0^\infty e^{-|x-y|}\, G(y,x')\,dy = \delta(x-x'), \quad x > 0,$$

$$G(0,x') = 0,$$

$$G(x,x') \to 0, \quad x \to \infty,$$

and show that it is related to the solution of Problem 8 by[6]

$$G(x,x') = -\int_0^{x'} \phi(s)\,\phi(s+x-x')\,ds$$
$$= \int_0^{\infty} \phi(s+x)\,\phi(s+x')\,ds + G_0(x-x'),$$

where G_0 is the Green's function for the infinite problem, i.e.,

$$\{\partial^2/\partial x^2 + \alpha^2\}\,G_0(x-x')$$
$$+ \int_{-\infty}^{\infty} e^{-|x-y|}\,G_0(y-x')\,dy = \delta(x-x'),\quad -\infty < x < \infty,$$
$$G_0(x-x') \to 0,\quad |x| \to \infty.$$

10. Derive the factorization

$$\pi z \coth(\pi z) = K_+(z)\,K_-(z),$$

$$K_+(z) = \pi^{1/2}\,(-iz)!/(-iz-\tfrac{1}{2})!,$$

$$K_-(z) = K_+(-z).$$

11. If $\phi(x,y)$ is determined by

$$\nabla^2\phi - \phi_x = 0,$$
$$\phi(x,y) \to 0,\quad x^2 + y^2 \to \infty,$$

$$\phi(x,0) = e^{-ax},\quad x \geq 0,$$

then show that

$$\Phi(\omega,y) = \frac{\sqrt{1+a}\;e^{-i\pi/4}}{(a-i\omega)\,\sqrt{\omega-i}}\;e^{-|y|\sqrt{\omega^2-i\omega}}$$

while the inversion contour lies in the strip $0 < \text{Im}(\omega) < 1$.

12. Consider the infinite strip $-b \leq y \leq +b$, $-\infty < x < \infty$, along which a wave $\phi_i = \exp(ikx)$ is incident from $x = -\infty$. The total wave field ϕ, which consists of the incident traveling wave and waves diffracted by a semi-infinite strip at $y = 0$, $x \geq 0$, satisfies the equations

$$\{\nabla^2 + k^2\}\, \phi(x,y) = 0,$$

$$\phi_y(x,\pm b) = 0, \quad -\infty < x < \infty,$$

$$\phi(x,0) = 0, \quad 0 \leq x < \infty.$$

Find explicit expressions for $\phi(x,y)$.

13. Solve the previous problem with the boundary condition on the strip replaced by $\phi_y(x,0) = 0$, $0 \leq x < \infty$.

Footnotes

1. The Wiener-Hopf technique is mentioned in a number of books; for a comprehensive review of the method see NOBLE (1958).

2. This problem may also be solved using the Kontorovich-Lebedev transform: see Sections 17.4-5.

3. One form of Liouville's theorem is as follows: if $E(z)$ is an entire function, and if

$$E(z) \sim \mathcal{O}(z^s), \quad |z| \to \infty,$$

then $E(z)$ is a polynomial of degree n, where n is an integer less than or equal to $\mathrm{Re}(s)$. See AHLFORS (1966), pp. 122-123.

4. See Section 10.4 for another example of this transformation, which is also discussed at some length in NOBLE (1958), p. 31ff.

5. See NOBLE (1958), p. 93ff. for a list and some references. In addition to problems in one complex variable, Kraut has considered mixed boundary value problems which may be resolved using a Wiener-Hopf type of decomposition in two complex variables. See E. A. Kraut, Proc. Amer. Math. Soc. (1969), 23, 24, and further references given there.

6. This relationship holds for a wide class of kernels, of which $\exp(-|x-y|)$ is the simplest. See G. A. Baraff, J. Math. Phys. (1970), 11, 1938.

§19. METHODS BASED ON CAUCHY INTEGRALS

19.1. Wiener-Hopf Decomposition by Contour Integration

The major difficulty in using the Wiener-Hopf technique is the problem of constructing a suitable factorization. We consider here a method based on contour integration which leads by natural extensions to the use of Cauchy integrals in the solution of mixed boundary-value problems. Suppose then that the function $\Phi(z)$ is analytic in the strip $\alpha < \mathrm{Im}(z) < \beta$, and that we wish to find functions $\Phi_+(z)$ and $\Phi_-(z)$ analytic in the half planes $\mathrm{Im}(z) > \alpha$ and $\mathrm{Im}(z) < \beta$ respectively, such that

$$\Phi(z) = \Phi_+(z) + \Phi_-(z), \quad \alpha < \mathrm{Im}(z) < \beta. \tag{1}$$

First choose z to lie inside the contour shown in Figure 1; then Cauchy's integral formula gives

$$\Phi(z) = \frac{1}{2\pi i} \int \frac{\Phi(\zeta)}{\zeta - z}\, d\zeta. \tag{2}$$

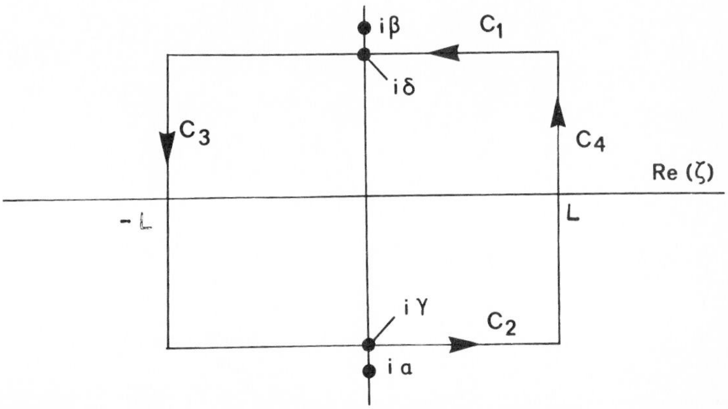

Figure 19.1

We consider only functions for which the integrals along C_3 and C_4 become zero as $L \to \infty$. Taking this limit, we obtain the result

$$\Phi(z) = \frac{1}{2\pi i}\int_{i\gamma-\infty}^{i\gamma+\infty} \frac{\Phi(\zeta)}{\zeta-z}d\zeta - \frac{1}{2\pi i}\int_{i\delta-\infty}^{i\delta+\infty} \frac{\Phi(\zeta)}{\zeta-z}\,d\zeta. \tag{3}$$

The first integral defines a function which is analytic for $Im(z) > \alpha$ [since for any such z we can always choose γ by $\alpha < \gamma < Re(z)$], and the second integral defines a function which is analytic for $Im(z) < \beta$, so that (3) gives the desired decomposition.

Equation (3) gives an additive splitting; however, the same trick may suffice to give a multiplicative splitting, that is, a factorization of the type

$$\Phi(z) = \frac{K_+(z)}{K_-(z)}\ . \tag{4}$$

If $K_+(z) \neq 0$ for $Im(z) > \alpha$ and $K_-(z) \neq 0$ for $Im(z) < \beta$, then we can reduce this problem to the previous one by either of two methods. The first is to take logarithms, writing

$$\ell n\ \Phi(z) = \ell n\ K_+(z) - \ell n\ K_-(z). \tag{5}$$

Alternatively, we may differentiate to write

$$\frac{\Phi'(z)}{\Phi(z)} = \frac{K_+'(z)}{K_+(z)} - \frac{K_-'(z)}{K_-(z)}\ . \tag{6}$$

Both approaches depend on $K_+(z)$ and $K_-(z)$ not having zeros, and both have been used extensively in the literature.[1]

Milne's Equation: On taking the Fourier transform of equation (18.32) we are led to the problem of factoring

$$\left(\frac{\arctan\,\omega}{\omega} - 1\right)\Phi_+(\omega) = \Psi_-(\omega). \tag{7}$$

The function $(\omega^{-1} \arctan \omega)-1$ has branch points at $\omega = \pm i$, so the strip in which we must operate is $-1 < \mathrm{Im}(\omega) < 1$. It is not difficult to show that in this strip the only zero of this function is a double root at $\omega = 0$. We must remove this zero in order to apply a contour integral method; therefore we consider the problem of finding the factorization

$$\left(\frac{\omega^2+1}{\omega^2}\right)\left(\frac{\arctan \omega}{\omega} - 1\right) = \frac{G_+(\omega)}{G_-(\omega)} . \tag{8}$$

This is suitable for use in the contour integral method since the logarithm of the function tends to zero as $|\omega| \to \infty$ in the strip $-1 < \mathrm{Im}(\omega) < 1$, and the double zero has been removed. Assuming that $\ln G_+(\omega)$ and $\ln G_-(\omega)$ have been found by contour integration, (7) now admits the factorization

$$\left(\frac{\omega}{\omega+i}\right) G_+(\omega)\ \Phi_+(\omega) = \left(\frac{\omega-i}{\omega}\right) G_-(\omega)\ \Psi_-(\omega), \tag{9}$$

and the functions $\Phi_+(\omega)$ and $\Psi_-(\omega)$ may be found by the usual arguments. For details and numerical computations the reader is referred to the literature.[2]

19.2. Cauchy Integrals[3]

Equation (3) leads us to consider functions of the type

$$F(z) = \frac{1}{2\pi i} \int_C \frac{f(\zeta)}{\zeta - z}\, d\zeta, \tag{10}$$

which are usually known as Cauchy integrals. We restrict ourselves to contours C which are piecewise smooth, and functions $f(\zeta)$ which satisfy the Hölder condition[4] wherever the contour is smooth. Restrictions on $f(\zeta)$ near a corner or an end of C will be specified below. It is a standard result that, with the restrictions applied to (10), the integral defines a function $F(z)$ which is analytic in the entire complex plane excluding C. Of particular interest is the value which $F(z)$ approaches as z approaches an arbitrary point of C. Since the definition of C includes a specification of the direction in which the contour is to be traversed, we may define the meaning of the left-hand and right-hand side of the contour. It is conventional to refer to the left-hand side as positive and the right-hand side as negative, and to define the functions

$$F_{\pm}(\zeta) = \lim_{z\to\zeta} F(z), \tag{11}$$

where the point ζ on C is approached from the side which is indicated by the suffix. We also define the principal value by

$$F_p(\zeta) = \lim_{\varepsilon\to 0} \frac{1}{2\pi i} \int_{C'} \frac{f(u)\,du}{u-\zeta}, \tag{12}$$

where C' is that part of C satisfying $|u-\zeta| \geq \varepsilon$.

<u>Plemelj Formulas</u>: We now derive some very important results in the case that $f(\zeta)$ is an analytic function. These results are in fact true for functions $f(\zeta)$ which satisfy a Hölder condition, although the proofs are more intricate in that case.[5] Consider first a point ζ on C which is not an end point or a corner; then, by some elementary considerations from the theory of contour integrals, we have

$$F_+(\zeta) = \frac{1}{2\pi i}\int_{C'+C_1} \frac{f(u)}{u-\zeta}\,du,$$

$$F_-(\zeta) = \frac{1}{2\pi i}\int_{C'+C_2} \frac{f(u)}{u-\zeta}\,du, \tag{13}$$

$$f(\zeta) = \frac{-1}{\pi i}\int_{C_1} \frac{f(u)}{u-\zeta}\,du$$

$$= \frac{1}{\pi i}\int_{C_2} \frac{f(u)}{u-\zeta}\,du,$$

where the various contours are shown in Figure 2. On taking the limit $\varepsilon \to 0$, we recover the Plemelj formulas

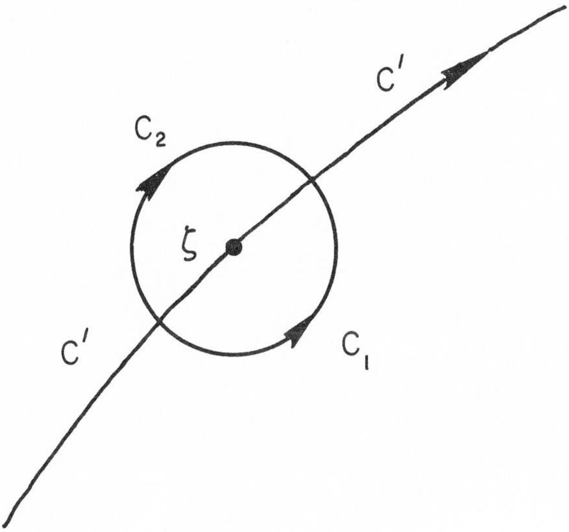

Figure 19.2

$$F_+(\zeta) = F_p(\zeta) + \frac{1}{2} f(\zeta),$$

$$F_-(\zeta) = F_p(\zeta) - \frac{1}{2} f(\zeta). \tag{14}$$

Near a corner of included angle α (see Figure 3), we have

$$f(\zeta) = -\frac{1}{\alpha i}\int_{C_1}\frac{f(u)}{u-\zeta}\,du = \frac{1}{(2\pi-\alpha)i}\int_{C_2}\frac{f(u)}{u-\zeta}\,du, \tag{15}$$

and hence the Plemelj formulas are changed to

$$F_+(\zeta) = F_p(\zeta) + \left(1 - \frac{\alpha}{2\pi}\right) f(\zeta),$$
$$F_-(\zeta) = F_p(\zeta) - \frac{\alpha}{2\pi} f(\zeta). \tag{16}$$

In either case, subtraction of one result from the other leads to the important common result that

$$F_+(\zeta) - F_-(\zeta) = f(\zeta), \tag{17}$$

so that the Cauchy integral gives a construction for an analytic function having a prescribed discontinuity across C.

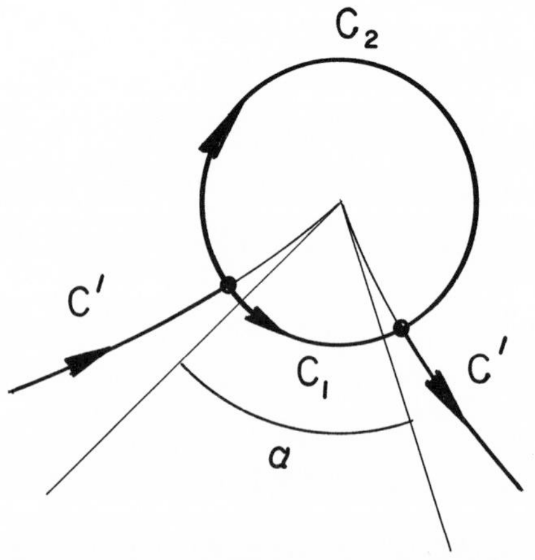

Figure 19.3

Behavior Near End Points: For simplicity we will consider a smooth arc C with end points a and b. We assume that $f(\zeta)$ satisfies a Hölder condition at every interior point of C, and examine the behavior of $F(z)$ near $z = a$. If $f(\zeta)$ satisfies the Hölder condition at $z = a$, then we can write (10) as

$$F(z) = \frac{1}{2\pi i}\int_a^b \frac{f(\zeta)-f(a)+f(a)}{\zeta - z}\, d\zeta$$
$$= \frac{1}{2\pi i} f(a)\, \ln\left(\frac{z-b}{z-a}\right) + \Omega(z), \tag{18}$$

where

$$\Omega(z) = \frac{1}{2\pi i}\int_a^b \frac{f(\zeta)-f(a)}{\zeta - z}\, d\zeta. \tag{19}$$

Since the integral defining $\Omega(z)$ converges when we set $z = a$, the singularity in $F(z)$ near $z = a$ is of the form

$$F(z) \sim \frac{-f(a)}{2\pi i} \ln (z-a). \tag{20}$$

Similarly, if $f(\zeta)$ satisfies the Hölder condition at $\zeta = b$, $F(z)$ has a singularity of the form

$$F(z) \sim \frac{f(b)}{2\pi i} \ln (z-b) \tag{21}$$

near that end point.

More generally, we may consider functions $f(\zeta)$ which are not finite at the end points, but for which there are constants γ_a and γ_b satisfying the inequalities

$$0 < \mathrm{Re}(\gamma_a) < 1,$$
$$0 < \mathrm{Re}(\gamma_b) < 1, \tag{22}$$

so that the functions $(\zeta-a)^{\gamma_a} f(\zeta)$ and $(\zeta-b)^{\gamma_b} f(\zeta)$ satisfy the Hölder condition at the end points a and b respectively. Then by a subtraction procedure similar to that employed in equations[6] (18) and (19), we may show that the singularities of $F(z)$ near the end points take the form

$$F(z) \sim \frac{e^{i\pi\gamma_a}\,\phi(a)}{2i\,\sin\,(\pi\gamma_a)\,(z-a)^{\gamma_a}}, \tag{23}$$

$$\phi(a) = \lim_{\zeta\to a}\,(\zeta-a)^{\gamma_a}\,f(\zeta)$$

near $z = a$, and

$$F(z) \sim -\frac{e^{i\pi\gamma_b}\,\phi(b)}{2i\,\sin\,(\pi\gamma_b)\,(z-b)^{\gamma_b}}, \tag{24}$$

$$\phi(b) = \lim_{\zeta\to b}\,(\zeta-b)^{\gamma_b}\,f(\zeta)$$

near $z = b$.

19.3. The Discontinuity Theorem

Cauchy integrals represent analytic functions which are discontinuous across an arc (or collection of arcs) C. The discontinuity theorem states that if C is a contour and $f(\zeta)$ a function defined on C, both satisfying the restrictions introduced in Section 19.2, then the only functions $\Phi(z)$ satisfying the three conditions

(i) $\Phi(z)$ analytic in the entire complex plane excluding C,

(ii) $\Phi_+(\zeta) - \Phi_-(\zeta) = f(\zeta)$, (25)

(iii) for any end point or corner,

$$|\Phi(z)| < A|z-c|^{\beta}, \quad \beta > -1 \tag{26}$$

are of the form

$$\Phi(z) = \frac{1}{2\pi i}\int_C \frac{f(\zeta)\,d\zeta}{\zeta - z} + E(z), \qquad (27)$$

where $E(z)$ is an entire function. The proof consists of using the properties of Cauchy integrals near end points (remembering that a corner is the adjunction of two end points) to show that the function $\Phi(z) - F(z)$ has no singularities, and is therefore entire. Details may be found in the literature; Muskhelishvili's book is a particularly comprehensive reference.

19.4. The Riemann-Hilbert Problem

The discontinuity theorem allows us to solve problems concerning analytic functions which have their properties specified on a contour C. The Riemann-Hilbert problem is to find a function $\Phi(z)$, analytic in the entire plane excluding C, which satisfies the condition

$$\Phi_+(\zeta) = g(\zeta)\,\Phi_-(\zeta) + f(\zeta) \qquad (28)$$

on C. To solve this problem, we first set $f(\zeta)$ to zero, and find functions $K_+(z)$ and $K_-(z)$, with the property

$$K_+(\zeta) = g(\zeta)\,K_-(\zeta), \quad \zeta \text{ on } C. \qquad (29)$$

Leaving aside the construction of such a pair of functions until later, we see that (29) reduces the Riemann-Hilbert problem to

$$\frac{\Phi_+(\zeta)}{K_+(\zeta)} - \frac{\Phi_-(\zeta)}{K_-(\zeta)} = \frac{f(\zeta)}{K_+(\zeta)}, \quad \zeta \text{ on } C. \qquad (30)$$

The discontinuity theorem gives a solution immediately, viz.

$$\frac{\Phi(z)}{K(z)} = \frac{1}{2\pi i}\int_C \frac{f(\zeta)}{(\zeta-z)\,K_+(\zeta)}\,d\zeta + E(z). \tag{31}$$

Construction of K_+ and K_-: Taking the logarithm of equation (29) we obtain

$$\ln K_+(\zeta) - \ln K_-(\zeta) = \ln g(\zeta), \tag{32}$$

and the discontinuity theorem will again provide a solution, provided we can handle any singularities near end points in such a way that the function $f(\zeta)/K_+(\zeta)$ satisfies the necessary restrictions for the application of the discontinuity theorem to equation (30). For applications there are two important cases which we now consider.

(i) C is not a closed contour, and does not cross itself. Then we write

$$\ln L(z) = \frac{1}{2\pi i}\int_C \frac{\ln g(\zeta)}{\zeta - z}\,d\zeta, \tag{33}$$

and using the properties of Cauchy integrals near an end point, we find that near the beginning $(z=a)$ of C the function $L(z)$ behaves like

$$\begin{aligned} L(z) &\sim (z-a)^{\gamma_a}, \\ \gamma_a &= -\frac{\ln g(a)}{2\pi i}. \end{aligned} \tag{34}$$

Now choose an integer k_a so that $0 \leq \mathrm{Re}(\gamma_a + k_a) < 1$, and multiply $L(z)$ by the factor $(z-a)^{k_a}$. The singularity of this new function at $z = a$ is sufficiently weak to allow its use in equation (31). The other end point is treated similarly, so that the factorization (29) is given by

$$K(z) = (z-a)^{k_a} (z-b)^{k_b} L(z). \tag{35}$$

Because the constants k_a and k_b are integers, $K(z)$ satisfies equation (32) whenever $L(z)$ does.

(ii) C is a closed contour which does not cross itself. Then the function $\ell n\, g(\zeta)$ will increase by $2\pi in$ in one circuit of the contour, so we introduce a new function $g_0(\zeta)$ by

$$g_0(\zeta) = (\zeta-a)^{-n} g(\zeta), \tag{36}$$

where a is an interior point of the region bounded by C. It is easy to check that the desired decomposition is given in terms of $g_0(\zeta)$ by

$$\begin{aligned} \ell n\, K_+(z) &= \frac{1}{2\pi i} \int_C \frac{\ell n\, g_0(\zeta)}{\zeta - z}\, d\zeta, \\ \ell n\, K_-(z) &= n\, \ell n(z-a) + \frac{1}{2\pi i} \int_C \frac{\ell n\, g_0(\zeta)}{\zeta - z}\, d\zeta. \end{aligned} \tag{37}$$

19.5. <u>Simple Applications</u>

(i) If C is a smooth arc joining the points $z = a$ and $z = b$, with the Riemann-Hilbert problem given as

$$\Phi_+(\zeta) + \Phi_-(\zeta) = f(\zeta), \tag{38}$$

then we first evaluate the integral

$$\ell n[L(z)] = \frac{1}{2\pi i} \int_a^b \frac{i\pi}{\zeta - z}\, d\zeta, \tag{39}$$

giving

$$L(z) = (z-b)^{1/2} (z-a)^{-1/2}. \tag{40}$$

The behavior of $L(z)$ at $z = b$ is inappropriate for the

next step; dividing by the factor $(z-b)$ yields the correct factorization

$$K(z) = (z-a)^{-1/2} (z-b)^{-1/2}. \tag{41}$$

Thus the solution of (38) is

$$\frac{\Phi(z)}{K(z)} = \frac{1}{2\pi i}\int_a^b \frac{f(\zeta)}{K_+(\zeta)(\zeta-z)}\, d\zeta + E(z). \tag{42}$$

For example, if $a = -1$, $b = 1$, and $f(\zeta) = 1$, then

$$\Phi(z) = \frac{1}{2} - \frac{z}{2\sqrt{z^2-1}} + \frac{E(z)}{\sqrt{z^2-1}}. \tag{43}$$

(ii) The Wiener-Hopf problem. The Riemann-Hilbert problem given in equation (18.26) is

$$A(\omega)\, \Phi_+(\omega) + B(\omega)\, \Phi_-(\omega) + C(\omega) = 0, \qquad \alpha < \mathrm{Im}(\omega) < \beta. \tag{44}$$

Now we no longer need a strip of overlap; indeed on choosing a constant γ such that $\alpha < \gamma < \beta$, we can take the contour C as the line[7] $\mathrm{Im}(z) = \gamma$. Application of the method of Cauchy integrals leads immediately to the formulas of Section 19.1.

19.6. Problems in Linear Transport Theory[8]

Fourier transform methods have been extensively applied to the linear transport equation[9]

$$\left(1 + \frac{\partial}{\partial t} + \underset{\sim}{v}\cdot\underset{\sim}{\nabla}\right) \psi(\underset{\sim}{r},\underset{\sim}{v},t) = c\int \sigma(\underset{\sim}{v},\underset{\sim}{v}')\, \psi(\underset{\sim}{r},\underset{\sim}{v}',t)\,d^3\underset{\sim}{v}' + q(\underset{\sim}{r},\underset{\sim}{v},t). \tag{45}$$

Here ψ, the phase-space density, is to be determined in a region V with the boundary condition on the surface S

given by

$$\psi(\underset{\sim}{r},\underset{\sim}{v},t) = \psi_s(\underset{\sim}{r},\underset{\sim}{v},t), \quad \underset{\sim}{v} \text{ inwards.} \tag{46}$$

The scattering kernel σ and source function q are, like ψ_s, presumed to be given. We shall consider in this section half-plane problems for which V is the region $z \geq 0$, and show that whenever the function σ has the separable form

$$\sigma(\underset{\sim}{v},\underset{\sim}{v}') = f(\underset{\sim}{v})\, g(\underset{\sim}{v}'), \tag{47}$$

the Fourier transform method may be used in conjunction with Cauchy integrals to give a general method of attack for the analytic solution of the problem. We confine our attention to time independent solutions and drop the reference to t in what follows.

The collisionless Green's function (see Problem 11.16 for the various properties used here)

$$G(\underset{\sim}{r}-\underset{\sim}{r}',\underset{\sim}{v}) = \frac{1}{(2\pi)^3}\int \frac{e^{-i\underset{\sim}{k}\cdot(\underset{\sim}{r}-\underset{\sim}{r}')}}{1 + i\underset{\sim}{k}\cdot\underset{\sim}{v}}\, d^3\underset{\sim}{k} \tag{48}$$

can be used to rewrite the basic equations as an integral equation, namely[10]

$$\begin{aligned}\psi(\underset{\sim}{r},\underset{\sim}{v}) = &\int_{z\geq 0} G(\underset{\sim}{r}'-\underset{\sim}{r},\underset{\sim}{v})[c\, p(\underset{\sim}{r}',\underset{\sim}{v}) + q(\underset{\sim}{r}',\underset{\sim}{v})]\, d^3\underset{\sim}{r}' \\ &+ v_z \int_{z=0} G(\underset{\sim}{r}_s-\underset{\sim}{r},\underset{\sim}{v})\, \psi_s(\underset{\sim}{r}_s,\underset{\sim}{v})\, d^2\underset{\sim}{r}_s,\end{aligned} \tag{49}$$

where

$$p(\underset{\sim}{r},\underset{\sim}{v}) = \int \sigma(\underset{\sim}{v},\underset{\sim}{v}')\, \psi(\underset{\sim}{r},\underset{\sim}{v}')\, d^3\underset{\sim}{v}', \tag{50}$$

and the subscript s refers to the restriction to the boundary $z = 0$. We now take the three-dimensional Fourier transform so as to make use of the convolution property. In so doing, we shall have to add to the left-hand side of the equation an unknown function which is given by the right-hand side for $z < 0$. Without any fear of confusion we denote this function by $\psi(\underset{\sim}{r},\underset{\sim}{v})$, introducing the conventions

$$\Psi_{+}(\underset{\sim}{k},\underset{\sim}{v}) = \int_{z\geq 0} e^{i\underset{\sim}{k}\cdot\underset{\sim}{r}}\psi(\underset{\sim}{r},\underset{\sim}{v})\, d^3\underset{\sim}{r},$$

$$\Psi_{-}(\underset{\sim}{k},\underset{\sim}{v}) = \int_{z\leq 0} e^{i\underset{\sim}{k}\cdot\underset{\sim}{r}}\psi(\underset{\sim}{r},\underset{\sim}{v})\, d^3\underset{\sim}{r}, \tag{51}$$

$$\Psi(\underset{\sim}{k},\underset{\sim}{v}) = \Psi_{+}(\underset{\sim}{k},\underset{\sim}{v}) + \Psi_{-}(\underset{\sim}{k},\underset{\sim}{v}).$$

After these manipulations, we obtain the integral equation

$$\Psi(\underset{\sim}{k},\underset{\sim}{v}) = \frac{c\, P_{+}(\underset{\sim}{k},\underset{\sim}{v}) + Q_{+}(\underset{\sim}{k},\underset{\sim}{v})}{1 - i\underset{\sim}{k}\cdot\underset{\sim}{v}} + \frac{v_z}{1 - i\underset{\sim}{k}\cdot\underset{\sim}{v}} \int_{z=0} e^{i\underset{\sim}{k}\cdot\underset{\sim}{r}}\psi_s(\underset{\sim}{r},\underset{\sim}{v})\, d^2\underset{\sim}{r}. \tag{52}$$

<u>Solution for Separable Kernels</u>: When σ has the form (47), we can introduce the integrated density

$$\rho(\underset{\sim}{r}) = \int g(\underset{\sim}{v})\, \psi(\underset{\sim}{r},\underset{\sim}{v})\, d^3\underset{\sim}{v}, \tag{53}$$

with Fourier transform $\tilde{\rho}(\underset{\sim}{k})$, and by multiplying by g and integrating over $\underset{\sim}{v}$ immediately get

$$\Lambda(\underset{\sim}{k})\, \tilde{\rho}_{+}(\underset{\sim}{k}) + \tilde{\rho}_{-}(\underset{\sim}{k}) = B(\underset{\sim}{k}), \tag{54}$$

where

$$\Lambda(\underset{\sim}{k}) = 1 - c\int \frac{f(\underset{\sim}{v})\, g(\underset{\sim}{v})}{1 - i\underset{\sim}{k}\cdot\underset{\sim}{v}}\, d^3\underset{\sim}{v} \tag{55}$$

is the dispersion function, and

$$B(\underset{\sim}{k}) = \int \frac{g(\underset{\sim}{v})}{1 - i\underset{\sim}{k}\cdot\underset{\sim}{v}} \left[Q_{+}(\underset{\sim}{k},\underset{\sim}{v}) + v_z \int_{z=0} e^{i\underset{\sim}{k}\cdot\underset{\sim}{r}}\psi_s(\underset{\sim}{r},\underset{\sim}{v})\, d^2\underset{\sim}{r}\right] d^3\underset{\sim}{v} \tag{56}$$

represents the known contributions from the sources q and the boundary function ψ_s.

For convenience we denote the z-component (normal component) of $\underset{\sim}{k}$ by k; dependence on the transverse components k_x, k_y will not be explicitly mentioned. It is evident from the above formulas that the real k-axis will separate regions of differing analytic behavior, and that equation (54) is a standard Riemann-Hilbert problem in the complex variable k. For suitably behaved functions $B(k)$, the solution of (54) will be given by the Cauchy integral

$$\tilde{\rho}_+(k)\,\Lambda_+(k) = \frac{1}{2\pi i}\int_{-\infty}^{\infty} \frac{B(k')\,\Lambda_-(k')}{k'-k}\,dk', \tag{57}$$

where the functions $\Lambda_+(k)$ and $\Lambda_-(k)$ factor the dispersion function $\Lambda(k)$ according to

$$\frac{\Lambda_+(k)}{\Lambda_-(k)} = \Lambda(k), \quad k \text{ real.} \tag{58}$$

<u>One-Speed Isotropic Scattering</u>, <u>$c < 1$</u>: In the particularly simple case that all the particles travel at the same speed and that collisions result in a random directional change, we can replace the velocity variable $\underset{\sim}{v}$ (which is three-dimensional) by a direction variable, $\underset{\sim}{\Omega}$, which is a unit vector. The dispersion function becomes simply

$$\begin{aligned}\Lambda(k) &= 1 - \frac{c}{4\pi}\int \frac{d\underset{\sim}{\Omega}}{1 - i\underset{\sim}{k}\cdot\underset{\sim}{\Omega}} \\ &= 1 + \frac{ic}{2\sqrt{k^2+K^2}}\,\ell n\left|\frac{1+i\sqrt{k^2+K^2}}{1-i\sqrt{k^2+K^2}}\right|,\end{aligned} \tag{59}$$

where $K^2 = k_x^2 + k_y^2$. The analytic structure of this function for $c < 1$ is indicated in Figure 4; there are two branch points at $k = \pm i\sqrt{1+K^2}$, and two zeros[11] at $k = \pm i\kappa_0$, where $\kappa_0 < \sqrt{1+K^2}$.

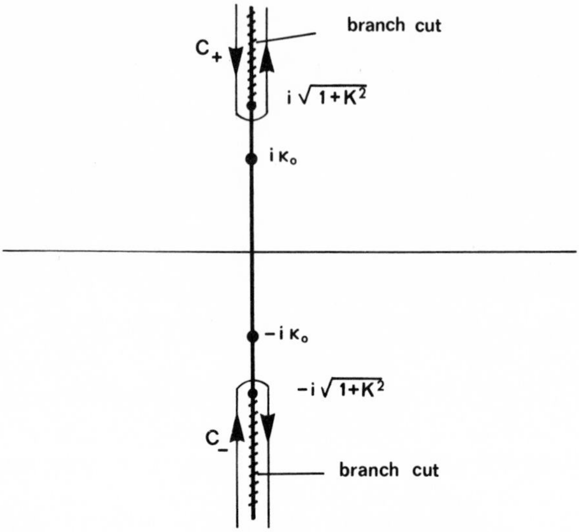

Figure 19.4

It is clear that the function

$$L(k) = \ln\left[\Lambda(k)\,\frac{(k^2+\kappa^2+1)}{(k^2+\kappa_0^2)}\right] \tag{60}$$

is analytic in the entire complex plane with the two branch cuts of Figure 4, and that it can be additively decomposed according to

$$L_\pm(k) = \frac{1}{2\pi i}\int_{-\infty}^{\infty}\frac{L(k')}{k'-k}\,dk'. \tag{61}$$

Now we deform the contours to $C_\pm$ of Figure 4, so that the factorization of the dispersion function can be written as

$$\Lambda_\pm(k) = \frac{k \pm i\kappa_0}{k \pm i\sqrt{1+K^2}}\,X_\pm(k),$$

$$X_\pm(k) = \exp\left[\pm\frac{1}{2\pi i}\int_{C_\mp}\frac{L(k')}{k'-k}\,dk'\right]. \tag{62}$$

This can now be substituted into equation (57) to solve particular problems.

19.7. The Albedo Problem

The Albedo problem is the problem of obtaining the (neutron) phase-space density in a source-free half-space $z \geq 0$, if a parallel beam is incident on the surface $z = 0$. We consider a beam which is uniform in the x and y variables,[12] which means that the solutions will be independent of these variables, and we can set k_x and k_y to zero. Then we have

$$\psi_s(x,y,0,\underset{\sim}{\Omega}) = \delta(\underset{\sim}{\Omega} - \underset{\sim}{\Omega}_0), \quad \underset{\sim}{\Omega} \text{ inward,} \tag{63}$$

where $\underset{\sim}{\Omega}_0$ is the direction of the incident beam, with $\Omega_{0z} > 0$. From this, using equations (56) and (57), and writing $\Omega_z = \mu$, $\Omega_{0z} = \mu_0$, it follows that

$$B(k) = \frac{i}{k + (i/\mu_0)} \tag{64}$$

and

$$\tilde{\rho}_+(k) = \frac{i\Lambda_-(-i/\mu_0)}{[k+(i/\mu_0)]\Lambda_+(k)} . \tag{65}$$

Substitution of this result into the inverse Fourier transform gives the result that $\rho(z) = 0$ for $z < 0$, while for $z \geq 0$, we can deform the contour into the lower half-plane, where the integrand has a simple pole and a branch cut, to obtain

$$\rho(z) = \frac{-1}{2\pi i}\int_{-\infty}^{\infty} \frac{e^{-ikz}\Lambda_-(-i/\mu_0)}{[k+(i/\mu_0)]\Lambda_+(k)}\, dk$$

$$= \frac{[(1/\mu_0)+\kappa_0](1-\kappa_0)\; X_-(-i/\mu_0)}{[(1/\mu_0)-\kappa_0][1+(1/\mu_0)]\; X_+(-i\kappa_0)}\, e^{-\kappa_0 z} \tag{66}$$

$$- \frac{1}{2\pi i}\frac{[(1/\mu_0)+\kappa_0]X_0(-i/\mu_0)}{[(1/\mu_0)+1]}\int_C \frac{e^{-ikz}(k+i)}{[k+(i/\mu_0)](k+i\kappa_0)X_+(k)}\, dk.$$

<u>Connection with Singular Eigenfunction Expansion</u>: We have chosen to write the Fourier inversion of $\rho(z)$ in some detail so that we can show, in a reasonably simple situation, the connection with the solution by singular eigenfunctions. In the notation of Case and Zweifel, the expression for $\rho(z)$ is

$$\rho(z) = -\frac{2\gamma(\mu_0)}{c\nu_0\; X(\nu_0)}\, e^{-z/\nu_0} + \int_0^1 A(\nu)\phi_\nu(\mu_0)d\nu,$$

$$A(\nu) = \frac{2(\nu_0-\mu_0)\;\gamma(\mu_0)}{c\nu(\nu_0-\nu)}\left\{\frac{1}{\hat{X}_\pm(\nu)\;\hat{\Lambda}_\mp(\nu)}\right\} e^{-x/\nu}, \tag{67}$$

where we have added a circumflex to their functions $\Lambda(\nu)$ and $X(\nu)$ to distinguish them from the functions appearing above. The eigenfunctions in (67) are

$$\phi_\nu(\mu) = \frac{c\nu}{2}\, P\, \frac{1}{\nu-\mu} + \lambda(\nu)\;\delta(\nu-\mu), \tag{68}$$

where the P refers to principal value integration. The meaning of that expansion is most easily expressed in terms of Cauchy integrals using

$$\int_0^1 A(\nu)\;\phi_\nu(\mu)\; d\nu = \left[\frac{1}{2\pi i}\int_0^1 \frac{\hat{\Lambda}_+(\nu)\; A(\nu)}{\nu-\mu}\, d\nu\right]_+$$

$$- \left[\frac{1}{2\pi i}\int_0^1 \frac{\hat{\Lambda}_-(\nu)\; A(\nu)}{\nu-\mu}\, d\nu\right]_- , \tag{69}$$

where μ is the complex variable to which the subscripts $\pm$ apply. With the help of these formulas, we can write $\rho(z)$ as

$$\rho(z) = -\frac{2\gamma(\mu_0)}{c\nu_0\, X(\nu_0)} e^{-x/\nu_0} + \frac{(\nu_0-\mu_0)\,\gamma(\mu_0)}{2\pi i}\int_C \frac{2\,e^{-z/\nu}}{c\nu(\nu_0-\nu)(\nu-\mu_0)\,\hat{X}(\nu)}\,d\nu, \tag{70}$$

where the contour is shown in Figure 5.

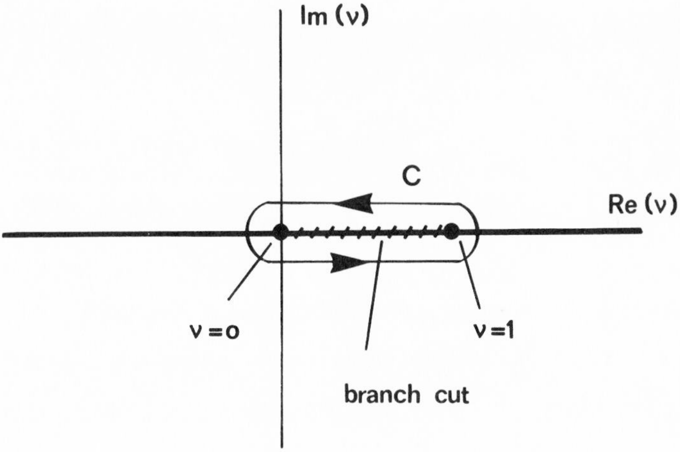

Figure 19.5

Under the variable change $\gamma = -i/k$, it is easily shown that we have the correspondences

$$\begin{aligned} \hat{\Lambda}(\nu) &= \Lambda(k), \\ \hat{X}(\nu) &= X_+(k), \\ \kappa_0 &= 1/\nu_0, \end{aligned} \tag{71}$$

so that the integral in equation (70) becomes

$$\frac{1}{2\pi i}\int_C \frac{2k_0\, e^{-ikz}}{c\mu_0(k+i\kappa_0)[k+(i/\mu_0)]\; X_+(k)}\,dk \tag{72}$$

with the contour of Figure 4. Thus the correspondence is established, although it may be seen, as pointed out by Case and Hazeltine,[8] that the Fourier transform method provides a more direct approach and avoids the difficulties inherent in the interpretation of singular eigenfunction methods as explicit formulas suitable for direct computation.

Angular Density of Emergent Particles: Using (65) in (52), we obtain the result

$$\begin{aligned}\psi(k,\underset{\sim}{\Omega}) &= \frac{i\,\delta(\underset{\sim}{\Omega}-\underset{\sim}{\Omega}_0)}{k+(i/\mu_0)} \\ &\quad - \frac{c}{4\pi}\,\frac{\Lambda_-(-i/\mu_0)}{\mu\,\Lambda_+(k)[k+(i/\mu_0)][k+(i/\mu)]}\,.\end{aligned} \tag{73}$$

This function is analytic in the upper half k-plane except for a pole at $k = -i/\mu$ when $\mu < 0$. Hence, for $z = 0$, the inverse Fourier transform involves only a simple residue calculation, and the result of this calculation is an explicit expression for the angular density of particles which escape from the surface. This expression is

$$\psi(x,y,0,\underset{\sim}{\Omega}) = \frac{c}{4\pi\mu}\,\frac{\Lambda_-(-i/\mu_0)}{[(1/\mu)-(1/\mu_0)]\;\Lambda_+(-i/\mu)}\,. \tag{74}$$

19.8. A Diffraction Problem[13]

Consider the diffraction of scalar waves in an infinite two-dimensional region by a strip of finite width. The differential equation is $(\nabla^2+\kappa^2)\phi = 0$, and there are boundary conditions on the strip, which we take to be $y = 0$, $|x| \le a$, plus the additional condition that the solution consists of incoming plane waves, ϕ_i, and outgoing diffracted waves. The details of the problem will depend on the particular boundary condition applied on the strip; here we take $\phi = 0$. Denoting by $G(\underset{\sim}{r},\underset{\sim}{r}')$ the free-space two-dimensional Green's function and applying Green's theorem to a region whose boundaries are a large circle centered at the origin, and the two sides of the strip, we readily deduce the formula

$$\phi(x,y) - \phi_i(x,y) = \frac{1}{4\pi}\int_{-a}^{a} G(x,y,x',0)\,\rho(x')\,dx', \tag{75}$$

where

$$\rho(x) = \phi_y(x,0+) - \phi_y(x,0-). \tag{76}$$

We recall[14] that the Green's function may be written as

$$G(x,y,x',y') = \int_{-\infty}^{\infty} \frac{e^{-ik(y-y') - |x-x'|\sqrt{k^2-\kappa^2}}}{\sqrt{k^2-\kappa^2}}\,dk, \tag{77}$$

where we take as the branch cuts the contours C_1 and C_2 of Figure 6. Setting $y = 0$, $|x| \le a$ in (75), we obtain the integral equation

$$\begin{aligned} -4\pi\,\phi_i(x,0) &= \int_{-a}^{a} K(x-x')\,\rho(x')\,dx', \\ K(x-x') &= \int_{-\infty}^{\infty} \frac{e^{-|x-x'|\sqrt{k^2-\kappa^2}}}{\sqrt{k^2-\kappa^2}}\,dk. \end{aligned} \tag{78}$$

<u>The Function K</u>: By deforming contours around the branch cuts and then shrinking them onto these cuts, we may deduce the representations

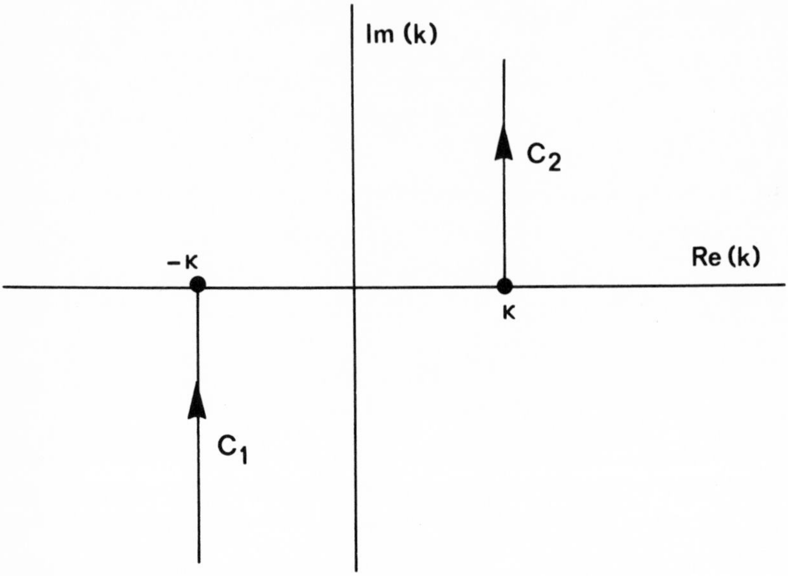

Figure 19.6

$$K(x-x') = \int_{C_1 \text{ or } C_2} e^{-ik(x-x')}\,\Delta(k)\,dk,$$

$$\Delta(k) = \eta_+(k) - \eta_-(k), \qquad (79)$$

$$\eta(k) = (k^2-\kappa^2)^{-1/2},$$

where, for convergence, C_1 is the contour if $x < x'$ and C_2 is the contour if $x > x'$. From this representation of $K(x-x')$, the following important property emerges:

$$\int_{-a}^{a} K(x-x')\,e^{-ik'x'}\,dx'$$

$$= \int_{C_1} \Delta(k)\,\frac{e^{-ikx-ia(k-k')} - e^{-ik'x}}{-i(k'-k)}\,dk \qquad (80)$$

$$+ \int_{C_2} \Delta(k)\,\frac{e^{-ik'x} - e^{ia(k-k')}}{i(k'-k)}\,dk.$$

We look for a solution of equation (78) having the form

$$\rho(x) = \sum_j A_j e^{-ik_j x} + \int_{C_1} \phi_1(k) e^{-ikx} dk + \int_{C_2} \phi_2(k) e^{-ikx} dk, \tag{81}$$

which can be substituted into the integral equation to convert it to

$$\begin{aligned} -4\pi\phi_i(x,0) = \sum_j A_j \Bigg\{ &\int_{C_1} \Delta(k) \left[\frac{e^{-ik_j x} - e^{-ikx+ia(k-k_j)}}{i(k_j-k)} \right] dk \\ &+ \int_{C_2} \Delta(k) \left[\frac{e^{-ikx-ia(k-k_j)} - e^{-ik_j x}}{i(k_j-k)} \right] dk \Bigg\} \\ &+ \left\{ \int_{C_1} \phi_1(k')\, dk' + \int_{C_2} \phi_2(k')\, dk' \right\} \\ &\times \Bigg\{ \int_{C_1} \Delta(k) \left[\frac{e^{-ik'x} - e^{-ikx+ia(k-k')}}{i(k'-k)} \right] dk \\ &+ \int_{C_2} \Delta(k) \left[\frac{e^{-ik'x-ia(k-k')} - e^{-ik'x}}{i(k'-k)} \right] dk \Bigg\}. \end{aligned} \tag{82}$$

<u>Reduction Using Cauchy Integrals</u>: None of the integrals in this expression is divergent, since the expressions in square braces are finite when their denominators are zero. What we do here is regard all of the integrals as principal value integrals, which allows us to separate the various terms which are presently grouped together to effect these cancellations. In doing this we need the following Plemelj formula:

$$\int_{C_2} \frac{\Delta(k')}{i(k'-k)} dk' - \int_{C_1} \frac{\Delta(k')}{i(k'-k)} dk' = \begin{cases} 2\pi\, \eta(k), & k \text{ not on } C_1 \text{ or } C_2 \\ 0 \quad , & k \text{ on } C_1 \text{ or } C_2. \end{cases} \tag{83}$$

With the aid of this formula, the original integral equation is reduced to

$$
\begin{aligned}
-4\pi\ \phi_i(x,0) &= \sum_j 2\pi\ A_j\ \eta(k_j) e^{-ik_j x} \\
&- \int_{C_1} \Delta(k)\ e^{ik(a-x)} \left\{ -\sum_j \frac{A_j\ e^{-iak_j}}{i(k-k_j)} \right. \\
&\left. + \int_{C_1} \frac{\phi_1(k') e^{-iak'}}{i(k'-k)}\, dk' + \int_{C_2} \frac{\phi_2(k')\ e^{-iak'}}{i(k'-k)}\, dk' \right\} dk \\
&- \int_{C_2} \Delta(k)\ e^{-ik(a+x)} \left\{ -\sum_j \frac{A_j\ e^{iak_j}}{i(k-k_j)} \right. \\
&\left. + \int_{C_1} \frac{\phi_1(k')\ e^{iak'}}{i(k'-k)}\, dk' + \int_{C_2} \frac{\phi_2(k')\ e^{iak'}}{i(k'-k)}\, dk' \right\} dk. \qquad (84)
\end{aligned}
$$

For the present problem, with incoming plane waves, $\phi_i(x,0)$ is simply one exponential term, and a solution of this last equation is achieved by simultaneously solving the three equations

$$-4\pi\ \phi_i(x,0) = 2\pi\ A\eta(k_1)\ e^{-ik_1 x}, \qquad (85)$$

$$\int_{C_1} \frac{\phi_1(k') e^{-iak'}}{i(k'-k)} dk' = \frac{A\ e^{-iak_1}}{i(k-k_1)} - \int_{C_2} \frac{\phi_2(k')\ e^{-iak'}}{i(k'-k)} dk', \qquad (86)$$

$$k \text{ on } C_1,$$

$$\int_{C_2} \frac{\phi_2(k') e^{iak'}}{i(k'-k)} dk' = \frac{A\ e^{iak_1}}{i(k-k_1)} - \int_{C_1} \frac{\phi_1(k')\ e^{iak'}}{i(k'-k)} dk', \qquad (87)$$

$$k \text{ on } C_2.$$

Reduction to Coupled Equations of Fredholm Type:[15] No explicit solution of these equations is known; however, we may easily reduce them to a much simpler form not involving any singular integrals. In the following we assume normal incidence, so that $k_1 = 0$, and the algebra is somewhat simplified. First introduce the functions

$$X_1(k) = (k+\kappa)^{-1/2}, \qquad X_2(k) = (k-\kappa)^{-1/2}, \tag{88}$$

which have the properties

$$X_{1+}(k) + X_{1-}(k) = 0 \quad \text{on} \quad C_1, \qquad X_{2+}(k) + X_{2-}(k) = 0 \quad \text{on} \quad C_2, \tag{89}$$

and the Cauchy integrals

$$\Phi_1(z) = \frac{1}{2\pi i}\int_{C_1}\frac{\phi_1(k)}{k-z}\,dk, \qquad \Phi_2(z) = \frac{1}{2\pi i}\int_{C_2}\frac{\phi_2(k)}{k-z}\,dk. \tag{90}$$

Then the Plemelj formulas enable us to express the functions $\phi_1(k)$ and $\phi_2(k)$ in terms of these Cauchy integrals as

$$\phi_1(k) = \frac{1}{2}[\Phi_{1+} - \Phi_{1-}] \quad \text{on} \quad C_1, \qquad \phi_2(k) = \frac{1}{2}[\Phi_{2+} - \Phi_{2-}] \quad \text{on} \quad C_2. \tag{91}$$

The equation for $\phi_1(k)$ now appears as

$$\Phi_{1+}X_{1+} - \Phi_{1-}X_{1-} = (X_{1+} - X_{1-})\,\psi(k),\ k \ \text{on} \ C_1, \tag{92}$$

where

$$\psi(k) = -\frac{2i\kappa}{k} - \int_{C_2}\frac{\phi_2(k')\,e^{-ik'a}}{k'-k}\,dk', \tag{93}$$

and the solution of this Riemann-Hilbert problem is

$$\Phi_1(z)\, X_1(z) = \frac{1}{2\pi i}\int_{C_1} \frac{X_{1+}(k) - X_{1-}(k)}{k-z}\psi(k)\, dk. \tag{94}$$

Introducing the expression for $\psi(k)$ into (94), the first term of $\psi(k)$ leads to the integral

$$\begin{aligned} &- \frac{\kappa}{\pi}\int_{C_1} \frac{[X_{1+}(k) - X_{1-}(k)]}{k(k-z)}\, dk \\ &= \frac{-\kappa}{\pi z}\int_{C_1} [X_{1+}(k) - X_{1-}(k)]\left[\frac{1}{k-z} - \frac{1}{k}\right] dk \qquad (95) \\ &= \frac{-2i\kappa}{z}\,[X_1(z) - X_1(0)], \end{aligned}$$

which therefore gives to $\Phi_1(z)$ the contribution

$$\frac{-2i\kappa}{z}\left\{1 - \frac{X_1(0)}{X_1(z)}\right\}, \tag{96}$$

and to $\phi_1(k)$ the contribution

$$- \frac{i\kappa}{z}\left\{\frac{X_1(0)}{X_{1-}(z)} - \frac{X_1(0)}{X_{1+}(z)}\right\} = \frac{2i\kappa}{z}\,\frac{X_1(0)}{X_{1+}(z)}\ . \tag{97}$$

The second term is dealt with using partial fractions in a similar manner, so that the singular integral equations (86) and (87) for the unknown functions $\phi_1(k)$ and $\phi_2(k)$ are reduced to the ordinary integral equations

$$\begin{aligned} i\pi X_{1+}(k)\,\phi_1(k) &= \frac{2i\kappa}{k}\,X_1(0) - \int_{C_2} \frac{\phi_2(k')X_1(k')\, e^{-ik'a}}{k'-k}\, dk', \\ &\qquad k \text{ on } C_1, \\ & \qquad\qquad\qquad\qquad\qquad\qquad\qquad\qquad (98) \\ i\pi X_{1-}(k)\,\phi_2(k) &= - \frac{2i\kappa}{k}\,X_2(0) + \int_{C_1} \frac{\phi_1(k')X_2(k')\, e^{ik'a}}{k'-k}\, dk', \\ &\qquad k \text{ on } C_2. \end{aligned}$$

We refer the reader to the literature, and particularly to the work of Wolfe,[16] for the solution of these equations by iteration and the relation of these solutions to other methods of treating this particular problem of diffraction.

Problems

1. Use Cauchy integrals to derive the factorization of (18.14).

2. Investigate the various solutions of (18.52) which may be obtained by choosing to take the Fourier transform along different contours, subsequently using this contour in the relevant Cauchy integrals.

3. Find an additive decomposition of the function

$$\Phi(z) = \frac{1}{\sqrt{z^2+k^2}}$$

by deforming the contours of the Cauchy integrals to turn them into loop integrals.

4. Find an additive decomposition of the function

$$\Phi(z) = \sqrt{z^2+k^2}.$$

5. Use Cauchy integrals to derive the decompositions needed in the solution of Problem 18.6.

Footnotes

1. See NOBLE (1958), p. 93ff. for details.

2. C. Mark, Phys. Rev. (1947), 72, 558; G. Placzek, Phys. Rev. (1947), 72, 556.

3. MUSKHELISHVILI (1953).

4. The function $f(\zeta)$ satisfies a Hölder condition on a contour C if there exist real positive constants A and μ such that

$$|f(\zeta_2) - f(\zeta_1)| < A|\zeta_2-\zeta_1|^{\mu}$$

for any two points ζ_1 and ζ_2 on C.

5. MUSKHELISHVILI (1953), Ch. 2.

6. MUSKHELISHVILI (1953), Ch. 4.

7. We could in fact use a Fourier inversion contour which is not a straight line parallel to the real axis, thus achieving a generalization of the Wiener-Hopf technique by using the Plemelj formula.

8. Based on work by K. M. Case and R. D. Hazeltine, J. Math. Phys. (1971), 12, 1970.

9. See CASE & ZWEIFEL (1967) for the derivation and interpretation of this equation.

10. In particular, we have used the property that

$$G(\underset{\sim}{r}_s-\underset{\sim}{r},\underset{\sim}{v}) = 0, \quad z > 0, \quad v_z < 0.$$

11. Proved in CASE & ZWEIFEL (1967), p. 62 ff.

12. See ref. 8 for another example.

13. Based on a paper by K. M. Case, Rev. Mod. Phys. (1964), 36, 669.

14. See Section 10.4.

15. The techniques used for the solution of these singular integral equations are quite standard; see MUSKHELISHVILI (1953).

16. P. Wolfe, SIAM J. Appl. Math. (1972), 23, 118.

§20. LAPLACE'S METHOD FOR ORDINARY DIFFERENTIAL EQUATIONS

20.1. Integral Transform Solutions

Transform methods are useful in finding solutions of ordinary differential equations far more complicated than those considered in Section 3. In fact, we have already seen in Section 3.4 that an explicit formula for the Bessel function $J_0(x)$, defined as the solution of an ordinary differential equation with variable coefficients, can be found with the Laplace transform. One advantage of the technique developed in this section over the simpler method for solution in terms of a power series expansion is that the transform method gives the solution required directly as an integral representation. In this compact form various properties of and relations between different solutions of an equation become quite clear, convenient asymptotic expansions can be obtained directly, and numerical computation may be facilitated. For applications, the analytic properties, asymptotic expansions, and ease of computation of a function are of primary interest.

Laplace's Method: We consider the differential equation

$$\sum_{k=0}^{n} (a_k + b_k x) y^{(k)}(x) = 0, \tag{1}$$

whose coefficients are linear polynomials. Laplace's method consists of assuming that the solutions have the integral representation

$$y(x) = \int_C S(p)\, e^{px}\, dp, \tag{2}$$

where the contour is to be chosen as part of the process of

solution. Equation (2) therefore represents a straightforward extension of the Laplace transform method, for which the contour is a vertical line in the p-plane. If we insert (2) into (1), and integrate by parts to remove the terms in x, we get

$$\begin{aligned} 0 &= \int_C [F(p) + x\, G(p)]\, S(p)\, e^{px}\, dp \\ &= \int_C e^{px} \left[F(p)\, S(p) - \frac{d}{dp} G(p)\, S(p) \right] dp \\ &\qquad + [G(p)\, S(p)\, e^{px}]_1^2, \end{aligned} \tag{3}$$

where

$$\begin{aligned} F(p) &= \sum_{k=0}^{n} a_k p^k, \\ G(p) &= \sum_{k=0}^{n} b_k p^k. \end{aligned} \tag{4}$$

The second term in (3) is evaluated at the end points of the contour, which are often at infinity. If we choose the contour so as to make this contribution vanish, then (2) will represent a solution to (1) provided the function $S(p)$ satisfies the differential equation

$$F(p)\, S(p) - \frac{d}{dp} G(p)\, S(p) = 0. \tag{5}$$

There is an immediate formal solution to (5), namely

$$S(p) = \frac{A}{G(p)} \exp\left[\int \frac{F(p)}{G(p)}\, dp \right], \tag{6}$$

where A is an arbitrary constant. Using (6), it is possible to determine suitable contours to complete the solution.

The simplification which allows us to write the explicit formula (6) arises from the fact that (5) is a first-order differential equation. This, in turn, is due to our requirement that the coefficients of the original equation be only linear polynomials.[1] If they were quadratic, the process of integrating by parts in (3) would lead to a second-order equation for $S(p)$. There is no general method for solving for $S(p)$ in this case, although if a solution can be found the method will still succeed.

20.2. Hermite Polynomials

As an illustration of Laplace's method, we consider Hermite's differential equation

$$w''(x) - 2x\,w'(x) + 2\nu w(x) = 0, \tag{7}$$

where ν is an arbitrary constant. We obtain solutions to this equation by using equations (2)-(6), which give

$$\begin{aligned} S(p) &= -\frac{1}{2p}\exp\left[\int \frac{p^2 + 2\nu}{-2p}\,dp\right] \\ &= -\frac{1}{2p}\exp\left[\frac{-p^2}{4} - \nu\,\ell n\,p\right] \\ &= -\frac{e^{-p^2/4}}{2p^{\nu+1}}, \end{aligned} \tag{8}$$

and this can be substituted into (2) to obtain an integral representation. In order to conform with established convention, we first introduce a new variable $s = p/2$, so that

$$w(x) = A\int_C \frac{e^{-s^2+2sx}}{s^{\nu+1}}\,ds, \tag{9}$$

where A is an arbitrary constant, and the path of integration must be chosen so that

$$\left[\frac{e^{-s^2+2sx}}{s^{\nu+1}}\right]_1^2 = 0, \tag{10}$$

where $[\phi]_1^2$ refers to the increment in ϕ as we traverse the contour C.

<u>Choice of Path</u>: In general a variety of paths will satisfy the condition that the integrated term vanish, and each path leads to a different solution. The particular path of interest is usually dictated by the physical or boundary conditions we impose. Possible paths which satisfy (10) are

(i) $-\infty < s < +\infty$ (avoiding $s = 0$),

(ii) $0 < s < \infty$ ($\mathrm{Re}(\nu) < 0$),

(iii) A contour which encircles the origin (ν an integer),

(iv) The path shown in Figure 1.

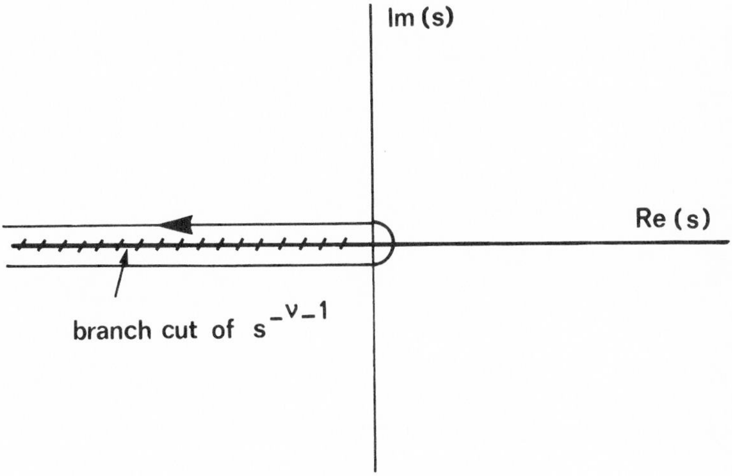

Figure 20.1

We will prove in Section 20.3 that Hermite functions have the behavior $w(x) \sim A \exp(x^2)$ either for $x \to \infty$ or $x \to -\infty$, or both, except when $\nu = 0,1,2,\ldots$. [This conclusion may also be reached using Watson's lemma for the paths (i), (ii) and (iv).]

Turning first to the bounded functions, for which ν must be an integer, we take C to be path (iii), and write

$$w_n(x) = A \int_C \frac{e^{-s^2+2sx}}{s^{n+1}}\, ds, \quad n = 0, 1, 2, \ldots. \tag{11}$$

The solutions (11), which will turn out to be polynomials, are normalized by the convention that the coefficient of x^n is 2^n; this gives

$$H_n(x) = \frac{n!}{2\pi i} \int_C \frac{e^{-s^2+2sx}}{s^{n+1}}\, ds, \tag{12}$$

where we use the symbol H_n as the (standard) notation for a Hermite polynomial.

Derivative Form: Another expression for $H_n(x)$ follows if we use the change of variables $u = s-x$ in (12) and apply the standard formula for evaluating the residue at a pole of order $(n+1)$. This gives

$$\begin{aligned} H_n(x) &= \frac{n!\, e^{x^2}}{2\pi i} \int_C \frac{e^{-u^2}}{(u+x)^{n+1}}\, du \\ &= (-1)^n\, e^{x^2} \frac{d^n}{dx^n}\, e^{-x^2}. \end{aligned} \tag{13}$$

Explicit Formula: From (12), we have

$$H_n(x) = n!\{\text{coefficient of } s^n \text{ in } e^{-s^2+2sx}\}, \tag{14}$$

giving the explicit formula

$$H_n(x) = \sum_{k=0}^{[n/2]} \frac{(-1)^k n!}{k!\,(n-2k)!} (2x)^{n-2k}, \tag{15}$$

where $[n/2]$ is the largest integer $\leq n/2$.

Generating Function: It follows immediately from (14) that the function $\exp(2xt-t^2)$ generates the Hermite polynomials in the sense that

$$e^{2xt-t^2} = \sum_{n=0}^{\infty} \frac{H_n(x)}{n!} t^n. \tag{16}$$

This formula plays an important role in the theory of Hermite polynomials.[2]

20.3. Hermite Functions

We return to the problem of finding solutions of Hermite's equation for arbitrary ν. Consider the integral

$$H_\nu(x) = \frac{\nu!}{2\pi i} \int_C \frac{e^{-p^2+2px}}{p^{\nu+1}} \, dp, \tag{17}$$

which is the Laplace integral solution to (7), with p replaced by $2p$. We take the contour shown in Figure 1, so that for $\nu = 0,1,2,\ldots$ (17) gives $H_n(x)$. When $\nu = -1, -2,\ldots,$ (17) is indeterminate. However, for $\mathrm{Re}(\nu) < 0$ we can obtain an alternative representation by 'shrinking' the contour around the branch cut; with the change of variables $p = t \exp(\pm i\pi)$, this gives

$$\begin{aligned} H_\nu(x) &= \frac{-\nu! \sin(\pi\nu)}{\pi} \int_0^\infty \frac{e^{-t^2-2tx}}{t^{\nu+1}} \, dt \\ &= \frac{1}{(-\nu-1)!} \int_0^\infty \frac{e^{-t^2-2tx}}{t^{\nu+1}} \, dt, \quad \mathrm{Re}(\nu) < 0. \end{aligned} \tag{18}$$

There are two important immediate consequences. The first is that $H_\nu(x)$ is an entire function of both x and ν; the second is that $H_\nu(0)$ can be evaluated. For, on putting $x = 0$ in (18), we have

$$H_\nu(0) = \frac{(-\frac{\nu}{2}-1)!}{2(-\nu-1)!}, \tag{19}$$

which holds for all ν by analytic continuation.

Recurrence Relations: If we multiply (17) by $2x$ and integrate by parts, we find that

$$\begin{aligned} 2xH_\nu(x) &= \frac{-\nu!}{2\pi i}\int_C e^{+2sx}\frac{d}{ds}\left[\frac{e^{-s^2}}{s^{\nu+1}}\right]ds \\ &= \frac{-\nu!}{2\pi i}\int_C \frac{e^{-s^2+2sx}}{s^{\nu+1}}\left[-2s-\frac{(\nu+1)}{s}\right]ds, \end{aligned} \tag{20}$$

and after using (17) again we see

$$H_{\nu+1}(x) - 2xH_\nu(x) + 2\nu H_{\nu-1}(x) = 0. \tag{21}$$

This recurrence relation (called a three-term relation because it connects Hermite polynomials of three different indices) can be used to tabulate the Hermite polynomials step by step, beginning with $H_0(x) = 1$.

Another recurrence relation may be obtained by differentiating (17) with respect to x. This gives

$$\begin{aligned} H_\nu'(x) &= \frac{\nu!}{2\pi i}\int_C \frac{2s\ e^{-s^2+2sx}}{s^{\nu+1}}\,ds \\ &= 2\nu\ H_{\nu-1}(x). \end{aligned} \tag{22}$$

The manipulations involved here are only valid for $\mathrm{Re}(\nu) > -1$, but the results are not restricted, because we can employ the principle of analytic continuation to extend them to all ν.

Independent Solutions: $H_\nu(x)$ is one solution of Hermite's equation, and it is trivial to show that $H_\nu(-x)$ is also a solution. The crucial question is whether this second solution is linearly independent of the first, and this depends on the Wronskian. For any two solutions of (7), we have[3]

$$W[u,w] = Ke^{x^2}. \tag{23}$$

In the present case, we find the constant by putting $x = 0$ and using (22), (19), and the properties of the factorial function. This gives

$$W\,[H_\nu(x),\, H_\nu(-x)] = \frac{2^{\nu+1}\sqrt{\pi}}{(-\nu-1)!}\, e^{x^2} \tag{24}$$

Consequently $H_\nu(x)$ and $H_\nu(-x)$ are linearly independent solutions provided $\nu \neq 0,1,2,\ldots$. In these exceptional cases, we can show that $H_n(x) = (-1)^n H_n(-x)$.

Other solutions can be found by noting that the functions $\exp(z^2)\, H_{-\nu-1}(\pm iz)$ also satisfy Hermite's equation. The Wronskians can again be obtained by direct calculation for $x = 0$. This gives

$$W\,[H_\nu(x),\, e^{x^2} H_{-\nu-1}(\pm ix)] = e^{x^2 \pm 1/2(\nu+1)\pi i}, \tag{25}$$

so that these new functions provide solutions independent of $H_\nu(x)$ for arbitrary ν. We return to this fact below, but first we wish to use these new solutions to obtain an integral representation similar to (18) for $\mathrm{Re}\,(\nu) > -1$. We commence by noting that any three solutions of a second-order differential equation must be linearly dependent; hence there are constants A, B, and C such that

$$AH_\nu(x) + e^{x^2}\,[BH_{-\nu-1}(ix) + CH_{-\nu-1}(-ix)] = 0. \tag{26}$$

By evaluating (26) and its derivative with respect to x at $x = 0$, we obtain the relation

$$H_\nu(x) = \frac{2^\nu \nu!}{\sqrt{\pi}}\, e^{x^2} \left[e^{i\pi\nu/2}\, H_{-\nu-1}(ix) + e^{-i\pi\nu/2} H_{-\nu-1}(-ix)\right]. \tag{27}$$

Substitution of (18) into this result yields the integral representation

$$H_\nu(x) = \frac{2^{\nu+1} e^{x^2}}{\sqrt{\pi}} \int_0^\infty e^{-t^2} t^\nu \cos(2xt - \frac{\nu\pi}{2}) dt, \ \mathrm{Re}(\nu) > -1. \quad (28)$$

Formulas (18) and (28) are useful in deriving asymptotic forms.

Asymptotic Forms: We commence with the integral (17), and replace the branch cut by the straight line $t = -p \exp(i\alpha)$, where α is chosen by the following considerations: (i) the integral must be unchanged and convergent, hence $-\pi/4 < \alpha < \pi/4$, and (ii) we want the factor $\exp(+2xp)$ to go to zero, which imposes the restriction $\mathrm{Re}(xp) < 0$. Denoting $\arg(x)$ by β, this implies $-\pi/2 < \alpha + \beta < \pi/2$ and thus $-3\pi/4 < \beta < 3\pi/4$. Now we replace $\exp(-t^2)$ by its Taylor series and apply Watson's lemma for loop integrals, which gives

$$H_\nu(x) \sim (2x)^\nu \sum_{k=0}^\infty \frac{(-1)^k (2k-\nu-1)!}{k!(-\nu-1)! (2x)^{2k}}, \ -3\pi/4 < \arg(x) < 3\pi/4. \quad (29)$$

We also need a formula to cover the region excluded from (29). Using (27) rearranged in the form

$$H_\nu(x) = e^{i\pi\nu} H_\nu(x e^{-i\pi}) + \frac{i\sqrt{\pi}}{2^\nu \nu!} e^{x^2} e^{i\pi\nu/2} H_{-\nu-1}(xe^{-i\pi/2}) \quad (30)$$

and applying (29) to each term, we get

$$H_\nu(x) \sim (2x)^\nu \sum_{k=0}^\infty \frac{(-1)^k (2k-\nu-1)!}{k!(-\nu-1)!(2x)^{2k}}$$

$$- \frac{\sqrt{\pi}\, e^{i\pi\nu}}{(-\nu-1)!} \frac{e^{x^2}}{x^{\nu+1}} \sum_{k=0}^\infty \frac{(2k+\nu)!}{k!\nu!(2x)^{2k}}, \quad (31)$$

$$\pi/4 < \arg(x) < 5\pi/4$$

with a similar formula for $-\frac{5\pi}{4} < \arg(x) < -\frac{\pi}{4}$. The essential difference between (31) and (29) is in the term $\exp(x^2)$; unless $\nu = 0,1,2, \ldots$ this causes the corresponding functions to be unbounded for $|\arg(x)| > 3\pi/4$, in support of the assertion made in Section 20.2.

20.4. Bessel Functions:[4] Integral Representations

The Hermite equation investigated above is typical of the simplest class of second-order equations whose general solutions are nontrivial. Slightly more difficult is the Bessel equation, which we consider here. Bessel functions satisfy the equation

$$f'' + \frac{1}{z} f' + \left(1 - \frac{\nu^2}{z^2}\right) f = 0. \tag{32}$$

Again, the solution is most easily effected if we remove the singularity at $z = 0$; this may be achieved by the substitution

$$f(z) = z^{\nu} u(z) \tag{33}$$

to obtain the equation

$$zu'' + (2\nu+1)u' + zu = 0, \tag{34}$$

which is amenable to solution by our method. Equation (34) is closely related to the confluent hypergeometric equation. However, it is worthwhile to discuss Bessel's equation in its own right, as there are many important properties and methods which are specific to Bessel functions.

The direct application of Laplace's method to (34) is straightforward. In the notation of Section 20.1 we have

$$
\begin{aligned}
F(p) &= (2\nu + 1)\, p, \\
G(p) &= p^2 + 1, \\
S(p) &= A\, (p^2 + 1)^{\nu - 1/2}.
\end{aligned} \tag{35}
$$

For large $|z|$, equation (34) may be approximated by $u'' + u = 0$, the solutions of which are periodic. It is therefore conventional to make the substitution $p = i\omega$ in the Laplace integral, so that the method yields

$$u(z) = \frac{A}{2\pi} \int_C (\omega^2 - 1)^{\nu - 1/2}\, e^{i\omega z}\, d\omega, \tag{36}$$

where the contour C and arbitrary constant A must be chosen. A variety of contours are suitable for defining various types of Bessel functions. These will be discussed as the occasion arises.

Another Representation: The integrand in (36) has two branch points at $\omega = \pm 1$, a feature which is characteristic of the confluent hypergeometric equation, and which causes some practical difficulties. An alternative approach, applicable only to Bessel functions, commences by replacing (33) by

$$f(z) = z^\nu\, g(z^2). \tag{37}$$

After writing $\xi = z^2$ and substituting into Bessel's equation (1), this yields the new differential equation[5] for $g(\xi)$

$$4\xi\, g'' + 4(\nu + 1) g' + g = 0. \tag{38}$$

Application of Laplace's method to this new equation gives

$$S(p) = \frac{1}{4p^2}\exp\left[\int \frac{4(\nu+1)p+1}{4p^2}\,dp\right] = \frac{1}{4}\,p^{\nu-1}\,e^{-1/4p}, \tag{39}$$

$$g(\xi) = \frac{A}{2\pi i}\int_C p^{\nu-1}\,e^{p\xi - \frac{1}{4p}}\,dp. \tag{40}$$

More symmetrical formulas may be obtained by a change of variables. If we substitute back into (37) after replacing p by $p/2z$, we obtain the representation

$$f(z) = \frac{A}{2\pi i}\int_C p^{\nu-1}\,e^{\frac{1}{2}z(p - \frac{1}{p})}\,dp, \tag{41}$$

where the constant A and contour C have yet to be chosen. Since the parameter ν occurs in Bessel's equation only as ν^2, its sign is undetermined, and another integral representation is given by

$$f(z) = \frac{A}{2\pi i}\int_C p^{-\nu-1}\,e^{\frac{1}{2}z(p - \frac{1}{p})}\,dp. \tag{42}$$

20.5. Bessel Functions of the First Kind

By choosing $A = 1$ and the contour shown in Figure 1, we obtain functions generally known as Bessel functions of the first kind, and denoted $J_\nu(z)$, i.e.,

$$J_\nu(z) = \frac{1}{2\pi i}\int_C p^{-\nu-1}\,e^{\frac{1}{2}z(p - \frac{1}{p})}\,dp, \quad \mathrm{Re}(z) > 0. \tag{43}$$

The restriction $\mathrm{Re}(z) > 0$ is necessary to make the integral converge.

Analytic Continuation: It is a simple matter to perform an analytic continuation of (43) to all z, and to elucidate the behavior of $J_\nu(z)$ about $z = 0$ at the same time. If we temporarily restrict z to be real and positive, then the change of variables $u = pz/2$ yields

$$J_\nu(z) = \frac{(z/2)^\nu}{2\pi i} \int_C u^{-\nu-1} e^{u-\frac{z^2}{4u}} du, \tag{44}$$

where the contour is unchanged since z is real. But the integral in (44) defines an entire function of z since it is single-valued and absolutely convergent for all z. Hence it is a valid representation for all z, and we see that $J_\nu(z)$ has a branch point at the origin, but that it has no other singularities. To complete the definition of $J_\nu(z)$ we must introduce a branch cut; the usual convention is to make this the negative real axis, specifying the branch by the restriction $-\pi < \arg(z) < \pi$.

<u>Series Expansion</u>: A series expansion for $J_\nu(z)$ can be obtained from (44) by replacing $\exp(-z^2/4u)$ by its Taylor series[6] and integrating term by term. Using (A8) we thus obtain

$$\begin{aligned} J_\nu(z) &= \frac{(z/2)^\nu}{2\pi i} \sum_{k=0}^{\infty} \frac{(-1)^k (z/2)^{2k}}{k!} \int_C u^{-(\nu+1+k)} e^u du \\ &= \sum_{k=0}^{\infty} \frac{(-1)^k (z/2)^{\nu+2k}}{(\nu+k)!\ k!} . \end{aligned} \tag{45}$$

<u>Recurrence Relations</u>: Recurrence relations can easily be obtained from (43) by temporarily assuming $\mathrm{Re}(z) > 0$ and then using analytic continuation to remove this restriction. If we differentiate under the integral sign, we have

$$\begin{aligned} J'_\nu(z) &= \frac{1}{2\pi i} \int_C \frac{1}{2}(p - \frac{1}{p})\ p^{-\nu-1}\ e^{\frac{1}{2}z(p-\frac{1}{p})}\ dp \\ &= \frac{1}{2}J_{\nu-1}(z) - \frac{1}{2}J_{\nu+1}(z). \end{aligned} \tag{46}$$

Alternatively, we can integrate by parts to get

$$\nu J_\nu(z) = \frac{-1}{2\pi i}\int_C (p^{-\nu})'\, e^{\frac{1}{2}z(p-\frac{1}{p})}\, dp$$

$$= \frac{1}{2\pi i}\int_C \frac{z}{2}(1+\frac{1}{p^2})\, p^{-\nu}\, e^{\frac{1}{2}z(p-\frac{1}{p})}\, dp \qquad (47)$$

$$= \frac{z}{2} J_{\nu+1}(z) + \frac{z}{2} J_{\nu-1}(z).$$

From these two relations a number of others can be deduced (see Problem 2).

Bessel's Integral: We modify the contour of Figure 1 to that shown in Figure 2 and let the straight line sections tend toward the negative real axis. The integral then splits up into two terms:

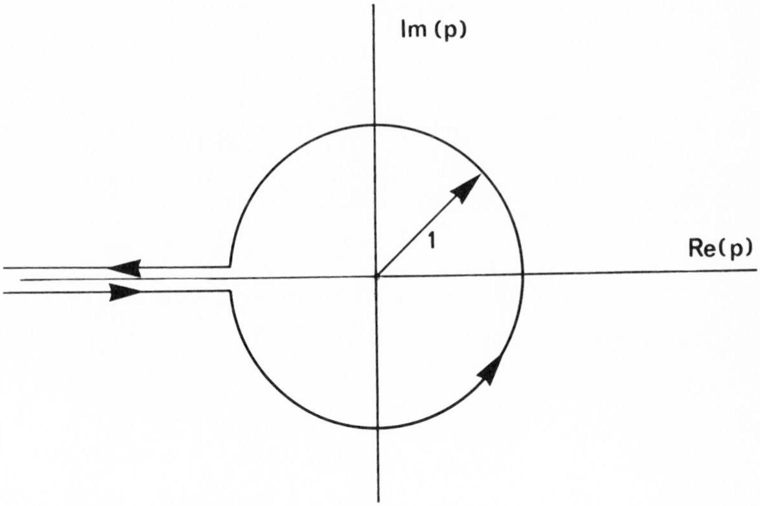

Figure 20.2

(i) Contribution from the circular path. If we write $p = \exp(i\theta)$, we have

$$\frac{1}{2\pi}\int_{-\pi}^{\pi} e^{-i\nu\theta+iz\sin\theta}\,d\theta = \frac{1}{\pi}\int_{0}^{\pi}\cos(\nu\theta - z\sin\theta)\,d\theta, \qquad (48)$$

which comes from writing the complex exponential in terms of sine and cosine functions, and using the fact that the former is an odd function and the latter an even function.

(ii) Contribution from straight paths. On the upper path we put $u = \exp(s - i\pi)$, on the lower path $u = \exp(s + i\pi)$. This yields the contribution

$$\begin{aligned}&\frac{1}{2\pi i}\int_{\infty}^{0} e^{-\nu s+i\pi\nu}\exp\left[-\tfrac{1}{2}z(e^{s}-e^{-s})\right]ds\\ &+\frac{1}{2\pi i}\int_{0}^{\infty} e^{-\nu s-i\pi\nu}\exp\left[-\tfrac{1}{2}z(e^{s}-e^{-s})\right]ds \qquad (49)\\ &= -\frac{\sin(\nu\pi)}{\pi}\int_{0}^{\infty}\exp(-z\sinh s-\nu s)\,ds.\end{aligned}$$

Adding (48) and (49) we have

$$\begin{aligned}J_\nu(z) &= \frac{1}{\pi}\int_0^{\pi}\cos(z\sin\theta-\nu\theta)\,d\theta\\ &\quad-\frac{\sin(\pi\nu)}{\pi}\int_0^{\infty}\exp(-z\sinh s-\nu s)\,ds,\quad \mathrm{Re}(z)>0.\end{aligned} \qquad (50)$$

For integer ν, the second integral gives no contribution. The first integral is known as Bessel's integral; the complete formula (50) which is valid for all ν is a generalization of Bessel's integral.

20.6. Functions of the Second and Third Kinds

For arbitrary ν the functions $J_\nu(z)$ and $J_{-\nu}(z)$ both satisfy Bessel's equation. To see if they are independent we evaluate the Wronskian. This is most easily done by first noting (see Problem 3) that for any pair of solutions of (32) the Wronskian must have the form

$$W\,[f_1,f_2] = \frac{A}{z}\,, \tag{51}$$

so we need only evaluate the constant A. Using (46), we have

$$\begin{aligned} A &= \tfrac{1}{2}zJ_\nu(z)\left[J_{-\nu-1}(z) - J_{-\nu+1}(z)\right] \\ &\quad - \tfrac{1}{2}zJ_{-\nu}(z)\left[J_{\nu-1}(z) - J_{\nu+1}(z)\right], \end{aligned} \tag{52}$$

and on inserting the power series (45) and considering the limit $z \to 0$,

$$\begin{aligned} A &= \frac{1}{\nu!(-\nu-1)!} - \frac{1}{(-\nu)!(\nu-1)!} \\ &= \frac{-2\sin(\pi\nu)}{\pi}\,. \end{aligned} \tag{53}$$

Hence $J_\nu(z)$ and $J_{-\nu}(z)$ are an independent pair of solutions provided ν is not an integer. If ν is an integer, then it may be shown (Problem 4) that $J_{-n}(z) = (-1)^n J_n(z)$.

Other solutions of Bessel's equation may be found from (42) by using a contour which has one end at the origin. Because of the essential singularity there, it is necessary to choose the contour so that it approaches the origin in the sector $|\arg(pz)| < \pi/2$, so that the integrand will become zero along this path. This essential singularity may be removed by the substitution $p = \exp(t)$, leading us to consider the functions

$$Z_\nu(z) = A\int_C e^{z\sinh t-\nu t}\,dt, \tag{54}$$

for suitable contours C. By repeating the arguments which led to (46) and (47), it can readily be shown that they satisfy the recurrence relations[7]

$$2Z'_\nu(z) = Z_{\nu-1}(z) - Z_{\nu+1}(z),$$
$$\frac{2\nu}{z} Z_\nu(z) = Z_{\nu-1}(z) + Z_{\nu+1}(z). \tag{55}$$

Choice of Path: The integrand in (54) is an entire function of t, consequently closed contours yield the trivial function $Z_\nu \equiv 0$. Furthermore, the integrand has no zeros, so both ends of the contour must approach infinity in such a way that $\mathrm{Re}(z \sinh t) \to -\infty$. If we restrict our attention to $\mathrm{Re}(z) > 0$, then these considerations impose the following restrictions:

(i) $\mathrm{Re}(t) \to \infty$, $\mathrm{Im}(t) \to (2n+1)\pi$,

(ii) $\mathrm{Re}(t) \to -\infty$, $\mathrm{Im}(t) \to 2n\pi$.

We also note that if two contours are related by the displacement $t \to t + 2\pi ni$, then the functions which they define are identical except for the constant multiplier $\exp(-2\pi\, i n\nu)$. We consider the four contours shown in Figure 3--any other suitable choices will give linear combinations of the functions which we obtain from these.

The Functions $J_{\pm\nu}(z)$: By a simple change of notation it can readily be shown that (54) reduces to the generalization of Bessel's integral (50) if $A = 1/2\pi i$. Hence

$$J_\nu(z) = \frac{1}{2\pi i}\int_{C_3} e^{z \sinh t - \nu t}\, dt, \quad \mathrm{Re}(z) > 0. \tag{56}$$

Next we consider the contour C_4. The substitution $t = i\pi - s$ maps the contour C_4 into C_3 and changes the

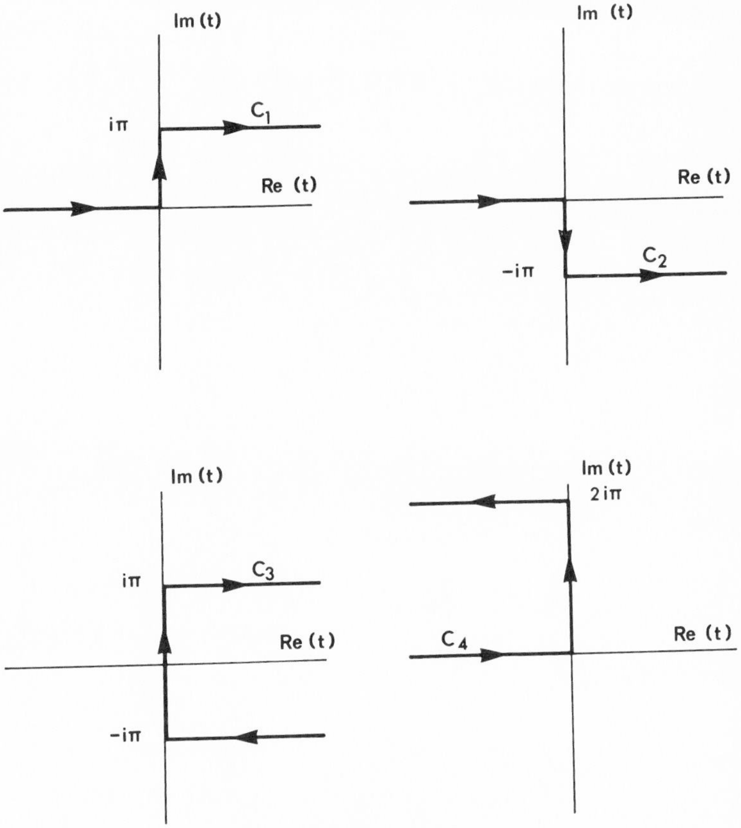

Figure 20.3

integrand to exp $[z \sinh s + \nu s - i\pi\nu]$. Hence, on choosing $A = -\exp(i\pi\nu)/2\pi i$, we obtain

$$J_{-\nu}(z) = \frac{-e^{i\pi\nu}}{2\pi i} \int_{C_4} e^{z \sinh t - \nu t}\, dt, \quad \mathrm{Re}(z) > 0. \quad (57)$$

Thus we recover functions of the first kind from the contours C_3 and C_4.

Hankel Functions: Functions of the third kind, named in honor of Hankel, are obtained from the contours C_1 and C_2. Explicitly, they are defined by

$$H_\nu^{(1)}(z) = \frac{1}{\pi i}\int_{C_1} e^{z\sinh t - \nu t}\,dt,$$
$$H_\nu^{(2)}(z) = -\frac{1}{\pi i}\int_{C_2} e^{z\sinh t - \nu t}\,dt, \quad \mathrm{Re}(z) > 0. \tag{58}$$

It follows immediately from (56) and (58) that

$$J_\nu(z) = \frac{1}{2}\left[H_\nu^{(1)}(z) + H_\nu^{(2)}(z)\right]. \tag{59}$$

Furthermore, if we translate the contour C_2 by the substitution $t \to t + 2\pi i$, we obtain from (57) and (58)

$$J_{-\nu}(z) = \frac{1}{2}\left[e^{i\pi\nu}H_\nu^{(1)}(z) + e^{-i\pi\nu}H_\nu^{(2)}(z)\right]. \tag{60}$$

These relations have been proved under the restriction $\mathrm{Re}(z) > 0$. However, analytic continuations of Hankel functions may readily be effected; if we introduce the negative real axis as the branch cut, (59) and (60) are valid in the entire cut plane.

Wronskian: The Wronskian of the two Hankel functions is conveniently evaluated by using (59) and (60). This gives

$$W[J_\nu, J_{-\nu}] = \frac{\sin(\pi\nu)}{2i}\,W[H_\nu^{(1)}, H_\nu^{(2)}], \tag{61}$$

and after substituting the value of $W[J_\nu, J_{-\nu}]$,

$$W[H_\nu^{(1)}(z), H_\nu^{(2)}(z)] = -\frac{4i}{\pi z}. \tag{62}$$

Thus the two Hankel functions are linearly independent for all ν.

Weber Functions: Functions of the second kind, named after Weber, are defined by

$$Y_\nu(z) = \frac{1}{2i}\left[H_\nu^{(1)}(z) - H_\nu^{(2)}(z)\right]. \tag{63}$$

The origin is again a branch point, and it is the usual convention to use the negative real axis as a branch cut. The pair of functions J_ν, Y_ν are linearly independent for all ν, as may be seen by evaluating the Wronskian

$$W\,[J_\nu(z),\, Y_\nu(z)] = \frac{2}{\pi z}\,. \tag{64}$$

Other properties of Weber functions are given in the problems.

Another Integral Representation: We conclude by deriving another integral representation for the Hankel functions which will prove useful later on. We commence by making the substitution $t \to s \pm i\pi$ in (58) so that

$$H_\nu^{(1)}(z) = \frac{e^{-i\pi\nu/2}}{\pi i}\int_{\Gamma_1} e^{iz\cosh s - \nu s}\, ds,$$

$$H_\nu^{(2)}(z) = -\frac{e^{i\pi\nu/2}}{\pi i}\int_{\Gamma_2} e^{-iz\cosh s - \nu s}\, ds, \tag{65}$$

$$\operatorname{Re}(z) > 0,$$

where the contours Γ_1 and Γ_2 are shown in Figure 4. If

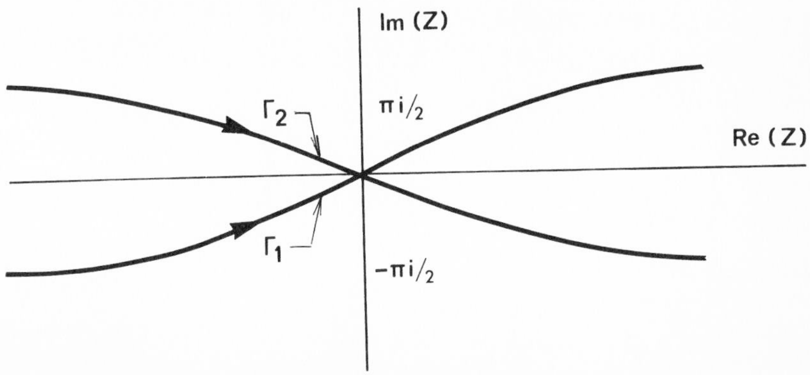

Figure 20.4

we further restrict z by $\mathrm{Im}(z) > 0$ for $H_\nu^{(1)}$, and $\mathrm{Im}(z) < 0$ for $H_\nu^{(2)}$, we can deform these contours in the s-plane to the real axis. Hence

$$H_\nu^{(1)}(z) = \frac{e^{-i\pi\nu/2}}{\pi i}\int_{-\infty}^{\infty} e^{iz\cosh s-\nu s}\,ds, \ \mathrm{Im}(z) > 0, \tag{66}$$

$$H_\nu^{(2)}(z) = -\frac{e^{i\pi\nu/2}}{\pi i}\int_{-\infty}^{\infty} e^{-iz\cosh s-\nu s}\,ds, \ \mathrm{Im}(z) < 0, \tag{67}$$

where we have removed the restriction $\mathrm{Re}(z) > 0$ since these formulas achieve analytic continuations of (58).

20.7. Poisson and Related Representations

In Section 20.4 we derived two distinct integral representations of Bessel functions, although we have only considered one of them (42) so far. We now examine the integral (36) with the contours shown in Figure 5. For the

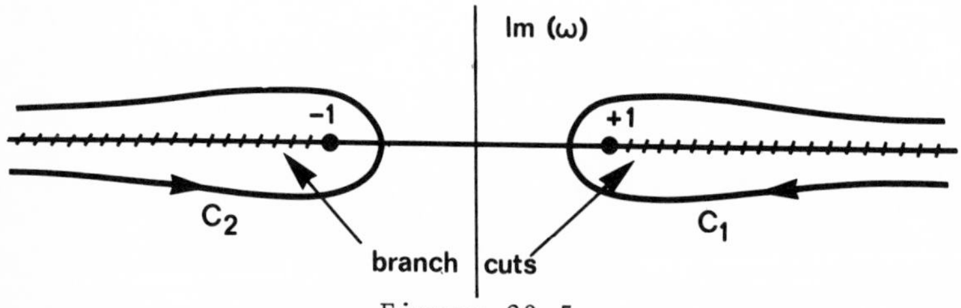

Figure 20.5

contour C_1 we temporarily assume that $z = i\xi$, with ξ real and positive, and consider the function defined by

$$\int_{C_1} e^{-\omega\xi}(\omega^2-1)^{\nu-\frac{1}{2}}\,d\omega. \tag{68}$$

We specify the branch of $(\omega^2-1)^{\nu-\frac{1}{2}}$ by requiring $\arg(\omega^2-1)$

to become zero as we move to large $|\omega|$ along the contour. Now we use the result (see Problem 1.17) that

$$\int_0^\infty e^{-\omega^2 t^2-(\xi/2t)^2}\,dt = \frac{\sqrt{\pi}}{2\omega}\,e^{-\omega\xi}, \qquad (69)$$

which is valid provided we require $|\arg(\omega)| < \pi/4$, which is consistent with ω lying on the contour C_1. We therefore use this integral to represent $\exp(-\omega\xi)$ in (68) and then interchange the orders of integration. This leads to the problem of evaluating

$$\int_{C_1} \omega e^{-\omega^2 t^2} (\omega^2-1)^{\nu-\frac{1}{2}}\,d\omega, \qquad (70)$$

which can be done by the substitution $\omega^2-1 = u\exp(-i\pi)$, giving

$$\frac{1}{2}e^{-i\pi(\nu+\frac{1}{2})}\, e^{-t^2}\int_{-\infty}^{0+} u^{\nu-\frac{1}{2}}\, e^{-ut^2}\,du. \qquad (71)$$

With the help of (A8) we can easily evaluate this new integral as $[2\pi i t^{-2\nu-1}/(-\nu-\frac{1}{2})!]$, and on using these results in (68) we have

$$\int_{C_1} e^{-\omega\xi}(\omega^2-1)^{\nu-\frac{1}{2}}\,d\omega$$

$$= \frac{2\sqrt{\pi}\,e^{-i\pi\nu}}{(-\nu-\frac{1}{2})!}\int_0^\infty e^{-t^2-(\xi/2t)^2}\,t^{-2\nu-1}\,dt \qquad (72)$$

$$= \frac{\sqrt{\pi}\,e^{-i\pi\nu/2}\,(z/2)^{-\nu}}{(-\nu-\frac{1}{2})!}\int_{-\infty}^{\infty} e^{iz\cosh s-\nu s}\,ds,$$

where the last step follows from the substitution $t^2 = (z/2)\exp(-s-\frac{1}{2}i\pi)$. Comparing this with (66) we see that we have recovered a Hankel function, and since (68) and (72) both define analytic functions for $\mathrm{Im}(z) > 0$, we may lift the restriction $z = i\xi$ to write

$$H_\nu^{(1)}(z) = \frac{(-\nu-\frac{1}{2})!\,(z/2)^\nu}{i\pi^{3/2}} \int_{C_1} e^{i\omega z}(\omega^2-1)^{\nu-\frac{1}{2}}\,d\omega, \quad \mathrm{Im}(z) > 0. \quad (73)$$

A similar argument leads to

$$H_\nu^{(2)}(z) = \frac{(-\nu-\frac{1}{2})!\,(z/2)^\nu}{i\pi^{3/2}} \int_{C_2} e^{i\omega z}(\omega^2-1)^{\nu-\frac{1}{2}}\,d\omega, \quad \mathrm{Im}(z) < 0, \quad (74)$$

where the phase of $(\omega^2-1)^{\nu-1/2}$ is again chosen so that it becomes zero as we move along the contour after circling the branch point.

20.8. Modified Bessel Functions

The differential equation

$$f'' + \frac{1}{z} f' - \left(1 + \frac{\nu^2}{z^2}\right) f = 0, \quad (75)$$

which is closely related to Bessel's equation, occurs frequently in applications. In principle, it may be solved by the substitution $z = \pm it$, which turns it into Bessel's equation, but because of its frequent occurrence, special terminology has grown up for it. This has been brought about particularly by the fact that Bessel functions are defined in the cut plane because the origin is a branch point.

Modified Bessel Functions of the First Kind: The functions

$$e^{\pm i\pi\nu/2}\, J_\nu(\mp iz) \quad (76)$$

are solutions of (75) which are finite as $z \to 0$ if $\mathrm{Re}(\nu) \geq 0$, and identical if $\mathrm{Re}(z) > 0$ (Problem 13). From them we may construct a solution of (75) in the region $-\pi < \arg(z) < \pi$ by

$$I_\nu(z) = \begin{cases} e^{-i\pi\nu/2} J_\nu(iz), & -\pi < \arg(z) < \pi/2 \\ \\ e^{i\pi\nu/2} J_\nu(-iz), & -\pi/2 < \arg(z) < \pi. \end{cases} \tag{77}$$

The function $I_\nu(z)$ thus defined is usually referred to as a modified Bessel function of the first kind. For real ν and z, it takes on real values.

Macdonald's Function: For a second independent solution of (75), it is often convenient to have a function which becomes small for large real z. Such a function cannot be obtained by replacing J_ν by Y_ν in (77). However, it may be verified that the functions

$$\begin{aligned} &\frac{\pi i}{2}\, e^{i\pi\nu/2}\, H_\nu^{(1)}(iz), \\ -&\frac{\pi i}{2}\, e^{-i\pi\nu/2}\, H_\nu^{(2)}(-iz), \end{aligned} \tag{78}$$

which are both solutions of (75), are equal for $\operatorname{Re}(z) > 0$ (Problem 13). Hence, it is conventional to define as the second solution to (75) the function $K_\nu(z)$, known as Macdonald's function, by

$$K_\nu(z) = \begin{cases} \frac{\pi i}{2}\, e^{i\pi\nu/2}\, H_\nu^{(1)}(iz), & -\pi < \arg(z) < \pi/2 \\ \\ -\frac{\pi i}{2}\, e^{-i\pi\nu/2}\, H_\nu^{(2)}(-iz), & -\pi/2 < \arg(z) < \pi. \end{cases} \tag{79}$$

Recurrence Relations: Recurrence relations corresponding to (55) are easy to derive from the basic definitions of $I_\nu(z)$ and $K_\nu(z)$ and (55). They are

$$
\begin{aligned}
2I'_\nu(z) &= I_{\nu-1}(z) + I_{\nu+1}(z), \\
-2K'_\nu(z) &= K_{\nu-1}(z) + K_{\nu+1}(z), \\
\frac{2\nu}{z} I_\nu(z) &= I_{\nu-1}(z) - I_{\nu+1}(z), \\
-\frac{2\nu}{z} K_\nu(z) &= K_{\nu-1}(z) - K_{\nu+1}(z).
\end{aligned}
\tag{80}
$$

Problems

1. Show that

$$H_n(x) = \frac{2n(-i)^n e^{x^2}}{\sqrt{\pi}} \int_{-\infty}^{\infty} e^{-t^2+2itx}\, t^n\, dt, \quad n = 1,2,3,\ldots .$$

2. Prove the recurrence relations

$$\frac{d}{dz}[z^\nu J_\nu(z)] = z^\nu J_{\nu-1}(z),$$

$$\frac{d}{dz}[z^{-\nu} J_\nu(z)] = -z^{-\nu} J_{\nu+1}(z),$$

$$J'_\nu(z) = J_{\nu-1}(z) - \frac{\nu}{z} J_\nu(z),$$

$$J'_\nu(z) = -J_{\nu+1}(z) + \frac{\nu}{z} J_\nu(z).$$

3. Show that the Wronskian of any two solutions of Bessel's equation is of the form A/z, where A is a constant. (Consider the differential equation satisfied by the Wronskian.)

4. Show that

$$J_{-n}(x) = (-1)^n J_n(x).$$

5. Show that

$$H^{(1)}_{-\nu}(z) = e^{i\pi\nu} H^{(1)}_{\nu}(z),$$

$$H^{(2)}_{-\nu}(z) = e^{-i\pi\nu} H^{(2)}_{\nu}(z).$$

6. Prove the relations

$$Y_\nu(z) = \frac{J_\nu(z)\cos(\pi\nu) - J_{-\nu}(z)}{\sin(\pi\nu)},$$

$$Y_n(z) = \frac{1}{\pi}\left[\frac{\partial J_\nu(z)}{\partial\nu}\right]_{\nu=n} - \frac{(-1)^n}{\pi}\left[\frac{\partial J_{-\nu}(z)}{\partial\nu}\right]_{\nu=n},$$

$$Y_{-n}(z) = (-1)^n\, Y_n(z).$$

7. Verify the expansion

$$Y_n(z) = -\frac{1}{\pi}\sum_{k=0}^{n-1}\frac{(n-k-1)!}{k!}\left(\tfrac{1}{2}z\right)^{2k-n}$$

$$+\frac{1}{\pi}\sum_{k=0}^{\infty}\frac{(-1)^k\left(\frac{1}{2}z\right)^{n+2k}}{k!\,(n+k)!}\left[2\ell n\left(\tfrac{1}{2}z\right)-\psi(k+1)-\psi(k+n+1)\right],$$

where the first sum is set equal to zero if $n = 0$, and $\psi(z) = d\,\ell n[(z-1)!]/dz$.

8. Show that

$$Y_0(z) \sim \frac{2}{\pi}\,\ell n\left(\tfrac{1}{2}z\right), \quad z \to 0,$$

$$Y_n(z) \sim -\frac{(n-1)!}{\pi}\left(\tfrac{1}{2}z\right)^{-n}, \quad z \to 0, \quad n \geq 1.$$

9. Prove the following properties governing the behavior of Bessel functions on the branch cut $\operatorname{Im}(z) = 0$, $\operatorname{Re}(z) \leq 0$:

$$J_\nu(-x+i0) - J_\nu(-x-i0) = 2i\sin(\pi\nu)\, J_\nu(x),$$

$$Y_\nu(-x+i0) - Y_\nu(-x-i0) = 2i[\cos(\pi\nu)\, J_\nu(x) + J_{-\nu}(x)],$$

$$H_\nu^{(1)}(-x+i0) - H_\nu^{(1)}(-x-i0) = -2[J_{-\nu}(x) + e^{-i\pi\nu}J_\nu(x)],$$

$$H_\nu^{(2)}(-x+i0) - H_\nu^{(2)}(-x-i0) = 2[J_{-\nu}(x) + e^{i\pi\nu}J_\nu(x)],$$

where $x > 0$.

10. Using Watson's lemma in conjunction with the Poisson integral representation, show that

$$H_\nu^{(1)}(z) \sim \left(\frac{2}{\pi z}\right)^{1/2} e^{-i[z-(\pi\nu/2)-(\pi/4)]} \sum_{k=0}^{\infty} \frac{(-1)^k(\nu+k-\frac{1}{2})!}{k!(\nu-k-\frac{1}{2})!(2iz)^k},$$

$$z \to \infty, \ -\pi/2 < \arg(z) < 3\pi/2,$$

$$H_\nu^{(2)}(z) \sim \left(\frac{2}{\pi z}\right)^{1/2} e^{i[z-(\pi\nu/2)-(\pi/4)]} \sum_{k=0}^{\infty} \frac{(\nu+k-\frac{1}{2})!}{k!(\nu-k-\frac{1}{2})!(2iz)^k},$$

$$z \to \infty, \ -3\pi/2 < \arg(z) < \pi/2.$$

11. Verify the following relations between Bessel functions and circular functions:

$$H_\nu^{(1)}(z) \sim \left(\frac{2}{\pi z}\right)^{1/2} e^{i[z-(\pi\nu/2)-(\pi/4)]},$$

$$H_\nu^{(2)}(z) \sim \left(\frac{2}{\pi z}\right)^{1/2} e^{-i[z-(\pi\nu/2)-(\pi/4)]},$$

$$J_\nu(z) \sim \left(\frac{2}{\pi z}\right)^{1/2} \cos[z-(\pi\nu/2)-(\pi/4)],$$

$$Y_\nu(z) \sim \left(\frac{2}{\pi z}\right)^{1/2} \sin[z-(\pi\nu/2)-(\pi/4)].$$

12. Verify the following formulas for Bessel functions of order integer plus one-half:

$$J_{1/2}(z) = \left(\frac{2}{\pi z}\right)^{1/2} \sin z,$$

$$Y_{1/2}(z) = -\left(\frac{2}{\pi z}\right)^{1/2} \cos z,$$

$$H^{(1)}_{1/2}(z) = -i\left(\frac{2}{\pi z}\right)^{1/2} e^{iz},$$

$$H^{(2)}_{1/2}(z) = i\left(\frac{2}{\pi z}\right)^{1/2} e^{-iz},$$

$$J_{n+\frac{1}{2}}(z) = (-1)^n \left(\frac{2}{\pi}\right)^{1/2} z^{n+1/2}\left(\frac{d}{zdz}\right)^n \frac{\sin z}{z},$$

$$Y_{n+\frac{1}{2}}(z) = (-1)^{n+1}\left(\frac{2}{\pi}\right)^{1/2} z^{n+1/2}\left(\frac{d}{zdz}\right)^n \frac{\cos z}{z}.$$

13. Prove that for real $x > 0$

$$e^{i\pi\nu/2} J_\nu(-ix) = e^{-i\pi\nu/2} J_\nu(ix),$$

$$e^{i\pi\nu/2} H^{(1)}_\nu(ix) = -e^{-i\pi\nu/2} H^{(2)}_\nu(-ix).$$

Prove the following integral representations.

14. $$I_\nu(z) = \frac{z^\nu}{2^\nu \pi^{1/2}(\nu-\frac{1}{2})!}\int_{-1}^{1} (\omega^2-1)^{\nu-\frac{1}{2}} e^{-\omega z}\, d\omega,$$

$$\operatorname{Re}(\nu) > -\frac{1}{2}$$

15. $$K_\nu(z) = \frac{\pi^{1/2} z^\nu}{2^\nu (\nu-\frac{1}{2})!}\int_0^\infty e^{-z\cosh t} \sinh^{2\nu} t\, dt,$$

$$\operatorname{Re}(z) > 0, \quad \operatorname{Re}(\nu) > -\frac{1}{2}$$

16. $$K_\nu(x) = \frac{2^\nu(\nu-\frac{1}{2})!}{x^\nu \pi^{1/2}}\int_0^\infty \frac{\cos(xt)}{(1+t^2)^{\nu+1/2}}\, dt, \quad x > 0, \quad \operatorname{Re}(\nu) > -\frac{1}{2}.$$

17. Show that

$$K_\nu(z) = \frac{\pi}{2}\,\frac{I_{-\nu}(z) - I_\nu(z)}{\sin(\pi\nu)}.$$

18. Verify that, for $x > 0$

$$I_\nu(-x+i0) - I_\nu(-x-i0) = 2i\,\sin(\pi\nu)\,I_\nu(x),$$

$$K_\nu(-x+i0) - K_\nu(-x-i0) = -i\pi[I_{-\nu}(x) + I_\nu(x)],$$

$$K_\nu(-x+i0) + K_\nu(-x-i0) = 2\cos(\pi\nu)\,K_\nu(x).$$

19. Show that

$$I_{1/2}(z) = \left(\frac{2}{\pi z}\right)^{1/2} \sinh z,$$

$$K_{1/2}(z) = \left(\frac{\pi}{2z}\right)^{1/2} e^{-z},$$

$$I_{n+1/2}(z) = \left(\frac{2}{\pi}\right)^{1/2} z^{n+1/2}\left(\frac{d}{zdz}\right)^n \frac{\sinh z}{z},$$

$$K_{n+1/2}(z) = (-1)^n \left(\frac{2}{\pi}\right)^{1/2} z^{n+1/2}\left(\frac{d}{zdz}\right)^n \frac{e^{-z}}{z}.$$

Integrals Involving Bessel Functions: There is an enormous amount of literature on the evaluation of integrals involving Bessel functions. WATSON (1958) is a primary reference on methods; extensive tables are also available. The two most important techniques are: (i) use an integral representation for one of the factors and interchange the order of integration, and (ii) expand one of the factors in a power series and integrate term by term. Verify the following:

20. $$\int_0^\infty x^\mu\, J_\nu(x)\,dx = \frac{2^\mu(\frac{1}{2}\mu + \frac{1}{2}\nu - \frac{1}{2})!}{(-\frac{1}{2}\mu + \frac{1}{2}\nu - \frac{1}{2})},$$

$$\mathrm{Re}(\mu) < -\frac{1}{2}, \quad \mathrm{Re}(\mu+\nu) > -1.$$

21. $$\int_0^\infty e^{-ax} J_\nu(bx)dx = \frac{[\sqrt{a^2+b^2} - a]^\nu}{b^\nu \sqrt{a^2+b^2}},$$

$$a > 0,\ b > 0,\ \operatorname{Re}(\nu) > -1.$$

[Use (43).]

22. $$\int_0^\infty e^{-a^2x^2} J_\nu(bx)x^{\nu+1}dx = \frac{b^\nu}{(2a^2)^{\nu+1}} e^{-b^2/4a^2}$$

(Expand $\exp(-a^2x^2)$ and use Problem 20.)

23. $$\int_0^\infty \frac{x^{\nu+1} J_\nu(bx)}{(x^2+a^2)^{\mu+1}} = \frac{a^{\nu-\mu}\, b^\mu}{2^\mu\, \mu!} K_{\nu-\mu}(ab),$$

$$a > 0,\quad b > 0,\ -1 < \operatorname{Re}(\nu) < 2\operatorname{Re}(\mu) + \tfrac{3}{2}.$$

[Use the representation

$$(x^2+a^2)^{-\mu-1} = \frac{1}{\mu!}\int_0^\infty e^{-(x^2+a^2)t}\, t^\mu\, dt.]$$

24. Using the convolution formula for Laplace transforms, obtain the result

$$\int_0^{\pi/2} J_\mu(\alpha \sin\theta)\, J_\nu(\beta\cos\theta)\sin^{\mu+1}\theta \cos^{\nu+1}\theta\, d\theta$$

$$= \frac{\alpha^\mu\beta^\nu}{(\alpha^2+\beta^2)^{\mu+\nu+1}} J_{\mu+\nu+1}(\sqrt{\alpha^2+\beta^2}).$$

Airy Functions:

25. Show that two independent solutions of Airy's equation

$$u'' - zu = 0$$

are

$$u_1 = \mathrm{Ai}(z) = \frac{z^{1/2}}{3}\,[I_{-1/3}(\xi) - I_{1/3}(\xi)]$$

$$u_2 = \mathrm{Bi}(z) = \left(\frac{z}{3}\right)^{1/2}\,[I_{-1/3}(\xi) + I_{1/3}(\xi)]$$

where $\xi = 2z^{3/2}/3$. These solutions are known as Airy functions of the first and second kinds, respectively.

26. By the application of Laplace's method, show that two solutions of Airy's equation are

$$f_1 = \frac{1}{\pi}\int_0^\infty \cos(\tfrac{1}{3}\,t^3+xt)\,dt,$$

$$f_2 = \frac{1}{\pi}\int_0^\infty \left[e^{-\frac{1}{3}t^3+xt} + \sin(\tfrac{1}{3}\,t^3+xt)\right]dt, \quad x \geq 0.$$

27. Prove that the solutions $f_1(x)$ and $f_2(x)$ of Problem 25 are the Airy functions $\mathrm{Ai}(x)$ and $\mathrm{Bi}(x)$, respectively.

28. Show that as $|z| \to \infty$

$$\mathrm{Ai}(z) \sim \frac{e^{-\xi}}{2\pi^{1/2}z^{1/4}}, \quad -\frac{2\pi}{3} < \arg(z) < \frac{2\pi}{3},$$

$$\mathrm{Bi}(z) \sim \frac{e^{-\xi}}{\pi^{1/2}z^{1/4}}, \quad -\frac{\pi}{3} < \arg(z) < \frac{\pi}{3},$$

where $\xi = \frac{2}{3}\,z^{3/2}$.

29. Show that as $x \to \infty$

$$\mathrm{Ai}(-x) \sim \frac{\cos[\xi-(\pi/4)]}{\pi^{1/4}\,x^{1/4}},$$

$$\mathrm{Bi}(-x) \sim -\frac{\sin[\xi-(\pi/4)]}{\pi^{1/2}\,x^{1/4}}, \quad x > 0,$$

where $\xi = \frac{2}{3}\,x^{3/2}$.

30. Show that

$$\mathrm{Ai}(x^2) = \frac{1}{2\pi} e^{-2x^3/3} \int_0^\infty e^{-xu} \cos(u^{3/2}/3) \frac{du}{\sqrt{u}},$$

and hence derive the asymptotic series

$$\mathrm{Ai}(z) \sim \frac{e^{-\xi}}{2\pi z^{1/4}} \sum_{k=0}^{\infty} \frac{(-1)^k (3k-\frac{1}{2})!}{3^{2k} (2k)! \xi^k},$$

where $\xi = \frac{2}{3} z^{3/2}$.

Footnotes

1. More complicated equations can sometimes be reduced to this form by suitable transformation.

2. For details beyond those given in this section see, for instance, ABRAMOWITZ & STEGUN (1965), Ch. 22, and LEBEDEV (1965), pp. 60ff.

3. Since on differentiating W and using (7) we have $W' = 2xW$, whose solution is (23).

4. The classic and monumental reference on Bessel functions is WATSON (1958).

5. Bessel's equation is a special case of the confluent hypergeometric equation; one of its distinguishing features is that under this transformation it remains an equation of the same form.

6. This is permissible even though the function has an essential singularity at $u = 0$.

7. Functions satisfying (55) are known as cylinder functions. They satisfy Bessel's equation as a consequence of (55).

§21. NUMERICAL INVERSION OF LAPLACE TRANSFORMS

There are many problems whose solutions may be found in terms of Laplace or Fourier transforms which are then too complicated for inversion using the techniques of complex analysis. In this section we discuss some of the methods which have been developed -- and in some cases are still being developed -- for the numerical evaluation of the Laplace inversion integral. We make no explicit reference to inverse Fourier transforms, although they may obviously be treated by similar methods because of the close relationship between the two transforms.

21.1. Gaussian Quadrature Formulas for the Laplace Inversion Integral

If $F(p)$ is a Laplace transform having the form

$$F(p) = p^{-s}\,\Phi(p^{-1}), \quad s \geq 1, \tag{1}$$

where Φ is a polynomial of degree N, then we can easily find the inverse function $f(t)$ numerically using n-point Gaussian quadrature formulas, which are exact whenever $2n-1 \geq N$. Many such rules have been derived for real integrals of the type

$$\int_a^b w(x)\,f(x)\,dx, \tag{2}$$

depending on the choice of weight function $w(x)$ and limits a and b. We give here the derivation of a similar rule for the evaluation of the Laplace inversion integral,[1] under the assumption that for some $s \geq 1$ the function $p^s F(p)$ can be approximated by polynomials in p^{-1}. Setting $u = pt$ and writing

$$F(p) = p^{-s}\, G(p), \tag{3}$$

our aim is to approximate the inversion integral

$$\begin{aligned} f(t) &= \frac{1}{2\pi i}\int_{c-i\infty}^{c+i\infty} F(p)\, e^{pt}\, dp \\ &= \frac{t^{s-1}}{2\pi i}\int_{ct-i\infty}^{ct+i\infty} e^{u} u^{-s}\, G(u/t)\, du \end{aligned} \tag{4}$$

by

$$f(t) \simeq t^{s-1} \sum_{k=1}^{N} A_k\, G(u_k/t) \tag{5}$$

in such a manner that the approximation is exact whenever $G(p)$ is a polynomial in p^{-1} of degree less than or equal to $2N-1$.

It is well known[2] that a Gaussian quadrature formula for (2) can be constructed once we have a set of polynomials $P_N(x)$ of degree N, each of which satisfies the orthogonality relations

$$\int_a^b w(x)\, x^k P_N(x)\, dx = 0, \quad k = 0,1,2, \ldots, N-1. \tag{6}$$

In the present case, we need to find polynomials $P_{N,s}(u^{-1})$ with the orthogonality properties

$$\frac{1}{2\pi i}\int_{\gamma-i\infty}^{\gamma+i\infty} e^{u}\, u^{-s-k}\, P_{N,s}(u^{-1})\, du = 0, \tag{7}$$
$$k = 0,1,2, \ldots, N-1.$$

The problem is most easily solved using the convolution property of Laplace transforms, since the inverses of polynomials in p^{-1} must be polynomials in t. Therefore we define a set of functions $\phi_{N,s}(t)$ by

$$\mathcal{L}[\phi_{N,s}(t)] = p^{-1}\, P_{N,s}(p^{-1}), \tag{8}$$

and after noting that $\mathscr{L}[t^{s+k-2}] = (s+k-2)!/p^{s+k-1}$, we see that equation (7) is equivalent to

$$\int_0^1 (1-t)^{s+k-2}\ \phi_{N,s}(t)\ dt = 0,\ k = 0,1,2,\ \ldots,\ N-1. \quad (9)$$

The substitution $t \to (2t-1)$ converts this to a standard formula expressing the orthogonality of the Jacobi polynomials. Using the notation of hypergeometric functions, the solution is

$$\phi_{N,s}(t) = {}_2F_1(-N,N+s-1;1;t). \quad (10)$$

The polynomials $P_{N,s}(p^{-1})$ are readily found by taking the Laplace transform; they are

$$P_{N,s}(p^{-1}) = {}_2F_0(-N,N+s-1;;p^{-1}). \quad (11)$$

Therefore, by the standard techniques for the derivation of Gaussian quadrature rules,[2] the points u_k in (5) are the zeros of the polynomials $P_{N,s}(p^{-1})$, and the weights A_k are given by the formula

$$A_k = (-1)^{N-1}\ \frac{(N-1)!}{N(N+s-2)!}\ \left[\frac{2N+s-2}{u_k\ P_{N-1,s}(u_k^{-1})}\right]^2 . \quad (12)$$

Tables of u_k and A_k for various values of s and N are available.[3]

<u>Restrictions on Application</u>: Suppose that the function $G(p)$ has the asymptotic expansion

$$G(p) \sim \sum_{n=0}^{\infty} a_n p^{-n},\quad |p| \to \infty; \quad (13)$$

then the Heaviside series expansion gives for the inverse function $f(t)$ the asymptotic expansion

$$f(t) \sim \sum_{n=0}^{\infty} a_n \frac{t^{s+n-1}}{(s+n-1)!} . \tag{14}$$

Denoting now by $f_N(t)$ the approximation to $f(t)$ obtained from the N-point Gaussian rule, we can write

$$\begin{aligned} f_N(t) &= t^{s-1} \sum_{k=1}^{N} A_k \, G(u_k/t) \\ &\sim t^{s-1} \sum_{n=0}^{\infty} a_n \left[\sum_{k=1}^{N} A_k (t/u_k)^n \right] \\ &= \sum_{n=0}^{\infty} a_n \, \Gamma_{n,N} t^{n+s-1}, \\ \Gamma_{n,N} &= \sum_{k=1}^{N} A_k \, u_k^{-n}, \end{aligned} \tag{15}$$

where the interchange of order of summation is valid because the series are only asymptotic. Now $\Gamma_{n,N}$ is the approximate value of the inverse of the function u^{-s-n}, hence by the properties of the quadrature rule

$$\Gamma_{n,N} = \frac{1}{(s+n-1)!} , \qquad n \leq 2N-1, \tag{16}$$

and we see that the method recovers a sequence of functions whose asymptotic series match that of $f(t)$ more and more closely. Conversely, if $G(p)$ has the asymptotic expansion

$$\begin{aligned} G(p) &\sim \sum_{n=0}^{\infty} a_n \, p^{-\lambda_n}, \\ \lambda_0 &= 0 < \mathrm{Re}(\lambda_1) < \mathrm{Re}(\lambda_2) < \dots , \end{aligned} \tag{17}$$

where some of the λ_n are not integers, then the Gaussian formula will not reproduce the coefficients of the corresponding terms in the asymptotic expansion for any value of N. Hence the method must be restricted to functions for which (13) holds, and we have

$$f(t) = f_N(t) + \mathcal{O}(t^{s+2N}) \tag{18}$$

for such functions.

Convergence in N.[4] Assume that $G(p)$ is analytic at infinity, in which case (13) is a convergent series for $|p| > p_0$. It is well-known that Gaussian quadrature is equivalent to replacing the integrand by a Lagrangian polynomial interpolation formula, and then integrating this formula analytically. For the present case the Lagrangian formula is

$$\begin{aligned} L_N(u) &= \sum_{k=1}^{N} \frac{G(u_k/t)\ P_{N,s}(1/u)}{[(1/u)-(1/u_k)][P'_{N,s}(1/u)]_{u=u_k}} \\ &= \frac{1}{2\pi i}\int_{C_N} \frac{[P_{N,s}(w)-P_{N,s}(1/u)]}{[w-(1/p)]P_{N,s}(w)}\ G(1/tw)\ dw. \end{aligned} \tag{19}$$

The representation as a contour integral follows from the use of standard residue theory, and the contours C_N are circles centered at the origin and large enough to contain all the zeros of $P_{N,s}(u^{-1})$. Since $G(p)$ is analytic at infinity we can write

$$G(u/t) = \frac{1}{2\pi i}\int_{C_N} \frac{G(1/tw)}{w-(1/u)}\ dw, \tag{20}$$

and on substituting this into (19) we derive an expression for the error $E_n(t) = f(t) - f_N(t)$:

$$\begin{aligned} E_n(t) &= \frac{t^{s-1}}{2\pi i}\int_{ct-i\infty}^{ct+i\infty} e^u\ u^{-s-1}\ \{G(u/t)-L_N(u)\}\ du \\ &= \frac{t^{s-1}}{2\pi i}\int_{ct-i\infty}^{ct+i\infty} e^u\ u^{-s-1} P_{N,s}(u^{-1}) \left\{\frac{1}{2\pi i}\int_{C_N}\frac{G(1/tw)}{w-(1/u)}dw\right\} du. \end{aligned} \tag{21}$$

The orders of integration can be interchanged, which gives the result in the more useful form

$$E_n(t) = \frac{(-1)^n\, n!\, t^{s-1}}{2\pi i(2n+s-1)!} \int_{C_N} \frac{G(1/tw)\,{}_1F_1(n+1;2n+s;w^{-1})}{P_{N,s}(w)\, w^{N+1}} dw. \quad (22)$$

From this formula, it is not difficult to show that

$$\lim_{N\to\infty} E_N(t) = 0, \qquad 0 \le t < 2/p_0. \quad (23)$$

Thus the approximations $f_N(t)$ are shown to converge to $f(t)$ as N increases for a range of values of t which depends on the analytic behavior of $G(p)$. In fact, a function with the assumed properties of $G(p)$ would give an inverse such that $t^{-s+1}f(t)$ is analytic for all t, and it seems plausible that the functions $f_N(t)$ should converge to $f(t)$ for all t. No proof of such a property seems to be known.

21.2. Approximation of $F(p)$ by Chebyshev Polynomials[5] for Real p

The Gaussian method approximates $F(p)$ using a selection of values of $F(p)$ from various points in the complex p-plane. An alternative approach is to make the series expansion

$$F(p) = p^{-s} \sum_{n=0}^{\infty}{}' a_n T_n(1-bp^{-1}), \quad (24)$$

where T_n is the Chebyshev polynomial of degree n, b is a free constant to be chosen later, and the prime on the summation symbol means that the first term must be multiplied by one-half. If this series is truncated after N terms then we have a least squares approximation of $F(p)$ over the interval $b/2 \le p < \infty$ with weight function

$$p^{-3/2}\,[1-(b/2p)]^{-1/2}. \quad (25)$$

Inverting the series term by term we obtain

$$f(t) = \frac{t^{s-1}}{(s-1)!} \sum_{n=0}^{\infty}{}' \; a_n \; \phi_n(bt/2), \tag{26}$$

where $\phi_n(x)$ is a polynomial of degree k. The first three of these polynomials are readily found to be

$$\begin{aligned}
\phi_0(x) &= 1, \\
\phi_1(x) &= 1 - \frac{2x}{s}, \\
\phi_2(x) &= 1 - \frac{8x}{s} + \frac{8}{s(s+1)} x^2.
\end{aligned} \tag{27}$$

For $n > 3$, the direct evaluation of the $\phi_n(x)$ is a numerically unstable procedure owing to cancellation of large alternating terms; however, Piessens has shown that they obey the recurrence relations[6]

$$\begin{aligned}
\phi_n &= (A+Bx)\;\phi_{n-1} + (C+Dx)\;\phi_{n-2} + E\phi_{n-3}, \\
A &= 2n - \frac{(n-1)(2n-3)(s+n-2)}{(n-2)(s+n-1)}, \\
B &= \frac{-4}{n+s-1}, \\
C &= 1-A-E, \\
D &= -\frac{4(n-1)}{(n-2)(s+n-1)}, \\
E &= -\frac{(n-1)(s-n+2)}{(n-2)(s+n-1)},
\end{aligned} \tag{28}$$

and that these relations are numerically stable and thus suitable for automatic computation.

The coefficients a_n in the expansion may be expressed as definite integrals in the usual manner for orthogonal polynomial expansions. However, the Chebyshev polynomials enjoy some remarkable orthogonality properties over finite sets of of points, one of which results in the following property:[7]

if $f(x)$ is approximated by the Chebyshev expansion

$$f(x) = \sum_{n=0}^{N-1}{}' A_n T_n(x), \tag{29}$$

then the coefficients are obtained by the formula

$$A_n = \frac{2}{N} \sum_{j=1}^{N} f(\xi_j)\, T_n(\xi_j), \quad n = 0,1,\ldots,N-1, \tag{30}$$

where the ξ_j are the zeros of $T_N(x)$. Applied to the present problem, in which we have truncated the fundamental expansion (24), it gives the formula

$$a_n = \frac{2}{N} \sum_{k=0}^{N-1} \Psi\left[\cos \frac{(2k+1)}{N}\frac{\pi}{2}\right] \cos \frac{(2k+1)n}{N}\frac{\pi}{2}, \tag{31}$$

where

$$\Psi(u) = \left[\frac{b}{1-u}\right]^s F\left[\frac{b}{1-u}\right]. \tag{32}$$

It only remains to choose the parameter b. Clearly we need b sufficiently large so that $F(p)$ is analytic for $\mathrm{Re}(p) > b/2$, but apart from that the choice is arbitrary. The polynomials $\phi_n(x)$ have n real positive zeros, so that they are initially oscillatory, and then increase like x^n once the last zero has been passed. The argument of the polynomials in equation (26) is $bt/2$, thus we will not want $bt/2$ too large or the series will contain serious cancellations. On the other hand, we see from the formula for the coefficients a_n that a large value of b will make them rapidly converge to zero for large n. In fact, if we take the first neglected term in the truncated series (24) as an indication of the error, we will need to choose b large enough to make a_N small, and small enough to avoid having $\phi_N(bt/2)$ too large. Clearly there is a need for some experimentation in any particular calculation.

21.3. Approximation of $f(t)$ by Orthogonal Polynomials[8]

The relation

$$F(p) = \int_0^\infty f(t)\, e^{-pt}\, dt \tag{33}$$

may be regarded as an integral equation for the unknown function $f(t)$ and solved using one of the many methods for the numerical solution of such equations. We sketch here two possibilities based on the use of orthogonal polynomials; several other possibilities are found in the problems and in the literature.

Legendre Polynomials: Let σ be a positive constant, introduce the variable change

$$\sigma t = -\ln x, \tag{34}$$

and expand $f(t)$ in shifted Legendre functions:[9]

$$f(t) = \sum_{n=0}^{\infty} a_n P_n^*(e^{-\sigma t}). \tag{35}$$

To determine the coefficients a_n, we observe that $P_n^*(e^{-\sigma t})$ is a polynomial in $\exp(-\sigma t)$ of degree n, and thus has as its transform the function

$$\Phi_n(p) = \frac{N(p)}{p(p+\sigma)(p+2\sigma)\ \dots\ (p+n\sigma)}, \tag{36}$$

where $N(p)$ is a polynomial of degree less than n. The orthogonality property of the Legendre functions can be expressed as

$$\begin{aligned}\int_0^1 x^k P_n^*(x)\, dx \\ = \sigma \int_0^\infty e^{-(k+1)\sigma t} P_n^*(e^{-\sigma t})\, dt = 0, \quad k < n,\end{aligned} \tag{37}$$

which implies $\Phi_n(p)$ has the form

$$\Phi_n(p) = A \frac{(p-\sigma)(p-2\sigma)\cdots(p-n\sigma)}{p(p+\sigma)(p+2\sigma)\cdots(p+n\sigma)}. \tag{38}$$

Only the constant A remains to be determined; since $p\Phi_n(p) \to (-1)^n P_n^*(1)$ for large p, we see immediately that $A = (-1)^n$. Therefore the transform of our expansion (35) is

$$F(p) = \frac{a_0}{p} + \sum_{n=1}^{\infty} (-1)^n \frac{(p-\sigma)\cdots(p-n\sigma)}{p(p+\sigma)\cdots(p+n\sigma)} a_n, \tag{39}$$

and setting $p = \sigma, 2\sigma, 3\sigma, \ldots$ we obtain the relations

$$\sigma F(\sigma) = a_0,$$

$$\sigma F(2\sigma) = \frac{a_0}{2} - \frac{a_1}{2\cdot 3},$$

$$\vdots$$

$$\sigma F(n\sigma) = \frac{a_0}{n} - \frac{(n-1)a_1}{n(n+1)} + \cdots + (-1)^n \frac{(n-1)\cdots 2\cdot 1\cdot a_{n-1}}{n(n+1)\cdots(2n-1)},$$

which can be solved recursively for the unknowns a_n.

<u>Trigonometric Functions</u>: Defining the variable θ by

$$e^{-\sigma t} = \cos\theta, \tag{41}$$

the equation for the Laplace transform becomes

$$\sigma F(p) = \int_0^{\pi/2} (\cos\theta)^{(p/\sigma)-1} \sin\theta\, g(\theta)d\theta, \quad g(\theta) = f(t), \tag{42}$$

and hence with $p = (2k+1)\sigma$, $k = 0,1,2,\ldots$ we have

$$\sigma F[(2k+1)\sigma] = \int_0^{\pi/2} \cos^{2k}\theta \sin\theta\, g(\theta)\, d\theta. \tag{43}$$

If we expand $f(t)$ in the Fourier series

$$g(\theta) = \sum_{n=0}^{\infty} a_n \sin(2n+1)\theta, \tag{44}$$

then the coefficients a_n can be determined as follows.

We write

$$\cos^{2n}\theta \sin\theta = [\tfrac{1}{2}(e^{i\theta} + e^{-i\theta})]^{2n} \frac{1}{2i}(e^{i\theta} - e^{-i\theta}), \tag{45}$$

and, with the use of the binomial theorem, we obtain

$$\begin{aligned} 2^{2n} \cos^{2n}\theta \sin\theta &= \sin(2n+1)\theta + \dots \\ &+ \left[\binom{2n}{k} - \binom{2n}{k-1}\right] \sin[2(n-k)+1]\theta + \dots + \left[\binom{2n}{n} - \binom{2n}{n-1}\right] \sin\theta. \end{aligned} \tag{46}$$

This can be used in (43) to determine a system of equations for the unknown coefficients, namely

$$\begin{aligned} \frac{4}{\pi}\sigma F(\sigma) &= a_0, \\ 2^2 \frac{4}{\pi}\sigma F(3\sigma) &= a_0 + a_1, \\ &\vdots \\ 2^{2n} \frac{4}{\pi}\sigma F[(2n+1)\sigma] &= \left[\binom{2n}{n} - \binom{2n}{n-1}\right] a_0 + \dots \\ &+ \left[\binom{2n}{k} - \binom{2n}{k-1}\right] a_{n-k} + \dots + a_n, \end{aligned} \tag{47}$$

which may again be solved recursively to calculate the first N terms of the basic expansion (44).

The influence of σ on these calculations is to determine the range of p for which $F(p)$ is sampled. An appropriate choice must be made for any particular inversion, keeping in mind that for small t, $f(t)$ is significantly affected by values of $F(p)$ to quite large values of p, whereas for large t the properties of $F(p)$ need only be known quite close to $p = 0$. Thus for a given value of N we expect that a suitable value of σ will be proportional to t^{-1}, while the constant of proportionality will have to be determined by some form of numerical experimentation.

21.4. Padé Approximation

As a prelude to the discussion in Section 21.5, we give here a brief treatment of approximations using rational functions. Suppose that $f(x)$ has the representation

$$f(x) = \sum_{k=0}^{p+q} c_k x^k + \mathcal{O}(x^{p+q+1}); \tag{48}$$

then we call the rational function

$$\begin{aligned} F_{p,q}(x) &= \frac{f_p(x)}{g_q(x)}, \\ f_p(x) &= \sum_{k=0}^{p} \alpha_k x^k, \\ g_q(x) &= \sum_{k=0}^{q} \beta_k x^k, \quad \beta_0 = 1, \end{aligned} \tag{49}$$

a Padé approximation to $f(x)$ if $F_{p,q}$ has the same representation (48) up to, but not including, the remainder term. This is equivalent to the requirement that

$$\begin{aligned} g_q(x)f(x) - f_p(x) &= x^{p+q+1} h_{p,q}(x), \\ h_{p,q}(0) &\neq 0. \end{aligned} \tag{50}$$

The functions $F_{p,q}$ can be arranged as a matrix, and this is usually known as the Padé table for $f(x)$. Explicit formulas may be given for the elements of the Padé table for certain special functions $f(x)$; we refer the reader to specialized literature for details.[10]

In general, the coefficients of the polynomials may be obtained by equating powers of x up to x^{p+q} in (50), which gives

$$\begin{aligned} \alpha_k &= \sum_{j=0}^{k} c_j\,\beta_{k-j}, \quad k = 0,1,\ \ldots\ ,p; \\ 0 &= \sum_{j=0}^{k} c_j\,\beta_{k-j}, \quad k = p+1,\ p+2,\ \ldots\ ,\ p+q. \end{aligned} \tag{51}$$

This is a set of $p+q+1$ inhomogeneous equations; for practical purposes they are not very convenient, and we now show how they may be solved recursively.[11] The method uses the fact that the first row and the first column of the Padé table are easy to compute. For the first row, we have

$$\begin{aligned} f_p(x) &= \sum_{k=0}^{p} c_k\,x^k, \\ g_q(x) &= 1, \end{aligned} \tag{52}$$

and for the first column, equations (51) become

$$\begin{aligned} \alpha_0 &= c_0, \\ c_0\beta_k &= -\sum_{j=1}^{k} c_j\beta_{k-j}, \quad k = 1,2,\ \ldots\ ,\ q, \end{aligned} \tag{53}$$

which can be solved recursively. Suppose that we already have $F_{p,q-1}$ and $F_{p-1,q}$. We denote the corresponding coefficients by α_k^+, β_k^+ (for $F_{p,q-1}$) and α_k^*, β_k^* (for $F_{p-1,q}$). If we write out the equations corresponding to (50) for these two approximations, we have

$$\begin{aligned} f(x)\sum_{k=0}^{p} \alpha_k^+\,x^k - \sum_{k=0}^{q-1} \beta_k^+ x^k &= \mathcal{O}(x^{p+q}), \\ f(x)\sum_{k=0}^{p-1} \alpha_k^*\,x^k - \sum_{k=0}^{q} \beta_k^* x^k &= \mathcal{O}(x^{p+q}), \end{aligned} \tag{54}$$

and on subtracting the second relation from the first

$$f(x)\left[\sum_{k=0}^{p-1}(\alpha_k^+ - \alpha_k^*)x^k + \alpha_p^+ x^p\right] - \left[\sum_{k=0}^{q-1}(\beta_k^+ - \beta_k^*)x^k - \beta_q^* x^q\right] = \mathcal{O}(x^{p+q}). \tag{55}$$

Since $\beta_0 = 1$ and $\alpha_0 = c_0$ for any element of the Padé table, we can divide out a factor x in (55). This gives

$$f(x)\sum_{k=0}^{p-1} a_k x^k - \sum_{k=0}^{q-1} b_k x^k = \mathcal{O}(x^{p+q-1}),$$

$$a_k = \begin{cases}\alpha_{k+1}^+ - \alpha_{k+1}^*, & k = 0,1,\ldots, p-2\\ \alpha_p^+ , & k = p-1,\end{cases}$$

$$b_k = \begin{cases}\beta_{k+1}^+ - \beta_{k+1}^*, & k = 0,1,\ldots, q-2\\ -\beta_q^* , & k = q-1.\end{cases} \tag{56}$$

Equation (56) is almost identical to the defining relation (50) for $F_{p-1,q-1}$. If $\beta_1^+ \neq \beta_1^*$, then we can divide by the factor b_0 to obtain

$$\alpha_k = a_k/b_0,$$
$$\beta_k = b_k/b_0, \tag{57}$$

for the coefficients of the element $F_{p-1,q-1}$. If $b_0 = 0$, then we have $\alpha_1^+ = \alpha_1^*$, and (56) gives for $F_{p-1,q-1}$ a reducible rational function, a circumstance which needs special consideration.[12] Normally, however, $b_0 \neq 0$, so we have a recursive relation which enables a triangular portion of values $F_{p,q}$ for $p+q \leq r$ to be built up from a knowledge of the functions $F_{k,0}$ and $F_{0,k}$ for $k = 0,1, \ldots ,r$.

21.5. Rational Approximation of F(p)

Writing, as usual,

$$F(p) = p^{-s}\, G(p), \quad s \geq 1, \tag{58}$$

where s is chosen so that $G(p) \to$ const. as $p \to \infty$, we can approximate $G(p)$ by rational functions, using a Padé table, and then invert these rational functions using standard methods[13] (see Section 2). For relatively low-order approximations, it will be possible to determine the position of the poles and calculate the corresponding residues. More often, however, particularly in conjunction with high-order approximations, it is necessary to find some method of inversion which does not rely on first computing the zeros of polynomials.[14] We discuss such a method here, assuming that $s = 1$ so that we want to invert

$$F(p) = \frac{A(p)}{B(p)}, \tag{59}$$

where

$$\begin{aligned} A(p) &= \sum_{i=0}^{m} a_i p^i, \\ B(p) &= \sum_{i=0}^{n} b_i p^i, \quad n > m. \end{aligned} \tag{60}$$

If we assume that the inversion is carried out by decomposing $F(p)$ into partial fractions and then inverting each term, and that the roots $\alpha_1, \alpha_2, \ldots, \alpha_n$ of $B(p)$ are distinct, then we can write immediately

$$\begin{aligned} F(p) &= \sum_{i=1}^{n} \frac{A(\alpha_i)}{(p-\alpha_i)\, B'(\alpha_i)}, \\ f(t) &= \sum_{i=1}^{n} \frac{A(\alpha_i)}{B'(\alpha_i)}\, e^{\alpha_i t}, \end{aligned} \tag{61}$$

with similar but more complicated results if one or more roots is repeated. The problem is to replace these formulas by an algorithm which does not employ the α_i.

Special Case: Suppose that $A(p) = 1$ and that all the roots of $B(p)$ are distinct. Since equation (61) for $f(t)$ is a linear combination of a finite number of exponentials, we may expand each exponential as a Taylor series and reverse the order of summation. This gives

$$\begin{aligned} f(t) &= \sum_{i=1}^{n} \frac{1}{B'(\alpha_i)} \sum_{k=0}^{\infty} \frac{\alpha_i^k t^k}{k!} \\ &= \sum_{k=0}^{\infty} u_k \frac{t^k}{k!} \end{aligned} \tag{62}$$

where

$$u_k = \sum_{i=1}^{n} \frac{\alpha_i^k}{B'(\alpha_i)} . \tag{63}$$

It is easily shown (Problem 6) that $u_k = 0$ for $k \le n-2$, and that $u_{n-1} = b_n^{-1}$. For $k \ge n$, we rearrange the definition of u_k using the fact that $B(\alpha_i) = 0$, to give

$$\begin{aligned} b_n u_k &= \sum_{i=1}^{n} \frac{\alpha_i^{k-n}}{B'(\alpha_i)} b_n \alpha_i^n \\ &= - \sum_{i=1}^{n} \frac{\alpha_i^{k-n}}{B'(\alpha_i)} \sum_{j=0}^{n-1} b_j \alpha_i^j \\ &= - \sum_{j=0}^{n-1} b_j u_{k-n+j} . \end{aligned} \tag{64}$$

Hence we may calculate the coefficients u_k which we need in (62) recursively, starting with $u_n = -b_{n-1}/b_n^2$, $u_{n+1} = (b_{n-1}^2/b_n^3) - (b_{n-2}/b_n^2)$, etc.

As a simple example, consider the function

$$F(p) = \frac{1}{p^2+1} \,. \tag{65}$$

Here we have $b_2 = 1$, $b_1 = 0$, $b_0 = 1$, and hence

$$u_k = -u_{k-2}, \quad k \geq 2. \tag{66}$$

Equation (62) now gives

$$f(t) = \sum_{\ell=0}^{\infty} \frac{(-1)^{\ell}\, t^{2\ell+1}}{(2\ell+1)!} = \sin t. \tag{67}$$

General Case, Distinct Roots: We now relax the condition $A(p) = 1$, while still assuming that the roots of $B(p)$ are distinct. Then we must replace equations (62) and (63) by

$$f(t) = \sum_{k=0}^{\infty} \frac{v_k t^k}{k!} \,, \tag{68}$$

where

$$\begin{aligned} v_k &= \sum_{i=1}^{n} \frac{A(\alpha_i)\, \alpha_i^k}{B'(\alpha_i)} \\ &= \sum_{i=1}^{n} \frac{\alpha_i^k}{B'(\alpha_i)} \sum_{j=0}^{m} a_j\, \alpha^j \\ &= \sum_{j=0}^{m} a_j\, u_{k+j}. \end{aligned} \tag{69}$$

Hence we may again calculate the coefficients of the Taylor series (68) without calculating the roots of $B(p)$, but we need a second recursion relation (69) to determine the coefficients v_k from the coefficients u_k. As a simple example, we consider

$$F(p) = \frac{p+1}{p^2+1} \,. \tag{70}$$

Then we have $a_0 = a_1 = 1$, and hence

$$v_k = u_{k+1} + u_k. \tag{71}$$

For $f(t)$, we obtain

$$f(t) = \sum_{\ell=0}^{\infty} \frac{(-1)^{\ell}\, t^{2\ell}}{(2\ell)!} + \sum_{\ell=0}^{\infty} \frac{(-1)^{\ell}\, t^{2\ell+1}}{(2\ell+1)}$$
$$= \cos t + \sin t. \tag{72}$$

Repeated Roots: If some of the roots of $B(p)$ are repeated, we may still write the series representation (68) for $f(t)$. Now, however, the derivations given for the recursion relations (64) and (69) break down, since some of the $B'(\alpha_i)$ are zero. Nevertheless it may be shown that (68) still holds with the coefficients u_k and v_k determined from these recursion relations. In a sense, this result is not surprising, since any method for determining $f(t)$ without a knowledge of the roots α_i ought not to break down if the roots happen to have a particular property. As a simple example, consider the function

$$F(p) = \frac{1}{(p-1)^2} . \tag{73}$$

It is easily shown that $u_k = k$ for $k \geq 0$ and hence

$$f(t) = \sum_{k=0}^{\infty} \frac{kt^k}{k!}$$
$$= t\, e^t. \tag{74}$$

More difficult (and realistic) examples have been given by Longman.[15]

Problems

1. Show that if (35) is truncated after N terms, the use of a Gaussian quadrature formula gives

$$\sum_{i=1}^{N} A_i x_i^k g(x_i) = \sigma F[(k+1)\sigma], \quad k = 0,1,\ldots,N-1,$$

where

$$g(x) = f(t),$$
$$P_N^*(x_i) = 0,$$

and the coefficients A_i are the appropriate Gaussian weights.

2. Let

$$f_j(x) = \frac{P_N^*(x)}{(x-x_j)\, P_N^{*\prime}(x_j)} = \sum_{k=0}^{N-1} c_{jk}\, x^k.$$

Use the coefficients c_{jk} in conjunction with the previous problem to show that[16]

$$\sigma \sum_{k=1}^{N} c_{jk}\, F[(k+1)\sigma] = A_j\, g(x_j), \quad j = 1,2,\ldots,N.$$

3. Derive a scheme similar to the result of the previous problem when (35) is replaced by

$$f(t) \simeq \sum_{n=0}^{N-1}{}' a_n T_n(e^{-\sigma t}),$$

where the T_n are Chebyshev polynomials.

4. Let

$$e^{-\frac{1}{2}\sigma t} = \sin \tfrac{1}{2}\theta\ ,$$
$$\phi(\theta) = \cot \tfrac{1}{2}\theta \sin^{2a} \tfrac{1}{2}\theta\, f(t) = \sum_{n=0}^{\infty}{}' a_n \cos n\theta.$$

Derive a recursive formula for the evaluation of the coefficients a_n from the Laplace transform $F(p)$.

5. Let

$$f(t) \simeq e^{-\sigma t} \sum_n a_n L_n(t)$$

where the L_n are the Laguerre polynomials

$$L_n(t) = \sum_{k=0}^{n} \binom{n}{k} \frac{(-t)^k}{k!} .$$

Show that

$$F(p) \simeq \sum_n a_n \frac{(p+\sigma-1)^n}{(p+\sigma)^{n+1}}$$

and that

$$a_n = \sum_j \binom{n}{j} b_{n-j},$$

where the coefficients b_k are obtained from the Taylor series expansion[17]

$$G(s) = F(s-\sigma+1)$$

$$= \sum_k b_k s^k.$$

6. In Problem 5, make the substitution[18]

$$z = \frac{p + \sigma - 1}{p + \sigma}$$

and show that the expansion for $F(p)$ becomes

$$\frac{F(p)}{1-z} \simeq \sum_{n=1}^{N} a_n z^n .$$

By setting $z_j = \exp(2\pi i(j+\frac{1}{2})/N)$, $j = 1,2,\ldots,N$, show that the coefficients a_n may be computed from

$$a_n = \frac{1}{N} \sum_{j=1}^{N} G(z_j) \exp(-2\pi i(j+\tfrac{1}{2})/N)$$

where

$$G(z) = F(p)/(1-z).$$

7. Using residues, write a contour integral representation for the coefficients u_k in (63). By letting the contours tend to infinity, show that

$$u_k = 0, \quad k \leq n-2.$$

Footnotes

1. Based on H. E. Salzer, Math. Tables Aids Comp. (1955), 9, 164; J. Maths. Phys. (1958), 37, 89.

2. For example, see STROUD (1974), pp. 135ff.

3. H. E. Salzer, J. Maths. Phys. (1961), 40, 72; STROUD & SECREST (1966), pp. 307ff.

4. This argument is given in LUKE (1969), vol. II, pp. 253ff.

5. Based on R. Piessens, J. Inst. Math. Appl., (1972), 10, 185. In the original paper, Piessens writes

$$F(p) = p^{-s} \sum_{n=0}^{\infty} a_n P_n^{(\alpha,\beta)}(1-bp^{-1})$$

where the $P_n^{(\alpha,\beta)}$ are Jacobi polynomials. We consider only the special case $\alpha = \beta = -1/2$, which forms the main body of Piessens' paper.

6. We have corrected Piessens' formulas for the coefficients to remove some errors.

7. RIVLIN (1974), p. 47.

8. Based on A. Papoulis, Q. Appl. Math. (1956), 14, 405.

9. ABRAMOWITZ & STEGUN (1965), Ch. 22.

10. A very thorough treatment may be found in LUKE (1969), vol. II, Ch. 10.

11. I. M. Longman, Int. J. Comp. Math. B (1971), 3, 53.

12. Obviously such a curcumstance would cause peculiar difficulties.

13. Some other possibilities for the use of Padé approximation are discussed in LUKE (1969), vol. II, pp. 255ff.

14. See I. M. Longman & M. Sharir, Geophys. J. Roy. Astr. Soc. (1971), 25, 299.

15. I. M. Longman, J. Comp. Phys. (1972), 10, 224.

16. This method is the subject of BELLMAN, KALABA & LOCKETT (1966).

17. LUKE (1969), vol. II, pp. 247-251.

18. W. T. Weeks, J. Ass. Comp. Mach. (1966), 13, 419; R. A. Spinelli, SIAM J. Num. Anal. (1966), 3, 636. See also R. Piessens & M. Branders, Proc. I.R.E.E. (1971), 118, 1517 for a generalization.

Appendix A: The Factorial Function

Definition: The factorial function is defined by

$$\alpha! = \int_0^\infty x^\alpha e^{-x}\,dx, \quad \mathrm{Re}(\alpha) > -1, \tag{1}$$

since it has the value $n!$ when $\alpha = n$. Analytic continuation can be effected by splitting the integral, i.e., by writing

$$\begin{aligned}\alpha! &= \int_0^1 x^\alpha e^{-x}\,dx + \int_1^\infty x^\alpha e^{-x}\,dx \\ &= \int_0^1 \sum_{n=0}^\infty \frac{(-1)^n\, x^{n+\alpha}}{n!}\,dx + \Omega(\alpha) \\ &= \sum_{n=0}^\infty \frac{(-1)^n}{n!\,(\alpha+n+1)} + \Omega(\alpha)\end{aligned} \tag{2}$$

where $\Omega(\alpha)$ is an entire function. We see that the function has simple poles at $\alpha = -(n+1)$, with residues $(-1)^n$, $n = 0, 1, 2, \ldots$.

Functional Relationships: The factorial function satisfies a number of important functional relationships. The most important three are

$$\alpha! = \alpha(\alpha-1)! \tag{3}$$

$$\alpha!(-\alpha-1)! = -\pi/\sin(\pi\alpha) \tag{4}$$

$$\alpha!(\alpha+\tfrac{1}{2})! = 2^{-(2\alpha+1)}\,\pi^{1/2}(2\alpha+1)! \tag{5}$$

Derivations of all these results can be obtained by manipulation of the appropriate integrals, and may be found in many places.

Hankel's Integral Representation: Consider the loop integral

$$I(z) = \int_{-\infty}^{0+} t^z e^t \, dt. \tag{6}$$

where the contour is as in (6.22). For $\mathrm{Re}(z) > -1$, we can shrink the contour to encircle the branch cut, giving

$$\begin{aligned} I(z) &= -\int_0^{\infty} (r\, e^{-i\pi})^z e^{-r} \, dr + \int_0^{\infty} (r\, e^{i\pi})^z e^{-r} \, dr \\ &= -\, 2i \sin(\pi z)\, z! \end{aligned} \tag{7}$$

By analytic continuation, the restriction $\mathrm{Re}(z) > -1$ can be removed, and, after using equation (4) together with the change of variables $z \to -(z+1)$, we have Hankel's integral representation

$$\frac{1}{z!} = \frac{1}{2\pi i} \int_{-\infty}^{0+} t^{-(z+1)} e^t \, dt. \tag{8}$$

This result shows that $1/z!$ is an entire function, so that $z!$ has no zeros.

Asymptotic Form: An application of the method of steepest descents to (8) gives the important result

$$\ell n(\alpha!) \sim (\alpha + \tfrac{1}{2})\, \ell n\, \alpha - \alpha + \tfrac{1}{2}\, \ell n(2\pi), \tag{9}$$

$$\alpha \to \infty, \quad |\arg(\alpha)| < \pi.$$

The complete asymptotic expansion, of which we have written down the first three terms, is derived in (13.33) using the Mellin transform.

Beta Function: Related to the factorial function, and often occuring in applications, is the function

$$B(p,q) = \int_0^1 x^{p-1} (1-x)^{q-1} dx, \tag{10}$$

$$\mathrm{Re}(p) > 0, \quad \mathrm{Re}(q) > 0.$$

Another common form, related by a variable change, is

$$B(p,q) = \int_0^\infty \frac{x^{p-1} dx}{(1+x)^{p+q}} \tag{11}$$

In terms of the factorial function, the Beta function has the value

$$B(p,q) = \frac{(p-1)!\ (q-1)!}{(p+q-1)!} \tag{12}$$

a result whose derivation may be found in many places. (See also Problem 1.18.)

Appendix B: Rieman's Zeta Function

Definition: The zeta function may be defined by

$$\zeta(s) = \sum_{n=1}^{\infty} n^{-s}, \quad \mathrm{Re}(s) > 1. \tag{1}$$

An integral representation is obtained quite easily by observing that

$$\int_0^{\infty} x^{s-1} e^{-nx}\, dx = n^{-s}(s-1)!, \quad \mathrm{Re}(s) > 0. \tag{2}$$

After summing both sides and reversing the order of summation and integration, this gives

$$(s-1)!\, \zeta(s) = \int_0^{\infty} \frac{x^{s-1} dx}{e^x - 1}, \quad \mathrm{Re}(s) > 1. \tag{3}$$

Analytic Continuation: Consider the loop integral

$$\frac{1}{2\pi i} \int_{-\infty}^{0+} \frac{z^{s-1}}{e^{-z}-1}\, dz. \tag{4}$$

For $\mathrm{Re}(s)$ 1 we can shrink the contour about the branch cut to get

$$\begin{aligned} &- \frac{1}{\pi} \sin[\pi(s-1)] \int_0^{\infty} \frac{r^{s-1} dr}{e^r - 1} \\ &= - \frac{1}{\pi} \sin[\pi(s-1)]\ (s-1)!\ \zeta(s). \end{aligned} \tag{5}$$

Using (A4) this gives

$$\zeta(s) = \frac{(-s)!}{2\pi i} \int_{-\infty}^{0+} \frac{z^{s-1}}{e^{-z}-1}\, dz. \tag{6}$$

This shows that $\zeta(s)$ is analytic except possibly where $(-s)!$ has poles.

Integer s: When s is an integer, the integrand is single-valued, and we may replace the contour by a small circle about the origin and use residue theory. It follows immediately that the integral is zero for $s = 2, 3, 4, \ldots$, cancelling out the poles of $(-s)!$ there. For $s \leq 1$, we recall the Taylor series

$$\frac{z}{2} \cot \frac{z}{2} = 1 - \sum_{n=1}^{\infty} \frac{B_n z^{2n}}{(2n)!}, \tag{7}$$

where B_n are the Bernoulli numbers. Using this in the present case to construct a Laurent expansion of the integrand of (6), we find that when $s = 1-m$, $m = 0, 1, 2, \ldots$, we have

$$\frac{z^{-m}}{e^{-z}-1} = z^{-(m+1)} \left[-1 - \frac{z}{2} + \sum_{n=1}^{\infty} \frac{(-1)^n B_n z^{2n}}{(2n)!} \right]. \tag{8}$$

Excluding for the moment the case $m = 0$, we can evaluate (6) immediately to give

$$\begin{aligned} \zeta(0) &= -1/2, \\ \zeta(-2n) &= 0, \quad n = 1, 2, 3, \ldots, \\ \zeta(1-2n) &= (-1)^n B_n/2n, \quad n = 1, 2, 3, \ldots . \end{aligned} \tag{9}$$

When $s = 1$ $(m = 0)$, the value of the integral in (6) is -1, and since $(-s)!$ has a simple pole with residue -1 at $s = 1$, we conclude that $\zeta(s)$ has a simple pole with residue $+1$ at $s = 1$. Apart from this, it is analytic for all s. Some other special values are

$$\begin{aligned} \zeta(2) &= \pi^2/6, \\ \zeta(4) &= \pi^4/90, \\ \zeta'(0) &= -\frac{1}{2} \ln 2\pi. \end{aligned} \tag{10}$$

<u>Riemann's Functional Relationship</u>: By further arguments involving the integral (6), which we shall not reproduce here, it can be shown that

$$\pi^s \zeta(1-s) = 2^{1-s}(s-1)! \cos(\pi s/2)\zeta(s). \tag{11}$$

<u>Asymptotic Forms</u>: It is important to know the asymptotic behavior of $\zeta(s)$ for large s. For $\mathrm{Re}(s) > 1$, (3) can be rearranged as follows:

$$\begin{aligned} \zeta(s) &= \frac{1}{(s-1)!}\int_0^\infty \frac{e^{-x}x^{s-1}}{1-e^{-x}}\{1 - e^{-x} + e^{-x}\}\,dx \\ &= 1 + \frac{1}{(s-1)!}\int_0^\infty e^{-2x}x^{s-1}\left\{\frac{1}{x} + \frac{1}{1-e^{-x}} - \frac{1}{x}\right\}dx \\ &= 1 + \frac{2^{1-s}}{s-1} \\ &\quad + \frac{1}{(s-1)!}\int_0^\infty \left\{\frac{1}{1-e^{-x}} - \frac{1}{x}\right\} e^{-2x}\,x^{s-1}\,dx. \end{aligned} \tag{12}$$

This last formula is an analytic continuation to $\mathrm{Re}(s) > 0$. It is easy to bound the integral using the fact that the expression in the brackets is less than unity, and this gives the asymptotic information

$$\zeta(s) \sim 1, \quad s \to \infty, \quad \mathrm{Re}(s) > 0. \tag{13}$$

The picture is completed by the use of Riemann's functional relationship, giving

$$\zeta(1-s) \sim (2\pi)^{-s} \cos(\pi s)(s-1)!\,, \qquad s \to \infty, \quad \mathrm{Re}(s) > 0. \tag{14}$$

Appendix C: The Exponential Integral

The exponential integral is defined by

$$E_1(z) = \int_z^\infty \frac{e^{-t}}{t}\,dt. \tag{1}$$

It is a multi-valued function, since its value along a closed contour encircling the origin increases by $2\pi i$ for each counter-clockwise circuit, due to the simple pole in the integrand. It is conventional to take the negative real axis as a branch cut, restricting the contour appropriately. If we split up the integral as follows:

$$\begin{aligned} E_1(z) = &\int_1^\infty \frac{e^{-t}}{t}\,dt + \int_0^1 \frac{e^{-t}-1}{t}\,dt \\ &- \int_0^z \frac{e^{-t}-1}{t}\,dt - \int_z^1 \frac{dt}{t}, \end{aligned} \tag{2}$$

then the sum of the first two integrals is a constant; on substituting $t = 1/u$ in the first integral and $t = u$ in the second, we find that this constant is

$$-\gamma = -\int_0^1 \frac{1-e^{-u}-e^{-1/u}}{u}\,du, \tag{3}$$

which is one expression for Euler's constant, $\gamma = 0.5772157\ldots$. Thus we have

$$E_1(z) = -\ln(z) - \gamma - \int_0^z \frac{e^{-t}-1}{t}\,dt \tag{4}$$

where $\ln(-z)$ is the principal branch. Since the remaining integral is an entire function, we see the behavior at the branch point explicitly; also by using a Taylor series expansion and integrating we obtain the series representation

$$E_1(z) = -\ln z - \gamma - \sum_{n=1}^{\infty} \frac{(-1)^n z^n}{n\; n!} \tag{5}$$

Asymptotic forms for large z are derived in Section 14.2.

Bibliography

Abramowitz, M., & Stegun, I.A., 1965, "Handbook of Mathematical Functions", National Bureau of Standards, Applied Mathematics Series 55.

Ahlfors, L.V., 1966, "Complex Analysis", McGraw-Hill.

Apostol, T.M., 1957, "Mathematical Analysis", Addison-Wesley.

Bellman, R.E., Kalaba, R.E., & Lockett, J.A., 1966, "Numerical Inversion of the Laplace Transform", American Elsevier.

Berg, L., 1967, "Introduction to the Operational Calculus", North Holland.

Carrier, G.F., Krook, M., & Pearson, C.E., 1966, "Functions of a Complex Variable", McGraw-Hill.

Carslaw, H.S., & Jaeger, J.C., 1941, "Operational Methods in Applied Mathematics", Oxford University Press.

Case, K.M., & Zweifel, P.F., 1967, "Linear Transport Theory", Addison-Wesley.

Dingle, R.B., 1973, "Asymptotic Expansions: Their Derivation and Interpretation", Academic Press.

Ditkin, V.A. & Prudnikov, A.P., 1962, "Operational Calculus in Two Variables and its Applications", Pergamon Press.

----, 1965, "Integral Transforms and Operational Calculus", Pergamon Press.

Doetsch, G., 1971, "Guide to the Application of the Laplace and Z Transforms", Van Nostrand.

Erdelyi, A., 1962, "Operational Calculus and Generalized Functions", Holt, Rinehart, and Winston.

Erdelyi, A., Magnus, W., Oberhettinger, F., & Tricomi, F.G., 1954, "Tables of Integral Transforms", (2 volumes) McGraw-Hill.

Friedmann, B., 1956, "Principles and Techniques of Applied Mathematics", Wiley.

Gelfand, I.M. & Shilov, G.E., 1964, "Generalized Functions" vol. 1, Academic Press.

Goldberg, R.R., 1961, "Fourier Transforms" (Cambridge Tract No. 52) Cambridge University Press.

Kaplan, W., 1962, "Operational Methods for Linear Systems", Addison-Wesley.

Krebbe, G., 1970, "Operational Calculus", Springer-Verlag.

Lebedev, N.N., 1965, "Special Functions and Their Applications", Prentice-Hall.

Luke, Y.L., 1969, "The Special Functions and Their Approximations", (2 volumes), Academic Press.

Miles, J.W., 1961, "Integral Transforms in Applied Mathematics", Cambridge University Press.

Morse, P.M., & Feshbach, H., 1953, "Methods of Theoretical Physics", McGraw-Hill.

Murnaghan, F.D., 1962, "The Laplace Transform", Spartan Books, Washington.

Muskhelishvili, N.I., 1953, "Singular Integral Equations", Noordhoff.

Noble, B., 1958, "Methods Based on the Wiener Hopf Technique for the Solution of Partial Differential Equations", Pergamon Press.

Oberhettinger, F., 1971, "Tables of Bessel Transforms", Springer-Verlag.

----, & Badd, L., 1973, "Tables of Laplace Transforms", Springer-Verlag.

----, 1974, "Tables of Mellin Transforms", Springer-Verlag.

Olver, F.W.J., 1974, "Asymptotics and Special Functions", Academic Press.

Papoulis, A., 1963, "The Fourier Integral and its Applications", McGraw-Hill.

Rivlin, T.J., 1974, "The Chebyshev Polynomials", Wiley-Interscience.

Smith, M.G., 1966, "Laplace Transform Theory", Princeton.

Sneddon, I.N., 1966, "Mixed Boundary Value Problems in Potential Theory", North Holland.

----, 1972, "The Use of Integral Transforms", McGraw-Hill.

Stakgold, I., 1968, "Boundary-value Problems in Mathematical Physics", (2 volumes), Macmillan, New York.

Stoker, J.J., 1957, "Water Waves", Interscience.

Stroud, A.H., 1974, "Numerical Quadrature and Solution of Ordinary Differential Equations", Springer-Verlag.

----, & Secrest, D., 1966, "Gaussian Quadrature Formulas", Prentice-Hall.

Szego, G., 1959, "Orthogonal Polynomials", American Mathematical Society Colloquium Publications, vol. 23.

Thompson, W.T., 1950, "Laplace Transformation", Longmans, Green & Co.

Titchmarsh, E.C., 1948, "An Introduction to the Theory of Fourier Integrals", 2nd. ed., Oxford University Press.

----, 1953, "Eigenfunction Expansions Associated with Second-Order Differential Equations", Oxford: Clarendon Press.

Tranter, C. J., 1956, "Integral Transforms in Mathematical Physics", Methuen.

Van der Pol, B. & Bremmer, H., 1955, "Operational Calculus Based on the Two-sided Laplace Transform", Cambridge University Press.

Watson, G.N., 1958, "A Treatise on the Theory of Bessel Functions", 2nd ed., Cambridge University Press.

Whittaker, E.T. & Watson, G.N., 1963, "A Course of Modern Analysis", Cambridge University Press.

Widder, D.V., 1944, "The Laplace Transform", Princeton.

----, 1971, "An Introduction to Transform Theory", Academic Press.

Zemanian, A.H., 1965, "Distribution Theory and Transform Analysis", McGraw-Hill.

----, 1968, "Generalized Integral Transformations", Interscience.

Index

3741